Theory and Analysis
of
Sample Survey Designs

Theory and Analysis
of
Sample Survey Designs

DAROGA SINGH, M.A., Ph.D.
Ex-Director
Indian Agricultural Statistics Research Institute
New Delhi, India
(Formerly Senior Statistical Adviser, F.A.O.,
Amman, Jordan)

F.S. CHAUDHARY, M.Sc., Ph.D.
Deputy Director (Statistics)
Haryana Agricultural University,
Hisar, India

A HALSTED PRESS BOOK

JOHN WILEY & SONS
New York   Chichester   Brisbane   Toronto   Singapore

Copyright © 1986, WILEY EASTERN LIMITED
New Delhi

Published in the Western Hemisphere by
Halsted Press, A Division of
John Wiley & Sons, Inc., New York

Library of Congress Cataloging in Publication Data

Singh, Daroga.
  Theory and analysis of sample survey design.

  "A Halsted Press book."
  Includes bibliographies.
  1. Sampling (Statistics) I. Chaudhary, Fauran S.
II. Title.
QA276.6.C44   1986   519.5′2   85-24928

ISBN 0-470-20266-1

Printed in India at Maharani Printers, Delhi.

Preface

This book is a collocation of the *Hand Book on Sample Surveys*, written by the first author in collaboration with Padam Singh and others, and lecture notes prepared for advanced-level under-graduates and graduate students by the second author. The *Hand Book* by the first author was written as a study-aid for students and professional statisticians and did not include the theory of analysis. Suggestions, particularly emphasising that the book should be enlarged and theory and techniques be included with the analysis of survey data so that survey statisticians may use it as a ready reference in formulating their projects, were received. To meet these demands, this book has been prepared as a ready reference manual for statisticians for project planning and data analysis and as a text book for graduate students offering a course on sample surveys. Besides, covering basic theory, this book also contains related corollaries and illustrative examples. As sampling theory is meant to be used in practice, the application of sampling methods has been demonstrated with the help of actual sets of data from different sample surveys conducted in India.

Some important general features of the book are:
  (i) To give a systematic exposition of existing theory;
  (ii) To extend its application to field data;
  (iii) To provide further references of work done in the recent past.

An attempt has been made to include all modern developments which are considered important in survey work. A unique feature of this book is that some problems with real sets of data from various fields are included, either as illustrative examples to demonstrate the method of analysis, or as unsolved problems to be attempted by the reader to make ideas much clearer. The theory presented here has been illustrated with examples of actual agricultural surveys which were mainly conducted in developing countries where sampling expertise is not easily available.

This book is divided into 13 chapters. Chapter 1 is devoted to basic concepts and some mathematical preliminaries needed for developing proofs in subsequent chapters. The next four chapters present the basic method of simple random sampling, stratified random sampling, systematic sampling and varying probability sampling. The use of auxiliary information has been explained in Chapters 6 and 7. The subsequent four chapters are devoted to sampling and sub-sampling of clusters, double sampling and successive surveys. We have also included sequential sampling in one chapter, a development which has not been included so far in other textbook. In the last chapter, non-sampling errors are briefly discussed so that the survey statisticians should bear it in mind when they plan the survey and analysis data in their presence. This book also covers a number of new results which are published in text form for the first time.

The restriction of keeping the book within a predetermined size has compelled us to avoid more detailed descriptions of sampling theory and techniques. In spite of this, the aim of presenting theory and analysis in a straightforward manner has been achieved. It is assumed that the reader has a good working knowledge of college algebra, calculus, probability, statistical methods and elementary estimation theory as a pre-requisite for a proper understanding of the treatment given in this text.

The excellent works of Cochran, Sukhatme and Sukhatme, Deming, Des Raj, Hansen, Hurwitz and Madow, Murthy, Yates, and Zarkovich, and several other writers, on sampling theory technique and analysis, have been used as sources of the material included in this book. All these are gratefully acknowledged. Some of the illustrations have been taken from the *Hand Book on Sample Surveys*, to which a reference has already been made. We have also included material from a large number of research papers, cited in references at the end of each chapter, which are an added attraction to this book. We thankfully acknowledge them all. The material for illustrative examples, and unsolved problems was collected from various textbooks, journal, laboratory records, and examination papers. We have referred to the original source whenever we could locate it. Since the material was derived from secondary sources, it is quite possible that in some ins-tances we might not have recorded proper credit and we offer our regrets and apologies to the authors.

We shall appreciate it greatly if readers will bring to our attention any errors which they detect or suggestions for further improvement in this work.

DAROGA SINGH
F.S. CHAUDHARY

Contents

Basic Concept of
Sample Surveys

But I keep no log of my daily grog.
For what's the use O' being bothered ?
I drink a little more when the wind's off shore,
And most when the wind's from the north'ard

Arthur Macy

1.1 INTRODUCTION

As Deming (1950) so succinctly puts it:

"Sampling is not mere substitution of a partial coverage for a total coverage. Sampling is the science and art of controlling and measuring reliability of useful statistical information through the theory of probability."

The enumeration of population by sampling methods, proposed by Laplace in 1783, came into widespread use only by the mid-thirties of this century. From the outset, some basic questions arose:

(i) How should the observations be made?

(ii) How many observations should be made?

(iii) How should the total sample be made?

(iv) How should the data thus obtained be analysed?

The answers to these questions were sought and, in the process, a number of different techniques and methods were developed. These methods were tested to determine whether the above mentioned questions were adequately answered or not. In course of time, the concept of generalization through the introdcution of inductive logic, i.e. proceeding from the *part* to the *whole*, caught the attention of statisticians engaged in developing suitable techniques of sampling. In this context a related question cropped up, viz. how should this generalization be made? The answer to this question was sought with the help of the technique of probability. Thus the concept of probability sampling originated. The use of probability in sampling theory came to be recognised as a reliable tool in drawing inferences about the populations, whether finite or not.

In traditional applications the experimenter assumes some kind of probability structure with the observations, while in sample surveys he introduces the probability element by adopting the technique of randomization. The idea of probability structure in planned experiments was originally given by Fisher (1935) who showed that deliberately introduced randomization in the selection of a part from the whole provides a valid method of obtaining an estimate of the amount of error committed. He demonstrated practically that the randomization not only gives a procedure for valid selection of the part from the whole, but also gives an expression to the amount of risk committed in doing so. Thus, the problem was reduced to determining the methods for selection and estimation, which would minimize the risk involved. Mahalanobis (1944) introduced another important concept—that of cost function. The problem was considered to be one of finding the combinations of selection and estimation procedures which would minimize the cost function.

1.2 FIELDS OF APPLICATION OF SAMPLING TECHNIQUES AND LIMITATIONS

The main objective of this work is to present the theory and techniques of sample surveys with their application in different types of problems in the field. Sample surveys are to be widely used as a means of collecting information to meet a definite need in government, industry and trade, physical and life sciences and technology, social, educational and economical problems, etc. The sample survey technique is now commonly used for obtaining information on various social and economic activities of society. All walks of life are covered by sample surveys. It is not possible to include the full range of applications in this work.

However, some specific situations in which sampling techniques can successfully be employed have been given.

(i) When results with maximum accuracy or reliability with a fixed budget, or with the minimum number of units with a specified degree of reliability are required.

(ii) When the units under investigation show considerable variation for the characteristic under study.

(iii) When a total count of the population is not possible or is very costly or destructive.

(iv) When the scope of the investigation is very wide and the population is not completely known.

(v) When time, money and other resources are limited.

Sampling theory has its own limitations which may be briefly outlined as follows:

(i) In spite of the fact that a proper choice of design is employed, a sample does not fully cover the parent population and consequently results are not exact.

(ii) Sampling theory and its application in the field need the services of trained and qualified personnel without whom results of sample surveys are not dependable.

(iii) The planning and execution of sample surveys should be done very carefully, or the data may provide misleading results.

1.3 DEFINITIONS AND PRELIMINARIES

The following definitions and basic concepts are developed here before we discuss the details of sampling methods:

We are usually faced with a collection (called *population*):

$$\mathcal{U} = (u_1, \ldots, u_N)$$

of the objects, $u_1, \ldots, u_N$ (called *units* or *elements*), of which some property (called *characteristic*) y_i is defined for every unit u_i.

Sampling theory is mainly concerned with ways of obtaining samples, i.e. sequences or sets of units taken from $\mathcal{U}$, in order to estimate parameters such as Y, $\overline{Y}$, σ^2, R, etc. A useful theory can be developed on the basis of the theory of probability (it is assumed that the reader has some knowledge of it). We shall consider a random experiment, the outcome of which depends on *chance*. The results of a random experiment will be called *sample points* and the totality of all sample points consistent with the method of sampling adopted will be called the *sample space*. Every outcome of the experiment is described by one, and only one, sample point. The method of sampling must also define the probability $P(s)$ that a particular sample s is drawn such that $P(s) \geqslant 0$, and $\sum P(s) = 1$.

We shall take as proved a number of standard results in the probability theory, which will be used in the discussion of sampling theory. Let there be a random variate X taking the values x_i with probability p_i $(i = 1, \ldots, N)$. The expected value of X is defined as

$$E(X) = \sum_{i=1}^{N} p_i x_i = \sum_{i}^{N} p_i x_i = \bar{X}$$

(We shall be using the notation $\sum_{i}^{N}$ for $\sum_{i=1}^{N}$ throughout the text, unless mentioned otherwise.)

In general, if $\phi(X)$ be a function of X, then

$$E\{\phi(X)\} = \sum_{i}^{N} p_i \, \phi(x_i)$$

We shall be using the results contained in the following theorems:

THEOREM 1.1   If X and Y are two random variates,

$$E(X + Y) = E(X) + E(Y) \tag{1.3.1}$$

In a generalized form of k random variates,

$$E\left(\sum_{i}^{k} X_i\right) = \sum_{i}^{k} E(X_i) \tag{1.3.2}$$

A more generalized form can be written as

$$E\left(\sum_{i}^{k} a_i X_i\right) = \sum_{i}^{k} a_i E(X_i) \tag{1.3.3}$$

where a_i $(i = 1, \ldots, k)$ are the constants.

THEOREM 1.2   If X and Y are independent,

$$E(X \cdot Y) = E(X) \cdot E(Y) \tag{1.3.4}$$

THEOREM 1.3   The variance of a random variate

$$Z = aX + b$$

where a and b are constants, is given by

$$V(Z) = a^2 V(X) \tag{1.3.5}$$

THEOREM 1.4   The covariance of X and Y is given by

$$\left. \begin{array}{l} \text{Cov } (X, Y) = E(XY) - E(X) \cdot E(Y) \\ \qquad\qquad = \rho[V(X).V(Y)]^{1/2} \end{array} \right\} \tag{1.3.6}$$

where ρ stands for the correlation coefficient of X and Y.

THEOREM 1.5   If X_i and a_i $(i = 1, \ldots, k)$ are k random variates and constants respectively, then

$$V\left(\sum_i^k a_i X_i\right) = \sum_i^k \sum_j^k a_i a_j \text{ Cov } (X_i, X_j) \tag{1.3.7}$$

THEOREM 1.6   If X and Y are independent random variates, then

$$V(XY) = V(X) V(Y) + [E(X)]^2 V(Y) + [E(Y)]^2 V(X) \tag{1.3.8}$$

THEOREM 1.7   The expected value of a random variate X is given by

$$E(X) = E_1 E_2(X|Y) \tag{1.3.9}$$

where E_2 stands for the conditional expectation of X for a given Y, and E_1 stands for the expectation over the space of Y.

THEOREM 1.8   The variance of a random variate X is the sum of the expected value of the conditional variance and the variance of the conditional expected value. Symbolically,

$$V(X) = E_1 V_2 (X) + V_1 E_2 (X) \tag{1.3.10}$$

We assume that it is either not feasible or is too time consuming or expensive to measure X on each unit u_i, or if performed on all elements, measurements would be too inexact. We, therefore, select some of the elements (a *sample*). Thus, a *sampling unit* may be taken as a well-defined and identifiable element or group of elements on which obervations can be made. A collection of such units is usually called *population*. A population is said to be *finite* if the number of units contained in it is finite. Usually a list has to be prepared or a serial order has to be given to all units in the population. This creates the *sampling frame*. A part or fraction of the population is said to constitute a *sample*. The number of units, not necessarily distinct, included in the sample is known as the *sample size*. The number of distinct units in the sample is termed as the *effective sample size*.

Any function of sample values is called *statistic*. If it is used to estimate any parameter it will be called *estimator*. An estimator is a random variate and may take different values from sample to sample. The value the estimator takes on in any particular sample is then its *estimate*. The difference between the estimator (t) and the parameter (θ) is called *error*. An estimator (t) is said to be *unbiased* estimator for the parameter (θ) if

$$E(t) = \theta,$$

otherwise biased. Thus bias is given by

$$E(t - \theta) = B(t)$$

A relative measure of bias is $B(t)/\theta$. The mean of squares of error taken from θ is called *mean-square error* (MSE). Symbolically,

$$\text{MSE}(t) = E(t - \theta)^2$$

The *sampling variance* of t is defined by

$$V(t) = E[t - E(t)]^2$$

Obviously,

$$\begin{aligned}
\text{MSE}(t) &= E(t - \theta)^2 \\
&= E[\{t - E(t)\} + \{E(t) - \theta\}]^2 \\
&= E[t - E(t)]^2 + [E(t) - \theta]^2 \\
&= V(t) + B^2(t)
\end{aligned}$$

The positive square root of the variance is termed as the *standard error* of the estimator. The ratio of the standard error of the estimator to the expected value of the estimator is known as the *relative standard error*.

Given two estimators t_1 and t_2 of a parameter, the estimator t_1 is said to be more *efficient* than t_2 if the mean square error of t_1 is less than the mean square error of t_2. The *relative efficiency* of t_1 as compared to t_2, which differs in respect of sample size or sampling method or both, may be defined as the reciprocal of the ratio of the sampling variances of the estimator given by both techniques when the same number of units are taken. The relative efficiency may also be obtained as the ratio of the reciprocals of sample sizes for both sampling methods when the same type of sampling units are taken.

In fact, the question here is how to measure *accuracy*, i.e., to measure the expected difference between the estimate and the true value. It is usually measured by *efficiency* which is inversely proportional to the mean-square error. Since the true value of the parameter is generally unknown, the efficiency may be measured in terms of *precision*, i.e. the expected difference of the estimate from its expected value. Thus, precision of an estimator is inversely proportional to its sampling variance.

1.4 CENSUS AND SAMPLE SURVEYS, ADVANTAGES AND DISADVANTAGES

The total count of all units of the population for a certain characteristic is known as *complete enumeration,* also termed *census survey.* The money, manpower and time required for carrying out complete enumeration will generally be large and there are many situations with limited means where complete enumeration will not be possible. There are also instances where it is not practicable to enumerate all units due to their perishable nature where recourse to selection of a few units will be helpful. When only a part, called a sample, is selected from the population and examined, it is called *sample enumeration* or *sample survey.*

A sample survey will usually be less expensive than a census survey and the desired information will be obtained in less time. This does not imply that economy is the only consideration in conducting a sample survey. It is most important that a degree of accuracy of results is also maintained. Occasionally, the technique of sample survey is applied to verify the results obtained from a census survey. It has been a well-established fact that in many situations a well-conducted sample survey can provide much more precise results than a census survey. The relative merits and demerits of sample surveys vis-a-vis census surveys have been discussed by Mahalanobis (1950), Yates (1953), Zarkovich (1961) and Lahiri (1963). Cochran (1977) has very lucidly shown the advantages of sample surveys over census surveys. In brief, these are:

   (i) reduced cost of survey,
   (ii) greater speed of getting results,
  (iii) greater accuracy of results,
  (iv) greater scope, and
   (v) adaptability.

Fisher (1950) sums up the merits of sample surveys over census surveys as follows:

"I have made four claims for the sampling procedure. About the first three, adaptability, speed and economy, I need say nothing further.

Too many examples are already available to show how much the method has to give in these ways. But, why do I say that it is more scientific than the only procedure with which it may sometimes be in competition, complete enumeration? The answer, in my view, lies in the primary process of designing and planning an enquiry by sampling. Rooted as it is in the mathematical theory of the errors of random sampling, the idea of precision is from the first in the forefront. The director of the survey plans from the first for a predetermined and known level of precision; it is a consideration of which he never loses sight, and precision actually attained, subject to well-understood precautions, is manifest from the results of the enquiry."

Despite the above advantages, sample surveys are not always preferred to census surveys. Sampling theory has its own limitations and the advantages of sampling over complete enumeration can be derived only if
  (i) the units are drawn in a scientific manner,
 (ii) an appropriate sampling technique is used, and
(iii) the size of units selected in the sample is adequate. If information is required for each unit, census is the only answer.

1.5 PRINCIPLES OF SAMPLING THEORY

The main aim of sampling theory is to make sampling more effective so that the answer to a particular question is given in a valid, efficient and economical way. The theory of sampling is based on three important basic principles:
  (i) Principle of Validity,
 (ii) Principle of Statistical Regularity, and
(iii) Principle of Optimization.

Principle of Validity

This principle states that the sampling design provides valid estimates about population parameters. By valid we mean that the sample should be so selected that the estimates could be interpreted objectively and in terms of probability. Thus, the principle ensures that there is some definite and pre-assigned probability for each individual in the sampling design.

Principle of Statistical Regularity

The principle of statistical regularity, which has its origin in the theory of probability, can be explained in these words:

> "A moderately large number of items chosen at random from a large group are almost sure on the average to possess the characteristics of the large group."

This principle stresses upon the desirability and importance of selecting sample designs where inclusion of sampling units in the sample is based on probability theory.

Principle of Optimization

This principle takes into account the desirability of obtaining a sampling design which gives optimum results. In other words, optimization is meant to develop methods of sample selection and of estimation that provide: (i) a given level of efficiency with the minimum possible resources or (ii) a given value of cost with the maximum possible efficiency.

Thus, the principle of optimization minimises the risk or loss of sampling design, i.e. the principle stresses upon obtaining optimum results with minimization of the total loss in terms of cost and mean square error.

1.6   PRINCIPAL STEPS IN A SAMPLE SURVEY

Sample survey techniques have now come to be used widely as an organized and established fact-finding instrument and it is, therefore, essential to describe briefly the main steps which are involved in a sample survey. Some of the main steps to be included are given as follows:

  (i) Statement of objectives
  (ii) Definition of population to be studied
 (iii) Determination of sampling frame and sampling units
 (iv) Selection of proper sampling design
  (v) Organization of field work
 (vi) Summary and analysis of data

Statement of Objectives

In a sample survey, the first step is to lay down a clear statement of objectives of the survey. The user should ensure that these objectives are commensurate with available resources in terms of money, manpower and the time limit of the survey.

Definition of Population

The population from which the sample is to be drawn should be defined in clear and unambiguous terms. For example, to estimate the average yield per plot for a crop, it is necessary to define the size of the plot in clear terms. The *sampled population* (population to be sampled) should coincide with the *target population* (population about which information is required). The demographic, geographical, administrative and other boundaries of the population must be specified so that there remains no ambiguity regarding the coverage of the survey.

Determination of Sampling Frame and Sampling Units

The main requirement of sample surveys is to fix up the sampling frame, i.e., the list of all sampling units with reference to which relevant data are to be collected. It is the sampling frame which determines the sampling structure of a survey. A sampling frame is the key note around which the selection and estimation procedures revolve. The population should be capable of division into units which are distinct, unambiguous and non-overlapping and cover the entire population.

Selection of Proper Sampling Design

If an appropriate sampling design is selected, the final estimates will be quite reliable. The size of the sample, procedure of selection and estimation of parameters along with the amount of risk involved are some of the important statistical aspects which should receive careful attention. If a number of sampling designs for taking a sample is available, then the total risk, i.e. the cost and precision, should be considered before making a final selection of the sampling design.

Organization of Field Work

The achievement of the aims of a sample survey depends to a large extent on reliable field work. If field work is done honestly, sincerely and according to the instructions laid down and if there is careful supervision of field staff, there remains no doubt about achieving the aims of the survey. It is, therefore, necessary to make provisions for adequate supervisory staff for inspection of field work.

Summary and Analysis of Data

In a sample survey, the final step of the analysis and drawing inferences from a sample to a population is a very vital and fascinating issue. Since the results of the survey are the basis for policy making, it is the most essential part of the sample survey and should be handled carefully.

The analysis of the data collected in a survey may be broadly classified as follows:
  (i) Scrutiny and editing of the data
  (ii) Tabulation of data
  (iii) Statistical analysis
  (iv) Reporting and conclusions

Finally, a report of the findings of the survey, suggesting possible action to be taken, should be written.

1.7 PROBABILITY AND NON-PROBABILITY SAMPLING

The technique of selecting a sample is of fundamental importance in sampling theory and usually depends upon the nature of the investigation. The sampling procedures which are commonly used may be broadly classified under the following heads:
  (i) Probability sampling
  (ii) Non-probability sampling

Probability Sampling

This is the method of selecting samples according to certain laws of probability in which each unit of the population has some definite probability of being selected in the sample. It is to be noted here that there are a number of samples of a specified types $S_1, S_2, \ldots, S_k$ that can be formed by grouping units of a given population, and each possible sample S_i has, assigned to it, a known probability of selection p_i. A clear specification of all possible samples of a given type along with their corresponding probabilities of selection is said to constitute a *sampling design*. In subsequent chapters of this book we shall consider only the procedures of probability sampling.

Non-Probability Sampling

This is the method of selecting samples, in which the choice of selection of sampling units depends entirely on the discretion or judgement of the sampler. This method is also sometimes called *purposive* or *judgement* sampling. In this procedure, the sampler inspects the entire population and selects a sample of typical units which he considers close to the average of the population.

This sampling method is mainly used for opinion surveys, but cannot be recommended for general use as it is subject to the drawbacks of prejudice and bias of the sampler. However, if the sampler is experienced and an expert, it is possible that judgement sampling may yield useful results. It, however, suffers from a serious defect that it is not possible to compute the degree of precision of the estimate from the sample values.

1.8 SAMPLING UNIT

A sampling unit has already been defined in the previous section, but it needs further description as it forms the basis of sampling procedure. These units may be natural units of the population such as individuals in a locality, or natural aggregates of such units such as family, or they may be artificial units such as a farm, etc. Before selecting the sample, the population must be divided into parts which are distinct, unambiguous and non-overlapping, such that every element (smallest component part in which a population can be divided) of the population belongs to one and only one sampling unit. Since the collection of all sampling units of a specified type constitutes a population, the sampling units should be so specified that each and every element in the population occurs just in one sampling unit. Otherwise, some of the elements will not be included in any sample. For example, if the sampling unit is a family, it should be so defined that an individual does not belong to two different families nor should it leave out any individuals belonging to it.

1.9 SAMPLING FRAME

As defined earlier, a complete list of sampling units which represents the population to be covered is called the sampling frame popularly known as *frame*. The construction of a sampling frame is sometimes one of the major practical problems. Generally, it is assumed that a frame is perfect if it is exhaustive, complete, and up-to-date in respect of sampling units and character structures. So the frame should always be made up-to-date and free from errors of omission and duplication of sampling units. Different aspects of sampling frames have been discussed by Mahalanobis (1944), Yates (1960), Seal (1962), Hansen, Hurwitz, and Jabine (1963), Singh (1978) and others. A sampling frame is subject to several types of defect which may be broadly classified as follows:

(i) *A frame may be incomplete*   When some sampling units of the population are either completely omitted or included more than once.

(ii) *A frame may be inaccurate*   When some of the sampling units of the population are listed inaccurately or some units which do not actually exist are included in the list.

(iii) *A frame may be inadequate* When it does not include all classes of the population which are to be taken in the survey.

(iv) *A frame may be out-of-date*   When it has not been updated according to the exigencies of the occassion, although it was accurate, complete and adequate at the time of construction.

Lists which have been routinely collected for some purpose are generally found to be incomplete and inaccurate and often contain unknown amounts of duplication. Such lists should be carefully examined and scrutinised to ensure that they are free from all such defects. If they are not up-to-date they should be made so. A good frame is difficult to construct, but experience always helps in its construction.

1.10 DESIDERATA IN PLANNING OF SAMPLE SURVEYS

The main stages of a survey are planning, data collection and data processing. Some of the important aspects involved in the planning and execution of a sample survey may be classified under the following heads:

Specification of Data Requirements

When specifying the data requirements, the sampler should always include the following points:

(i) Statistical statement of the desired information
(ii) Clear specification of the domain of study
(iii) Form of data which are to be collected and limitations of budget
(iv) Degree of precision aimed at

Survey Reference and Reporting Periods

From the operational point of view, it is desirable to decide about these periods well in advance. The *survey period* is the time period during which the required data are collected. It is advisable to divide the survey period into shorter sub-periods to ensure an even representation of the sample. The *reference period* is the time period to which the data information should refer. It depends on the objective of the survey. For different items there may be different reference periods. The *reporting period* is the time period for which the information is collected for a unit and is determined by the nature and condition of the survey. For technical considerations there may be different reporting periods for different items, but from the operational point of view it is desirable to use the same reference period for all items as far as possible.

Preparation of Sampling Frame

Since the sampling frame decides the structure of the survey and its design, it is necessary to pay adequate attention to the preparation of an up-to-date and accurate sampling frame. Even if some resources are to be spent on such work, it will be worthwhile.

Choice of Sampling Design

A decision about an optimum sampling design, after taking the various technical, operational and risk factors into consideration, plays a very significant part. The principle of optimization should always be kept in mind, i.e. to achieve

(i) either a given degree of precision with a minimum cost, or
(ii) the maximum possible precision with a fixed cost.

Method of Data Collection

The planning and execution of a survey is influenced to a large extent by the method of data collection. After a very careful examination of the frame, design, budget and objectives of the survey, a decision should be taken regarding the choice of method of data collection, i.e. to collect primary data or to use secondary data. In case the primary data is to be collected, a clear-cut mode of collection should be given as to whether data are to be collected by personal interview, mail enquiry, physical measurement, etc. To maintain uniformity, it is necessary to give detailed field instructions.

Field Work and Training of Personnel

It is essential that the personnel should be thoroughly trained in locating sample units and the methods of collection of required data, before starting field work. The staff should be trained well, not only in the statistical aspects but also in the art of eliciting correct information from different sources.

Processing of Survey Data

Processing of the collected data in a survey may be broadly classified under the following heads:
  (i) Scrutiny and editing of data
  (ii) Tabulation of data
  (iii) Statistical analysis

It is, therefore, necessary to plan the survey work in such a way that the flow of work-material through various stages of data processing ensures the desired degree of precision in survey results.

Preparation of Reports

The guidelines, as formulated by the United Nations (1949), for preparation of sample-survey reports should be adequate for the purpose. The report may have sections such as objectives; scope; subject coverage; method of data collection; survey reference and reporting periods; sampling design and estimation procedure, tabulation procedure; presentation of results; accuracy; cost structure; responsibility; and references. It may be useful to give a summary of the main results which can be used by the financing agency for policy decisions.

1.11 SAMPLING AND NON-SAMPLING ERRORS

The errors involved in collection, processing and analysis of the data in a survey may be classified as:

    (i) Sampling error, and

    (ii) Non-sampling error

Sampling Error

The error which arises due to only a sample being used to estimate the population parameters is termed sampling error or sampling fluctuation. Whatever may be the degree of cautiousness in selecting a sample, there will always be a difference between the population value (parameter) and its corresponding estimate.

This error is inherent and unavoidable in any and every sampling scheme. A sample with the smallest sampling error will always be considered a good representative of the population. This error can be reduced by increasing the size of the sample. In fact, the decrease in sampling error is inversely proportional to the square root of the sample size and the relationship can be examined graphically as shown in Fig. 1.1. When the sample survey becomes a census survey, the sampling error becomes zero.

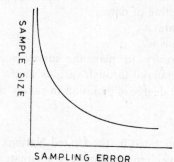

Fig. 1.1. Relationship of sampling error with sample size.

Non-Sampling Error

Besides sampling error, the sample estimate may be subject to other errors which, grouped together, are termed non-sampling errors. The main sources of non-sampling errors are:

    (i) Failure to measure some of the units in the selected sample.

    (ii) Observational errors due to defective measurement technique.

    (iii) Errors introduced in editing, coding and tobulating the results.

    In practice, the census survey results may suffer from non-sampling errors although these may be free from sampling error. The non-sampling error is likely to increase with increase in sample size, while sampling error decreases with increase in sample size.

SET OF PROBLEMS

1.1 Describe the advantages of a sample survey in comparison with a census survey. Write the circumstances under which census surveys are preferred to sample surveys.

1.2 Explain what you understand by probability sampling and non-probability sampling. What are their relative advantages and disadvantages?

1.3 Draw a schedule for enquiry into the state of employment, under-employment in the rural sector of India. Justify your definitions and concepts and elaborate a set of instructions to field workers.

1.4 What are the points which need special attention at the planning stages of sample surveys ? With reference to any suitable example (preferably surveys actually conducted in India), explain whether these have been covered or not.

1.5 Define population, sampling unit and sampling frame for conducting surveys on each of the following subjects. Mention other possible sampling units, if any, in each case and discuss their relative merits.

  (i) Popularity of family planning among families having more than two children.

 (ii) Monthly fish-catch along a given stretch of sea coast.

(iii) Election for a political office with adult franchise.

 (iv) Measurement of the volume of timber available in a forest.

  (v) Annual yield of apple fruit in a hilly district.

 (vi) Labour manpower in the urban area of a state.

(vii) Study of incidence of lung cancer and heart attacks among the rural inhabitants of a country.

(viii) Housing conditions in a rural area.

 (ix) Pre-harvest acreage under a specified crop in a region.

  (x) Study of birth-rate in a district.

 (xi) Study of nutrient contents of food consumed by the residents in a city.

  1.6 You are required to plan a sample survey to study the problem of indebtedness among the rural agricultural population in India. Suggest a suitable survey plan on the following points:

  (i) Sampling unit

 (ii) Sampling frame

(iii) Method of sampling

(iv) Method of data collection

Draft a suitable questionnaire that may be used in this regard.

REFERENCES

Cochran, W.G., *Sampling Technique*, John Wiley & Sons, New York, (1977).

Deming, W.E., *Some Theory of Sampling*, John Wiley & Sons, New York, (1950)

Fisher, R.A., *The Design of Experiments*, Oliver & Boyd, London (1935). "The Sub-Commission on Statistical Sampling of the United Nations," *Bull. Int. Statist. Inst.* **32**, 207-209, (1950).

Hansen, M.H., Hurwitz, W.N. and Jabine, T.B., "The use of imperfect lists for probability sampling at the U.S. Bureau of the Census," *Bull. Int. Statist. Inst.*, **40**, 497-517, (1963).

Lahiri, D.B., "Some thoughts on multi-subject sample survey system," *Contributions to Statistics*, 175-220 (Presented to Professor P.C. Mahalanobis on his 70th Birthday). Statistical Publishing Society, Calcutta, (1963).

Mahalanobis, P.C., "On large-scale sample surveys", *Phil. Trans. R. Soc*, **231**, 329-451, (1944). "Cost and accuracy of results in sampling and complete enumeration", *Bull. Int. Statist. Inst.*, **32**, 210-213, (1950).

Seal, K.C., "Use of outdated frames in large sample surveys," *Bull. Cal. Statist. Assoc.*, **11**, 68-84, (1962).

Singh, D., *Taking Agricultural Censuses*, FAO, Rome (1978).

United Nations, "Recommendations concerning the preparation of reports on sampling surveys." *Statistical Paper Series*, **C**, **1**, New York, revised in 1964, *Statistical Paper Series*, **C**, **1**, rev. 2, (1949).

Yates, F., *Sampling Methods for Censuses and Surveys*, Charles Griffin & Co., London, (1960).

Zarkovich, S.S., *Sampling Methods and Censuses*, Food and Agricultural Organization of the United Nations, Rome, (1961).

2

Basic Methods of Simple Random Sampling

To this end was I born for this cause came I into the world, that I should bear witness unto the truth ... Pilate saith unto him "What is truth?"

St. John

2.1 SIMPLE RANDOM SAMPLING

The simplest and most common method of sampling is simple random sampling in which the sample is drawn unit by unit, with equal probability of selection for each unit at each draw. Therefore, simple random sampling is a method of selecting n units out of a population of size N by giving equal probability to all units, or a sampling procedure in which all possible combinations of n units that may be formed from the population of N units have the same probability of selection. It is also sometimes referred to as *unrestricted random sampling*. If a unit is selected and noted and then returned to the population before the next drawing is made and this procedure repeated n times, it gives rise to a simple random sample of n units. This procedure is generally known as simple random sampling *with replacement* (wr). If this procedure is repeated till n distinct units are selected and all repetitions are ignored, it is called a simple random sampling *without replacement* (wor).

THEOREM 2.1.1 The probability that a specified unit of the population being selected at any given draw is equal to the probability of its being selected at the first draw.

Proof The probability that the specified unit is selected at the rth draw is clearly the product of (a) the probability that the specified unit is not selected in any of the previous $(r - 1)$ draws, and (b) the probability that it is selected at the rth draw with the condition that it is not selected in the previous $(r - 1)$ draws.

The probability under (a) is given by

$$\frac{N-1}{N} \cdot \frac{N-2}{N-1} \cdots \frac{N-r+1}{N-r+2} = \frac{N-r+1}{N}$$

The probability under (b) is given by $1/(N - r + 1)$. Hence the required probability is $1/N$ which is independent of the term r, i.e. draw number.

THEOREM 2.1.2 The probability of a specified unit being included in the sample is equal to n/N.

Proof Let n denote the sample size. Since the specified unit may be included in the sample at any of the n draws, the probability that a specified unit is included in the sample is the sum of the probabilities of n mutually exclusive events, viz. it is included in the sample at the first draw, second draw, . . ., nth draw. As shown in Theorem 2.1.1 the probability of each case is $1/N$. Thus the required probability is n/N.

COROLLARY 1 The probability of a specified sample of n units, ignoring order, is $1/\binom{N}{n}$.

COROLLARY 2 If in the population of N units, m units are deleted and m' are added, show that the probability of selection of any unit at a specified draw is $(N - m + m')^{-1}$.

2.2 PROCEDURES OF SELECTING A RANDOM SAMPLE

Since the theory of sampling is based on the assumption of random sampling, the technique of random sampling is of basic significance. Some of the procedures used for selecting a random sample are as follows:

(i) Lottery Method
(ii) Use of Random Number Tables

2.2.1 Lottery Method

In practice, a ticket/chit may be associated with each unit of the population. Thus, each sampling unit has its identification mark from 1 to *N*. The procedure of selecting an individual is simple. All the tickets/ chits are placed in a container, drum or metallic spherical device, in which a thorough mixing or reshuffling is possible, before each draw. Draws of tickets/chits may be continued until a sample of the required size is obtained.

This procedure of numbering units on tickets/chits and selecting one after reshuffling becomes cumbersome when the population size is large. It may be rather difficult to achieve a thorough shuffling in practice. Human bias and prejudice may also creep in this method.

2.2.2 Use of Random Number Tables

A random number table is an arrangement of digits 0 to 9, in either a linear or rectangular pattern, where each position is filled with one of these digits. A table of random numbers is so constructed that all numbers 0, 1, 2, . . . , 9 appear independent of each other. Some random number tables in common use are:

(i) Tippett's randon number tables
(ii) Fisher and Yates tables
(iii) Kendall and Smith tables
(iv) A million random digits

To ascertain whether these series of random numbers are really random, the following tests may be applied:

(i) Frequency test
(ii) Serial test
(iii) Gap test
(iv) Poker test

A practical method of selecting a random sample is to choose units one-by-one with the help of a table of random numbers. By considering two-digit numbers, we can obtain numbers from 00 to 99, all having the same frequency. Similarly, three or more digit numbers may be obtained by combining three or more rows or columns of these tables.

The simplest way of selecting a sample of the required size is by selecting a random number from 1 to N and then taking the unit bearing that number. This procedure involves a number of rejections since all numbers greater than N appearing in the table are not considered for selection. The use of random numbers is, therefore, modified and some of these modified procedures are:

  (i) Remainder approach
  (ii) Quotient approach
  (iii) Independent choice of digits

Remainder Approach

Let N be an r-digit number and let its r-digit highest multiple be N'. A random number k is chosen from 1 to N' and the unit with the serial number equal to the remainder obtained on dividing k by N is selected. If the remainder is zero, the last unit is selected. As an illustration, let $N = 123$, the highest three-digit multiple of 123 is 984. For selecting a unit, one random number from 001 to 984 has to be selected. Let the random number selected be 287. Dividing 287 by 123, the remainder is 41. Hence, the unit with serial number 41 is selected in the sample.

Quotient Approach

Let N be an r-digit number and let its r-digit highest multiple be N' such that $N'/N = q$. A random number k is chosen from 0 to $(N'-1)$. Dividing k by q the quotient r is obtained and the unit bearing the serial number $(r-1)$ is selected in the sample. As an illustration, let $N = 16$ and hence $N' = 96$ and $q = 96/16 = 6$. Let the two-digit random number chosen be 65 which lies between 0 and 95. Dividing 65 by 6, the quotient is 10 and hence the unit bearing number $(10-1) = 9$ is selected in the sample.

Independent Choice of Digits

This method, suggested by Mathai (1954), consists of the selection of two random numbers which are combined to form one random number. One random number is chosen according to the first digit and other according to the remaining digits of the population size. If the number chosen is 0 the last unit is chosen. But if the number made up is greater than or equal to N, the number is rejected and the operation srepeated.

EXAMPLE 2.1 Select a random sample of 11 households from a list of 112 households in a village.

(i) By using the 3-digit random numbers given in columns 1 to 3, 4 to 6 and so on of the random number table and rejecting numbers greater than 112 (also the number 000), we have for the sample bearing serial numbers 033, 051, 052, 099, 102, 081, 092, 013, 017, 076, and 079.

(ii) In the above procedure, a large number of random numbers is rejected. Hence, a commonly used device, i.e. remainder approach, is employed to avoid the rejection of such large numbers. The greatest three-digit multiple of 112 is 896. By using three-digit random numbers as above, the sample will comprise of households with serial numbers 086, 033, 049, 097, 051, 052, 066, 107, 015, 106 and 020.

(iii) In case the quotient approach is applied, the 3-digit multiple of 112 is 896 and $896/112 = 8$. Using the same random numbers and dividing them by 8, we have the sample of households with list numbers 025, 004, 020, 026, 006, 006, 092, 041, 085, 027 and 086 with the replacement method and with list numbers 025, 004, 020, 026, 006, 092, 041, 085, 027, 036 and 042 without the replacement method.

EXAMPLE 2.2 Ten orchards in a locality near Lucknow were having 125, 793, 970, 830, 1502, 864., 503, 106, 970, 312 fruit trees, respectively. Draw a random sample of 10 fruit trees by using random numbers.

Let us assume that in the first orchard the fruit trees bear serial numbers 1 to 125, in the second orchard from 126 to 918, and so on. Hence, the cumulative serial numbers may be written as, 125, 918, 1888, 2718, 4220, 5084, 5587, 5693, 6663, 6975. By using 4-digit random numbers of the above-said table with similar notions, we have 1983, 0330, 1614, 2096, 0511, 0524, 3311, 6874, 2183 and 6926.

With the first random number 1983, we select the fruit tree bearing serial number 95 in the 4th orchard. Similarly, with the second random number 0330, we select the tree bearing serial number 205 in the 2nd orchard, and so on. Working out this procedure further, we find that orchards included in the sample have serial numbers 2, 3, 4, 5 and 10. Thus, it is not necessary to give serial numbers to fruit trees in all the orchards, only five orchards of the above-shown numbers sufficing for the purpose of selecting a random sample of 10 fruit trees.

2.3 ESTIMATION OF POPULATION PARAMETERS

Let us assume that each unit u_i in the population is associated with a variate value Y_i for the character y. For parameters, let us designate

The population total, $Y = \sum_i^N y_i$

The population mean, $\bar{Y} = \sum_i^N Y_i / N$

The population variance, $\sigma^2 = \sum_i^N (Y_i - \bar{Y})^2 / N$

Let the n units in the sample be $u_1, u_2, \ldots, u_n$, with variate values $y_1, y_2, \ldots, y_n$, respectively. The estimators of population total and mean are given by

$$\hat{Y} = \frac{N}{n} \sum_i^n y_i = N \bar{y}$$

and

$$\hat{\bar{Y}} = \sum_i^n y_i / n = \bar{y}$$

The factor N/n by which the sample total is multiplied is sometimes called the *expansion* or *raising* or *inflation factor*. Its inverse n/N is called the *sampling fraction* and is denoted by the letter f in this text.

THEOREM 2.3.1 In simple random sampling, wor, the sample mean $\bar{y}$ is an unbiased estimator of $\bar{Y}$ and its sampling variance is given by

$$V(\bar{y}) = (1 - n/N) S^2/n = (1 - f) S^2/n \qquad (2.3.1)$$

where

$$S^2 = N \sigma^2 / (N - 1)$$

Proof We have

$$E(\bar{y}) = E \left(\sum_i^n y_i/n \right)$$

$$= \sum_i^n E(y_i)/n$$

By definition,

$$E(y_i) = \sum_i^N p_i Y_i = \sum_i^N \frac{Y_i}{N} = \bar{Y}$$

Hence,

$$E(\bar{y}) = \bar{Y}$$

Thus the sample mean is an unbiased estimator of the population mean.

The variance of $\bar{y}$ is given by

$$V(\bar{y}) = E(\bar{y} - \bar{Y})^2 = E\{\sum_i^n (y_i - \bar{Y})\}^2/n^2$$

$$= \frac{1}{n^2} E\{\sum_i^n (y_i - \bar{Y})^2 + \sum_{i \neq j}^n (y_i - \bar{Y})(y_j - \bar{Y})\}$$

$$= \frac{1}{n^2} \{\sum_i^n E(y_i - \bar{Y})^2 + \sum_{i \neq j}^n E(y_i - \bar{Y})(y_j - \bar{Y})\}$$

Since

$$E(y_i - \bar{Y})^2 = \sum_i^N (y_i - \bar{Y})^2/N = \sigma^2$$

and

$$\underset{i \neq j}{E}(y_i - \bar{Y})(y_j - \bar{Y}) = \sum_{i \neq j}(y_i - \bar{Y})(y_j - \bar{Y})/N(N-1)$$

$$= \frac{1}{N(N-1)} [\{\sum_i^N (y_i - \bar{Y})\}^2 - \sum_i^N (y_i - \bar{Y})^2]$$

$$= -\sigma^2/(N-1)$$

Thus, $V(\bar{y}) = \frac{1}{n^2} \{n \sigma^2 - n(n-1) \sigma^2/(N-1)\}$

$$= \frac{N-n}{N-1} \frac{\sigma^2}{n} = \frac{N-n}{N} \frac{S^2}{n} = (1-f) \frac{S^2}{n}$$

which proves the theorem.

COROLLARY 1   In simple random sampling, wor, the standard error of $\bar{y}$ is given by

$$\sigma_{\bar{y}} = S[(N-n)/nN]^{1/2} = S[(1-f)/n]^{1/2} \qquad (2.3.2)$$

COROLLARY 2   In simple random sampling, wor, the variance of $\hat{Y} = N\bar{y}$, as an estimator of the population total Y, is obtained as

$$V(\hat{Y}) = E(\hat{Y} - Y)^2 = N^2 S^2 (N-n)/nN$$

$$= (1-f) N^2 S^2/n \qquad (2.3.3)$$

COROLLARY 3   In simple random sampling, wor, the standard error of $\hat{Y}$ is given by

$$\sigma_{\hat{Y}} = NS \left[(N - n)/nN\right]^{1/2} = NS[(1 - f)/n]^{1/2} \qquad (2.3.4)$$

If sampling is with replacement, a random sample of size n gives sample mean $\bar{y}$ as an unbiased estimator. From the above results, it can be seen easily that terms $(N - n)/N$ for the variance and $\left[\dfrac{N-n}{N}\right]^{1/2}$ for the standard error are introduced due to finiteness of the population and these are called *finite population corrections* (fpc). In case of randon sampling, wr, the variance and standard error can be obtained by ignoring finite population correction.

COROLLARY 4   In random sampling, wr, the variance and standard error of $\bar{y}$ can be written as

$$\left.\begin{array}{r} V(\bar{y}) = \sigma^2/n \\ \sigma_{\bar{y}_i} = \sigma/\sqrt{n} \end{array}\right\} \qquad (2.3.5a)$$

If n is small as compared to N, the fpc will not differ much from unity and the sampling variance of the mean will be nearly equal to that of the mean of a sample drawn from an infinite population.

COROLLARY 5   In random sampling, wr, the variance and standard error of $\hat{Y} = N\bar{y}$ can be written as

$$\left.\begin{array}{r} V(\hat{Y}) = N^2 \sigma^2/n \\ \sigma_{\hat{Y}} = N\sigma/\sqrt{n} \end{array}\right\} \qquad (2.3.5b)$$

THEOREM 2.3.2   In simple random sampling,

$$s^2 = \sum_{i}^{n} (y_i - \bar{y})^2/(n - 1) \qquad (2.3.6)$$

is an unbiased estimator of $S^2 = N\sigma^2/(N - 1)$

Proof   We may write

$$s^2 = \sum_{i}^{n}[(y_i - \bar{Y}) - (\bar{y} - \bar{Y})]^2/(n - 1)$$

$$= [\sum_{i}^{n} (y_i - \bar{Y})^2 - n(\bar{y} - \bar{Y})^2]/(n - 1)$$

Therefore,

$$E(s^2) = E\left[\sum_i^n (y_i - \overline{Y})^2 - n(\bar{y} - \overline{Y})^2\right]/(n - 1)$$

$$= \left[\sum_i^n E(y_i - \overline{Y})^2 - nE(\bar{y} - \overline{Y})^2\right]/(n - 1)$$

$$= [n(N - 1) S^2/N - n (N - n) S^2/Nn]/(n - 1)$$

Hence s^2 is an unbiased estimator of S^2.

COROLLARY 1 An unbiased estimator of the variance of $\bar{y}$ in random sampling, wor, is given by

$$v(\bar{y}) = (N - n)s^2/Nn = (1 - f)s^2/n \qquad (2.3.7)$$

For an estimate of the standard error, we take

$$s_{\bar{y}} = s [(1 - f)/n]^{1/2}$$

COROLLARY 2   An unbiased estimator of the variance of $\hat{Y} = N\bar{y}$ in random sampling, wor, is given by

$$v(\hat{Y}) = N(N - n) s^2/n = (1 - f) N^2 s^2/n \qquad (2.3.8)$$

For an estimate of the standard error, we take

$$s_{\hat{Y}} = Ns [(1 - f)/n]^{1/2}$$

COROLLARY 3   An unbiased estimator of the variance of $\bar{y}$ in random sampling, wr, is given by

$$v(\bar{y}) = s^2/n \qquad (2.3.9)$$

For an estimate of the standard error, we take

$$s_{\bar{y}} = s/\sqrt{n}$$

COROLLARY 4   An unbiased estimator of the variance $\hat{Y} = N\bar{y}$ in random sampling, wr, is given by

$$v(\hat{Y}) = N^2 s^2/n \qquad (2.3.10)$$

For an estimate of the standard error, we take

$$s_{\hat{Y}} = Ns/\sqrt{n}$$

EXAMPLE 2.3 A random sample of $n = 2$ households was drawn from a small colony of 5 households (hypothetical population) having monthly income (in rupees) as follows:

Household ...	1	2	3	4	5
Income (in rupees)...	156	149	166	164	155

(i) Calculate population mean $\bar{Y}$, variance (σ^2) and mean square error (S^2).

(ii) Enumerate all possible samples of size 2 by the replacement method and show that

(a) the sample mean gives an unbiased estimate of the population mean and find its sampling variance;

(b) sample variance (s^2) is an unbiased estimate of the population variance σ^2; and

(c) $v(\bar{y}) = \dfrac{(y_1 - y_2)^2}{4}$ is an unbiased estimator of $V(\bar{y})$, i.e.

$$Ev(\bar{y}) = V(\bar{y}), = \sigma^2/2.$$

(iii) Enumerate all possible samples of size 2 by the without replacement method and show that

(a) the sample mean gives an unbiased estimate of the population mean and find its sampling variance;

(b) the sample variance (s^2) is an unbiased estimate of the population variance S^2; and

(c) $v(\bar{y}) = 3 (y_1 - y_2)^2/20$ is an unbiased estimator of $V(\bar{y})$, i.e.

$$Ev(\bar{y}) = V(\bar{y}) = \left(\frac{1}{2} - \frac{1}{5} \right) S^2 = \frac{3}{10} S^2.$$

(i) The population mean of 5 households

$$\bar{Y} = (156 + 149 + 166 + 164 + 155)/5$$
$$= \text{Rs } 158.00$$

The population variance

$$\sigma^2 = [(156^2 + 149^2 + \ldots + 155^2) - 5 \times 158^2]/5 = 38.80$$

Hence $S^2 = N\sigma^2/(N - 1) = 5 \times 38.8/4 = 48.50$

(ii) It can be seen that the total number of possible samples is 25 $(= 5^2)$. Also, each of the 25 possible samples has the same probability $(1/25)$ of being selected.

Table 2.3.1 All samples of 2 units from 5 units in simple random sampling with replacement

Sample no. (1)	Units in the sample (2)	Probability (3)	Sample observations y_1 (4)	Sample observations y_2 (5)	Sample mean $(\bar{y})$ (6)	Sampling error $(\bar{y}-\overline{Y})$ (7)	Sampling variance $\dfrac{(y_1-y_2)^2}{4}$ (8)
1	1, 1	1/25	156	156	156.0	−2.0	0
2	1, 2	1/25	156	149	152.5	−5.5	49/4
3	1, 3	1/25	156	166	161.0	3.0	100/4
4	1, 4	1/25	156	164	160.0	2.0	64/4
5	1, 5	1/25	156	155	155.5	−2.5	1/4
6	2, 1	1/25	149	156	152.5	−5.5	49/4
7	2, 2	1/25	149	149	149.0	−9.0	0
8	2, 3	1/25	149	166	157.5	−0.5	289/4
9	2, 4	1/25	149	164	156.5	−1.5	225/4
10	2, 5	1/25	149	155	152.0	−6.0	36/4
11	3, 1	1/25	166	156	161.0	3.0	100/4
12	3, 2	1/25	166	149	157.5	−0.5	289/4
13	3, 3	1/25	166	166	166.0	8.0	0
14	3, 4	1/25	166	164	165.0	7.0	4/4
15	3, 5	1/25	166	155	160.5	2.5	121/4
16	4, 1	1/25	164	156	160.0	2.0	64/4
17	4, 2	1/25	164	149	156.0	−1.5	225/4
18	4, 3	1/25	164	166	165.0	7.0	4/4
19	4, 4	1/25	164	164	164.0	6.0	0
20	4, 5	1/25	164	155	159.5	1.5	81/4
21	5, 1	1/25	155	156	155.5	−2.5	1/4
22	5, 2	1/25	155	149	152.0	−6.0	36/4
23	5, 3	1/25	155	166	160.5	2.5	121/4
24	5, 4	1/25	155	164	159.5	1.5	81/4
25	5, 5	1/25	155	155	155.0	−3.0	0
Average					158.0		19.40

(a) The expected value of $\bar{y}$ is given by the average value of column (6) which works out to be the population mean 158.0, thus verifying that the estimator is unbiased.

We find from columns (6) and (7) of Table 2.3.1 that these estimates differ, in general, from $\bar{Y}$ ($= 158.0$), and that the error usually varies from sample to sample.

Further, it is of interest to note from column (7) that each sample mean ($\bar{y}$) given by column (6) is an unbiased estimate of the population mean ($\bar{Y}$) as the average of column (7) is zero, which proves the results. The sampling variance is the mean of the squares of error column (7) which works out to 19.40 ($= 38.80/2$).

(b) The sample variance (s^2) is given by $(y_1 - y_2)^2/2$ which is twice the value given in column (8). Hence, $2 \times$ (8) shows that $E(s^2)$ is equal to 38.80 (σ^2), which shows that s^2 is an unbiased estimator for the population variance in simple random sampling, wr.

(c) An estimator of $V(\bar{y})$ is given by

$$v(\bar{y}) = (y_1 - y_2)^2/4$$

the values which are given in column (8) of Table 2.3.1 for possible samples. The expected value of $v(\bar{y})$ is the average of the values in column (8), showing that

$$E[v(\bar{y})] = V(\bar{y}) = \sigma^2/2$$

Thus the estimate $v(\bar{y})$ is unbiased.

Table 2.3.2 All samples of 2 units from 5 units in simple random sampling without replacement

Sample no.	Units in sample	Probability	Sample observations		Sample mean	Error	Sampling variance
			y_1	y_2	$(\bar{y})$	$(\bar{y}-\bar{Y})$	$\frac{3}{5} \times (y_1-y_2)^2/4$
(1)	(2)	(3)	(4)	(5)	(6)	(7)	(8)
1	1, 2	1/10	156	149	152.5	-5.5	7.35
2	1, 3	1/10	156	166	161.0	3.0	15.00
3	1, 4	1/10	156	164	160.0	2.0	9.60
4	1, 5	1/10	156	155	155.5	-2.5	0.15
5	2, 3	1/10	149	166	157.5	-0.5	43.35
6	2, 4	1/10	149	164	156.5	-1.5	33.75
7	2, 5	1/10	149	155	152.0	-6.0	5.40
8	3, 4	1/10	166	164	165.0	7.0	0.60
9	3, 5	1/10	166	155	160.5	2.5	18.15
10	4, 5	1/10	164	155	159.5	1.5	12.15
Average					158.0		14.55

(iii) In case of random sampling, wor, the number of possible samples is 10 $[= \binom{5}{2}]$. It can be seen that each of the 10 possible samples has the same probability (1/10) of being selected.

(a) The expected value of $\bar{y}$, which is given by the average of column (6) of Table 2.3.2, works out to be the population mean 158.0, thus verifying that $\bar{y}$ is an unbiased estimator of $\bar{Y}$. Further, the sampling variance which is obtained by averaging the squares of errors given in column (7) works out to be 14.55, verifying that $V(\bar{y}) = 3\sigma^2/8 \ (= 3S^2/10)$.

(b) Since the sample variance (s^2) is given by $(y_1 - y_2)^2/2$, which can be obtained easily, the average of 10 sample mean squares gives $E(s)^2$. Here, $$E(s^2) = 485/10 = 48.5$$
Also, the population mean square = 48.5
Thus, the sample mean square provides an unbiased estimator of the population mean square (S^2), verifying that
$$E(s^2) = S^2$$

(c) An estimator of $V(\bar{y})$ is given by
$$v(\bar{y}) = 3(y_1 - y_2)^2/20$$
the values which are given in column (8) of Table 2.3.2 and may be used here. The expected value of $v(\bar{y})$, which is the average of the values in the column (8), is 14.55, showing that the estimator is unbiased, i.e.
$$Ev(\bar{y}) = V(\bar{y}) = \frac{3S^2}{10}$$

2.4 ESTIMATION OF POPULATION PROPORTION

Sometimes, the units in the population are classified in two groups (i) having a particular characteristic and (ii) not having that characteristic. For example, a crop field may be irrigated or not irrigated. If it is irrigated, we say that it possesses the chararacteristic, 'irrigation'. If it is not irrigated, we say that it does not possess the particular characteristic of irrigation. If we are interested in estimating the proportion of irrigated fields, the population of N fields can be defined with variate y_i as having value 1 if the field is irrigated, otherwise zero. If the total number of irrigated fields be N_1 out of N

$$\sum_i^N y_i = N_1$$

Thus,

$$\overline{Y} = \frac{1}{N} \sum_i^N y_i = \frac{N_1}{N} = P = \text{proportion of irrigated fields}$$

and

$$\sum_i^N y_i^2 = N_1 = NP$$

Thus, the problem of estimating a population proportion becomes that of estimating a population mean by defining the variate as above. If n_1 units out of a random sample of size n possess that characteristic, the sample proportion is given by $p = n_1/n$. Thus

$$\sum_i^n y_i = n_1 = \sum_i^n y_i^2 = np$$

Hence, an unbiased estimator of P is given by

$$\hat{P} = (n_1/n) = p \tag{2.4.1}$$

THEOREM 2.4.1 In sampling, wor, the variance of p is given by

$$V(p) = (N-n)\, PQ/n\, (N-1) = (1-f)\, NPQ/n\, (N-1) \tag{2.4.2}$$

where $Q = 1 - P$. The proof is obvious.

COROLLARY 1 In sampling, wr, the variance of p is given by

$$V(p) = PQ/n \tag{2.4.3}$$

COROLLARY 2 The variance of $\hat{N_1} = Np$, the estimated total number of units with some desired characteristic, is given by

$$V(\hat{N_1}) = N^2\, (N-n)\, PQ/n\, (N-1) \tag{2.4.4}$$

THEOREM 2.4.2 In sampling, wor, an unbiased estimator of $V(p)$ is given by

$$v(p) = (1-f)\, pq/(n-1) \tag{2.4.5}$$

The proof is obvious.

COROLLARY 1 In sampling, wr, an unbiased estimator of $V(p)$ is given by

$$v(p) = pq/(n-1) \tag{2.4.6}$$

COROLLARY 2 An unbiased estimate of the variance of $\hat{N_1} = Np$ is given by

$$v(\hat{N_1}) = N(N-n)pq/(n-1) = (1-f)\, N^2\, pq/(n-1) \tag{2.4.7}$$

COROLLARY 3   The coefficient of variation of p is given by

$$CV = [PQ/n]^{1/2}/P = [Q/nP]^{1/2} \qquad (2.4.8)$$

EXAMPLE 2.4   A list of 3000 voters of a ward in a city was examined for measuring the accuracy of age of individuals. A random sample of 300 names was taken, which revealed that 51 citizens were shown with wrong ages. Estimate the total number of voters having a wrong description of age in the list and estimate the standard error. Here,

$$N = 3000, \; n{=}300, \; n_1 = 51, \; p = 0.17$$

The estimate of the total number of voters having a wrong description of age is obtained by

$$\hat{N}_1 = Np = (3000)(0.17) = 510$$

(i) If sampling, wr, is considered, the estimate of the standard error is given by

$$s_{\hat{N}_1} = N [pq/(n-1)]^{1/2}$$
$$= 3000 [(0.17)(0.83)/50]^{1/2}$$
$$= 159.3$$

(ii) If sampling, wor, is considered, the estimated standard error is given by

$$s_{\hat{N}_1} = N [(1-f) pq/(n-1)]^{1/2}$$
$$= 3000[(1-0.10)(0.17)(0.83)/50]^{1/2}$$
$$= 151.1$$

2.5   COMBINATION OF UNBIASED ESTIMATORS

There are situations where the estimates based on several samples will have to be pooled to get a combined estimate. If $t_i \; (i = 1, 2, \ldots, m)$ are unbiased estimators of a parameter θ, which are mutually independent, then the pooled estimator

$$\bar{t} = \sum_{i}^{m} t_i/m \qquad (2.5.1)$$

is also an unbiased estimator of θ. The variance of $\bar{t}$ is given by

$$V(\bar{t}) = \sum_{i}^{m} V(t_i)/m^2 \qquad (2.5.2)$$

and estimate of variance

$$v(\bar{t}) = \sum_i^m (t_i - \bar{t})^2 / m(m-1) \qquad (2.5.3)$$

We shall consider the problem for the following cases:
(i) Simple random sampling for variables
(ii) Simple random sampling for attributes

2.5.1 Simple Random Sampling for Variables

Let us suppose that $\bar{y}_1, \bar{y}_2, \ldots, \bar{y}_m$ are the sample means, each of which is drawn independently, with sizes $n_1, n_2, \ldots, n_m$. A pooled estimator of all samples is given by taking

(i) the arithmetic mean of m estimates

$$\bar{y}' = \sum_i^m \bar{y}_i / m \qquad (2.5.4)$$

(ii) the weighted mean of m estimates

$$\bar{y}'' = \sum_i^m n_i \bar{y}_i / n \qquad (2.5.5)$$

where

$$n = \sum_i^m n_i$$

When Sampling is With Replacement Method

The sampling variances of $\bar{y}'$ and $\bar{y}''$ are given by

$$V(\bar{y}') = \sum_i^m V(\bar{y}_i)/m^2 = \frac{\sigma^2}{m^2} \sum_i^m 1/n_i \qquad (2.5.6)$$

and

$$V(\bar{y}'') = \sum_i^m n_i^2 \, V(\bar{y}_i)/n^2 = \frac{\sigma^2}{n} \qquad (2.5.7)$$

Unbiased estimators of $V(\bar{y}')$ and $V(\bar{y}'')$ are provided by

$$v(\bar{y}') = \sum_i^m (\bar{y}_i - \bar{y}')^2 / m \, (m-1) \qquad (2.5.8)$$

and

$$v(\bar{y}'') = \sum_i^m \sum_j^{n_i} (y_{ij} - \bar{y}'')^2 / n \, (n-1) \qquad (2.5.9)$$

It should be noted that the estimate given in relation (2.5.5) is more efficient than that given in relation (2.5.4) and can be verified by comparing the variances in both the cases.

When Sampling is Without Replacement Method

The sampling variances of $\bar{y}'$ and $\bar{y}''$ are given by

$$V(\bar{y}') = \sum_i^m V(\bar{y}_i)/m^2 = \sum_i^m (1-f_i) S_i^2/m^2 n_i \qquad (2.5.10)$$

and

$$V(\bar{y}'') = \frac{S^2}{n} \left\{ 1 - \frac{1}{Nn} \sum_i^m n_i^2 \right\} \qquad (2.5.11)$$

Unbiased estimators of $V(\bar{y}')$ and $V(\bar{y}'')$ are provided by

$$v(\bar{y}') = \sum_i^m (1-f_i) s_i^2/m^2 n_i \qquad (2.5.12)$$

and

$$v(\bar{y}'') = \sum_i^m n_i (1-f_i) s_i^2/n^2 \qquad (2.5.13)$$

2.5.2 Simple Random Sampling for Attributes

The results obtained in the previous section can be applied to simple random sampling for attributes also. If $p_1, p_2, \ldots, p_m$ are the sample proportions based on m samples of size $n_1, n_2, \ldots, n_m$ each of which is drawn independently. A pooled estimator of all samples is given by taking

(i) the arithmetic mean of m estimates

$$\bar{p}' = \sum_i^m p_i/m \qquad (2.5.14)$$

(ii) The weighted mean of m estimates

$$\bar{p}'' = \sum_i^m n_i p_i/n \qquad (2.5.15)$$

where

$$n = \sum_i^m n_i$$

When Sampling is With Replacement Method

The sampling variances of $\bar{p}'$ and $\bar{p}''$ are given by

$$V(\bar{p}') = \sum_i^m V(p_i)/m^2 = \sum_i^m P(1 - P)/m^2 n_i \qquad (2.5.16)$$

and

$$V(\bar{p}'') = \sum_i^m n_i^2 \ V(p_i)/n^2 = P(1 - P)/n \qquad (2.5.17)$$

Unbiased estimators of $V(\bar{p}')$ and $V(\bar{p}'')$ are provided by

$$v(\bar{p}') = \sum_i^m (p_i - \bar{p}')^2/m \ (m - 1) \qquad (2.5.18)$$

and

$$v(\bar{p}'') = \sum_i^m n_i p_i \ (1 - p_i)/n^2 \ (n_i - 1) \qquad (2.5.19)$$

Another estimator is given by

$$v(\bar{p}'') = \bar{p} \ (1 - \bar{p})/n \qquad (2.5.20)$$

When Sampling is Without Replacement Method

The sampling variances of $\bar{p}'$ and $\bar{p}''$ are given by

$$V(\bar{p}') = \sum_i^m (N - n_i) \ P(1 - P)/n_i \ (N - 1)m^2 \qquad (2.5.21)$$

and

$$V(\bar{p}'') = \sum_i^m (N - n_i) \ P(1 - P)/n_i \ (N - 1) \qquad (2.5.22)$$

Unbiased estimators of $V(\bar{p}')$ and $V(\bar{p}'')$ are provided by

$$v(\bar{p}') = \sum_i^m (p_i - \bar{p}')^2/m \ (m - 1) \qquad (2.5.23)$$

and

$$v(\bar{p}'') = \sum_i^m (N - n_i) \ n_i^2 \ p_i \ (1 - p_i)/n^2 \ N \ (n_i - 1) \qquad (2.5.24)$$

2.6 CONFIDENCE LIMITS

After having the estimate of an unknown parameter, it becomes necessary to measure the reliability of these estimates and to construct some confidence limits with a given degree of confidence. If we assume that the estimator $\bar{y}$ is normally distributed about the population mean $\bar{Y}$, lower and upper confidence limits for the population mean $\bar{Y}$ are given by

$$\hat{\bar{Y}}_L = \bar{y} - t_{(\alpha,\, n-1)}\, s\, [(1 - f)/n]^{1/2} \tag{2.6.1}$$

and

$$\hat{\bar{Y}}_U = \bar{y} + t_{(\alpha,\, n-1)}\, s\, [(1 - f)/n]^{1/2} \tag{2.6.2}$$

where $t_{(\alpha,\, n-1)}$ stands for the value of student's t with $(n - 1)$ degrees of freedom at α level of significance. Similarly, the confidence limits for the population total may be written as

$$\hat{Y}_L = N\bar{y} - t_{(\alpha,\, n-1)}\, Ns\, [(1 - f)/n]^{1/2} \tag{2.6.3}$$

and

$$\hat{Y}_U = N\bar{y} + t_{(\alpha,\, n-1)}\, Ns\, [1 - f)/n]^{1/2} \tag{2.6.4}$$

EXAMPLE 2.5 Signatures to a petition were collected on 700 sheets. Each sheet was provided with space for 50 signatures, but the signatories put their signatures in erratic ways and the number of signatures per sheet was not definite. 12 sheets were spoiled in transit. Of the available sheets, a random sample of 50 was drawn and the numbers of signatures per sheet counted, which are shown as below:

Number of signatures (y_i) ..	52	51	46	42	40	37	32	29	27	15	14	10	8
Number of sheets (n_i) ..	1	2	21	8	7	2	2	1	1	2	1	1	1

Estimate the total number of signatures to the petition and calculate 95% confidence limits.

We have

$$N = 700 - 12 = 688, \; n = \sum n_i = 50$$

$$\sum n_i y_i = 1992, \; \sum n_i y_i^2 = 84820$$

Therefore, the estimate of total signatures is given by

$$\hat{Y} = N\bar{y} = (688)\,(1992)/50 = 27{,}410$$

and

$$s^2 = \sum n_i \, (y_i - \bar{y})^2/(n - 1)$$

$$= \frac{1}{49} \, [84820 - (1992)^2/50]$$

$$s = 10.55$$

Hence, the 95% confidence limits are

$$\hat{Y}_L = 27410 - 1.96 \times 688 \times 10.55 = 13,184$$

$$\hat{Y}_U = 27410 + 1.96 \times 688 \times 10.55 = 41,636$$

2.7 ESTIMATION OF SAMPLE SIZE

In planning a sample survey for estimating the population parameters, the important question is how to determine the size of the sample to be drawn. It can be done by specifying the degree of risk (or precision) in terms of permissible loss and the level of confidence. Before we discuss the problem for different sampling methods, let us approach a generalized solution to the problem of estimation of sample size.

Let z be the amount of error by taking the estimate and let $l\,(z)$ be the loss incurred by taking it. For a given sampling method, the theory will provide us the density function. Thus, the expected value of the loss for a given sample size is obtained by

$$L(n) = E\,[l\,(z)] \qquad (2.7.1)$$

Let us also consider the cost function for a sample of size n, denoted by

$$C\,(n) = a + cn \qquad (2.7.2)$$

where a is the over-head cost, and c is the cost per unit in the sampling method.

By combining relations (2.7.1) and (2.7.2), we get the total loss which is given by

$$\phi\,(n) = L\,(n) + \lambda C\,(n) \qquad (2.7.3)$$

where λ is some constant quantity.

Since the purpose in taking the sample is to minimize the total loss, n should be so chosen that relation (2.7.3) is minimized. By differentiating $\phi\,(n)$ with respect to n and equating $\partial\phi/\partial n = 0$, the optimum value of n can be determined.

EXAMPLE 2.6 If the loss function due to an error in $\bar{y}$ is proportional to $|\bar{y} - \overline{Y}|$ and if the total cost of the survey is $C = a + cn$, show that with simple random sampling, ignoring fpc, the optimum value of n is

$$(k\sigma/c\sqrt{2\pi})^{2/3}, \text{ where } k \text{ is a constant.}$$

Here

$$l(z) \propto |\bar{y} - \overline{Y}|$$

or

$$l(z) = k^1 |\bar{y} - \overline{Y}|$$

where k^1 is some constant.

Thus, it follows that

$$L(n) = k^1 E |\bar{y} - \overline{Y}| = k^1 \sqrt{\frac{2}{\pi}} \, \sigma/\sqrt{n}$$

[It is assumed that $\bar{y}$ is distributed $\mathcal{N}(\overline{Y}, \sigma/\sqrt{n})$]

Hence

$$\phi(n) = a + cn + \lambda k^1 \sqrt{\frac{2}{\pi}} \cdot \sigma/\sqrt{n}$$

$$= a + cn + k \sqrt{\frac{2}{\pi}} \cdot \sigma/\sqrt{n}$$

By differentiation and equating $\partial\phi/\partial n = 0$, we have

$$n = (k \, \sigma/c\sqrt{2\pi})^{2/3}$$

A similar treatment and analysis can be applied to any sampling method in which the loss function is inversely proportional to n and the cost function is also a function of n. A generalized discussion is given by Yates (1960), and Raiffa and Schlaifer (1961). Chaudhary (1977) and Chaudhary and Singh (1979) have discussed a line for sequential methods. For a classified discussion, a good account of methods is given by Nordin (1944), Blythe (1945), Deming (1950) and Tippett (1950).

Now let us elaborate the results for the simple random sampling for a characteristic which can be measured quantitatively. Let the marginal error permissible in the estimate be ϵ, and $(1 - \alpha)$ be the level of confidence. The sample mean $\bar{y}$ is assumed to be normally distributed with mean $\overline{Y}$ and variance

$$V(\bar{y}) \left(= \frac{(N - n)}{N} \frac{S^2}{n} \right).$$

Hence,

$$\epsilon = t_{(\alpha,\infty)} \left(\frac{N - n}{N} \cdot \frac{S^2}{n} \right)^{1/2} \tag{2.7.4}$$

where $t_{(\alpha, \infty)}$ is the value of the normal variate corresponding to given $(1 - \alpha)$, which gives

$$n = (S^2 t^2 / \epsilon^2) / (1 + t^2 S^2 / N\epsilon^2) \qquad (2.7.5)$$

(i) if fpc is ignored, we have

$$n_0 = \frac{t^2 S^2}{\epsilon^2} \qquad (2.7.6)$$

(ii) if fpc is not ignored, we can get the value of n by putting the value of n_0 in Eq. (2.7.5) and we have,

$$n_1 = n_0 / (1 + n_0 / N) \qquad (2.7.7)$$

EXAMPLE 2.7 A study of sampling methods was conducted in a population having 500 sampling units. By total count it was obtained that $\overline{Y} = 49$ and $S^2 = 44.6$. In a simple random sampling, how many sampling units should be chosen to estimate $\overline{Y}$ with a permissible marginal error of 10 per cent and 95 per cent confidence coefficient, when sampling is done (i) with replacement method (ii) without replacement method?

(i) *Sampling with replacement* In this case, we can ignore fpc and we have

$$n_0 = \frac{t^2 S^2}{\epsilon^2} = \frac{(1.96)^2 \times 44.6}{(4.9)^2} = 7.136 \cong 8$$

(ii) *Sampling without replacement* In this case, the fpc cannot be ignored and we take

$$n = \frac{n_0}{1 + n_0 / N} = 7.035 \cong 8$$

Similarly, we can also discuss these results when the sampling units are classified by the presence or absence of a characteristic. With similar notions, the sample proportion p may be assumed to be normally distributed with P and variance $(N - n) P (1 - P)/n (N - 1)$. Hence, the value of n with a pre-assigned level of precision can be estimated by

$$\epsilon = t_{(\alpha, \infty)} [(N - n) P (1 - P)/n (N - 1)]^{1/2} \qquad (2.7.8)$$

where $t_{(\alpha, \infty)}$ has the same meaning given in relation (2.7.4), which gives

$$n = \frac{Nt^2 P (1 - P)/\epsilon^2}{N + \left[\dfrac{t P(1-P)}{\epsilon^2} - \epsilon^2\right]} \qquad (2.7.9)$$

(i) if fpc is ignored, we have

$$n_0 = t^2 P (1 - P)/\epsilon^2 \qquad (2.7.10)$$

(ii) if fpc is not ignored, we get

$$n_1 = \frac{n_0}{1 + (n_0 - 1)/N} \qquad (2.7.11)$$

EXAMPLE 2.8   In a population of 4000 people who were called for casting their votes, 50 per cent returned to the polls. Estimate the sample size to estimate this proportion so that the marginal error is 5 per cent with 95 per cent confidence coefficient, when the sampling is done (i) with replacement and (ii) without replacement methods.

(i) *Sampling with replacement*   In this case, we can ignore fpc and have,

$$n_0 = t^2 P (1 - P)/\epsilon^2 = (1.96)^2 (0.5) (0.5)/0.0025$$

$$\cong 385$$

(ii) *Sampling without replacement*   In this case, fpc cannot be ignored and we have,

$$n_1 = \frac{n_0}{1 + (n_0 - 1)/N} \cong 352$$

SET OF PROBLEMS

2.1   Suppose, in a list of N factories serially numbered, m factories have gone out of existence and n new factories have been added to the list making the total number of factories $(N - m + n)$. Give a simple procedure for selecting one factory with equal probability from $(N - m + n)$ factories, avoiding renumbering of the original N factories and show that your procedure achieves equal probability for new factories.

2.2   With the help of random numbers, draw random samples, each of size 5, from the following:

(i) Cauchy's population:

$$f(x) = \frac{1}{\pi} \frac{\lambda}{\lambda^2 + (x - \mu)^2}$$

where $-\infty < x < \infty$

and $\lambda = 3.8$ cm and $\mu = -2.1$ cm

(ii) Normal population:

$$f(x) = \frac{1}{\sqrt{2\pi}\sigma} \exp\{-(x-\mu)^2\}/2\sigma^2$$

$$-\infty < x < \infty$$

$$\mu = 8 \text{ and } \sigma = 2$$

(iii) Bivariate normal population in which the means of two variates, x and y, are 68 cm and 170 kg, respectively; the s.d's of x and y are 3 cm and 7 kg, respectively; and the correlation coefficient ρ is

(i) $+1$ (ii) 0 and (iii) -1.

2.3 The following procedure has been used for selecting a sample of fields for crop-cutting experiments on paddy:

Against the name of each selected village are shown three random numbers smaller than the highest survey number. These random numbers correspond to three paddy fields for crop-cutting experiments. If the selected survey number does not grow paddy, select the next paddy growing survey number in its place.

Examine whether the above method will provide an equal chance of inclusion in the sample to all the paddy-growing survey numbers in the village, given the following:

   (i) Name of the village   . . .   Payagpur

   (ii) Total number of survey numbers   . . .   299

   (iii) Random numbers   . . .   28, 189, 269

   (iv) Paddy-growing survey numbers   . . .   39 to 88 and 189 to 299

Show that the survey number 39 has a chance of 39/299 of being included in the sample, the survey number 189 a chance of 101/299, while the remaining survey numbers have a chance of only 1/299 each.

(I.C.A.R., 1951)

2.4 A population contains N units, the variate value of one unit being known to be y_0. A random sample, wor, is drawn from the remaining $(N - 1)$ units. Show that the estimator $y_0 + (N - 1)\bar{y}$ has a smaller variance than $N\bar{y}$ based on a random sample, wor, of size n taken from the whole population.

2.5 (i) Define simple random sampling. Is the sample mean a consistent and unbiased estimator of the population mean? Give the variance of the sample mean and also unbiased estimate of this variance. What is the sample size required to estimate the population mean with a given standard error?

(ii) Also, if n_i of the units in the sample are of type A, give an unbiased estimate of the proportion of units of type A in the population and derive its sampling variance and estimate. If the sample size is sufficiently large, write 95% confidence limits for the unknown proportion of units of type A in the population.

2.6 Suppose ν distinct units occur in a sample of n units selected with equal probability with replacement from a population of N units. Show that the estimator $\bar{y}_\nu\ (= \sum_i^\nu y_i/\nu)$ is unbiased for the population mean.

Obtain an unbiased estimator of the variance of this estimator.

2.7 From a random sample of n units, a random sub-sample of m units is drawn without replacement and added to the original sample. Show that the mean based on $(n + m)$ units is an unbiased estimator of the population mean, and that ratio of its variance to that of the mean of the original n units is approximately $(1 + 3m/n)/(1 + m/n)^2$, assuming that the population size is large.

2.8 Derive the expression for variance of a simple random sample drawn without replacement from a finite population.

N balls, placed in a lot container, are drawn at random from a supply of Mp red and Mq white balls. Then a sample of n balls is drawn at random from the lot container and placed in a sample container. It is found that, out of these n balls, r are red. Find the $V(r)$ when the N balls are put into the lot container (i) with replacement and (ii) without replacement.

2.9 In simple random sampling, the variance of the sample mean is given by

$$V(\bar{y}) = (1/n - 1/N)(\sum_i^N y_i^2 - N\bar{y})^2/(N - 1).$$ Denoting an unbiased estimator of $V(\bar{y})$ by v and noting that

$$E\left(\frac{N}{n} \cdot \sum_i^n y_i^2\right) = \sum_i^N y_i^2, \qquad E(\bar{y}^2 - v) = \bar{Y}^2$$

Show that

$$E(v) = \left(\frac{1}{n} - \frac{1}{N}\right) \cdot \frac{1}{(N-1)} \cdot E\left[\frac{N}{n} \sum_i^n y_i^2 - N(\bar{y}^2 - v)\right]$$

2.10 The yields in quintals for wheat crop of 100 villages in a certain tehsil are given on page 44.

(The figures within brackets indicate the village numbers).

(i) Select simple random samples of size 20 and 25 units and estimate the average yield per plot along with their standard errors on the basis of selected units.

(ii) Set up 95 per cent confidence interval and comment.

(iii) Obtain the size of the sample required for estimating the mean yield with 5 per cent standard error.

(1) 20	(2) 21	(3) 32	(4) 41	(5) 55
(6) 22	(7) 64	(8) 42	(9) 28	(10) 35
(11) 25	(12) 25	(13) 24	(14) 32	(15) 75
(16) 28	(17) 29	(18) 38	(19) 19	(20) 19
(21) 16	(22) 28	(23) 30	(24) 29	(25) 29
(26) 19	(27) 37	(28) 34	(29) 31	(30) 35
(31) 29	(32) 19	(33) 27	(34) 42	(35) 39
(36) 11	(37) 26	(38) 21	(39) 45	(40) 61
(41) 16	(42) 29	(43) 32	(44) 40	(45) 63
(46) 30	(47) 21	(48) 35	(49) 28	(50) 18
(51) 24	(52) 32	(53) 23	(54) 8	(55) 35
(56) 27	(57) 35	(58) 25	(59) 29	(60) 29
(61) 25	(62) 31	(63) 38	(64) 31	(65) 43
(66) 21	(67) 36	(68) 30	(69) 37	(70) 47
(71) 15	(72) 19	(73) 32	(74) 19	(75) 50
(76) 10	(77) 27	(78) 36	(79) 28	(80) 43
(81) 28	(82) 25	(83) 31	(84) 6	(85) 4
(86) 22	(87) 24	(88) 39	(89) 71	(90) 44
(91) 24	(92) 34	(93) 18	(94) 28	(95) 10
(96) 70	(97) 20	(98) 32	(99) 42	(100) 47

2.11 Material for the construction of 5000 wells was issued during the year 1964 in a district as part of the Grow-More-Food Campaign in India. The list of cultivators to whom it was issued together with the proposed location of each well is available. It is proposed to estimate the proportion P of wells actually constructed and used for irrigation purpose. The sample is proposed to be selected by simple random sampling. Determine the size of the sample for values of P ranging from 0.5 to 0.9, if the permissible margin of error is 10 per cent and the degree of assurance desired is 95 per cent.

2.12 The data given below pertain to one complete lactation of milk yield (in 10 kg) of 250 cows in an organised dairy farm.

(i) Select a simple random sample of size 25.

(ii) Estimate the mean with its standard error.

(iii) Construct a 95 per cent confidence limit for the population mean.

230	293	163	290	200	173	194	322	169	230
297	151	248	271	259	214	167	207	240	286
184	248	327	338	165	177	270	177	202	155
155	293	190	172	150	319	151	118	213	114
186	167	129	185	231	199	265	306	173	276
291	231	205	220	246	239	186	299	233	208
265	204	300	195	239	173	237	282	221	218
197	215	213	290	146	232	305	184	149	267
188	219	171	99	329	199	180	225	257	202
189	207	792	327	201	300	206	199	299	153
175	287	277	230	258	137	174	301	260	282
211	212	284	214	283	139	223	212	207	224
207	111	272	192	127	303	221	187	309	263
203	176	233	239	176	218	193	243	236	275
288	198	241	219	167	193	234	179	126	173
279	178	275	260	191	174	235	338	242	238
211	187	184	189	305	221	253	225	327	203
195	158	156	185	170	271	160	188	165	218
312	143	267	298	196	139	205	298	238	217
145	201	313	230	185	166	147	223	271	133
155	230	287	329	265	150	286	271	268	198
214	231	163	335	198	270	187	174	163	201
192	247	247	297	178	240	290	234	170	227
230	353	170	159	236	181	230	240	212	242
151	158	253	179	263	158	250	226	246	301

2.13 The frequency distribution of 232 cities in a country, by population size ('000), is given below:

Population size class	No. of cities	Population size class	No. of cities	Population size class	No. of cities
50–75	81	500–550	2	1800–1850	1
75–100	45	550–600	3	1850–1950	0
100–150	42	600–650	1	1950–2000	1
150–200	14	650–700	1	2000–2050	0
200–250	9	700–750	0	2050–2100	1
250–300	5	750–800	1	2100–3600	0
300–350	6	800–850	2	3600–3650	1
350–400	5	850–900	1	3650–7850	0
400–450	5	900–950	2	7850–7900	1
450–500	2	950–1800	0		

Calculate the standard error of the estimator of the population mean when
  (i) a sample of 50 cities is selected with srswor, and
  (ii) the two largest cities are definitely included in the survey and only 48 cities are drawn from the remaining 230 cities with srswor.

2.14 In an agricultural survey, a sample of 36 holdings was selected with srswor, from a population of 432 holdings, in a village. Data relating to land holding size were recorded as follows:

S. No. of holding	Holding size in acres	S. No. of holding	Holding size in acres	S. No. of holding	Holding size in acres
1	21.04	13	8.29	25	22.13
2	12.59	14	7.27	26	1.68
3	20.30	15	1.47	27	49.58
4	16.16	16	1.12	28	1.68
5	23.82	17	10.67	29	4.80
6	1.79	18	5.94	30	12.72
7	26.91	19	3.15	31	6.31
8	7.41	20	4.84	32	14.18
9	7.68	21	9.07	33	22.19
10	66.55	22	3.69	34	5.50
11	141.80	23	14.61	35	25.29
12	28.12	24	1.10	36	20.99

Estimate along with the standard error, the proportions of holdings P_1, P_2, P_3, and P_4 in the four holding size classes 0–4.99, 5.00–9.99, 10.00–24.99 and 25 and above.

REFERENCES

Blyth, R.H., "The economics of sample size applied to the scaling of saw logs," *Bio. Bull.*, **1**, 67–70, (1945).

Chaudhary, F.S. *Sequential approach to sample surveys*, Ph. D. thesis, Meerut University, (1977).

Chaudhary, F.S. and D. Singh, 'Sequential estimation of population and sample sizes," (unpublished), (1979).

Deming, W.E., *Some theory of sampling*, John Wiley and Sons, New York, (1950).

I.C.A.R., "Sample surveys for the estimation of yield of food crops," *Bull*, **72**, New Delhi, (1951).

Mathai, A. "On selecting random numbers for large–scale sampling," *Sankhya*, **13**, 157–160, (1954).

Nordin, J.A., "Determining sample size," *J. Amer. Statist. Assoc.*, **39**, 497–506, (1944).

Raiffa, H. and R. Schlaifer, *Applied Statistical Decision Theory*, Harvard Business School, Boston, (1961).

Tippett, L.H.C., *Technological application of statistics*, John Wiley & Sons, New York, (1950).

Yates, F. *Sampling methods for censuses and surveys*, Charles Griffin and Co., London, (1960).

3

Stratified Random Sampling

Sir, In your otherwise beautiful poem ("*The Vision of Sin*") *there is a verse which reads* "*Every moment dies a man, every moment one is born.*"

Obviously, this cannot be true and I suggest that in the next edition you have it read "*Every moment dies a man, every moment* $1\frac{1}{16}$ *is born.*"

Even this value is slightly in error but should be sufficiently accurate for poetry.

(in a letter to Lord Tennyson)   Charles Babbage

3.1 INTRODUCTION

Of all the methods of sampling, the procedure most commonly used in surveys is stratified sampling. In stratified sampling, the population of N units is sub-divided into k sub-populations called *strata*, the ith sub-population having N_i units ($i=1, 2,\cdots, k$). These sub-populations are non-overlapping so that they comprise the whole population such that

$$N_1 + N_2 + \cdots + N_k = N \tag{3.1.1}$$

A sample is drawn from each stratum independently, the sample size within the ith stratum being n_i ($i=1, 2, \ldots, k$) such that

$$n_1 + n_2 + \cdots + n_k = n \tag{3.1.2}$$

The procedure of taking samples in this way is known as *stratified sampling*. If the sample is taken randomly from each stratum, the procedure is known as *stratified random sampling*.

The main objective of stratification is to give a better cross-section of the population so as to gain a higher degree of relative precision. To achieve this, the following points are to be examined carefully:

(i) Formation of strata

(ii) Number of strata to be made

(iii) Allocation of sample size within each stratum

(iv) Analysis of data from a stratified design

We shall discuss the first two points after examining the last two points relating to the theory of stratified sampling.

3.2 PRINCIPLES OF STRATIFICATION

The principles to be followed in stratifying a population are summarised below:

(i) The strata should be non-overlapping and should together comprise the whole population.

(ii) The stratification of population should be done in such a way that strata are homogeneous within themselves, with respect to the characteristic under study.

(iii) In many practical situations when it is difficult to stratify with respect to the characteristic under study, administrative convenience may be considered as the basis for stratification.

(iv) If the limit of precision for certain sub-populations is given, it will be better to treat each sub-population as a stratum.

3.3 ADVANTAGES OF STRATIFICATION

Stratification serves many useful purposes. The principal ones are the following:

(i) Stratification may be desired for administrative convenience. The agency conducting the survey work can establish its field offices in various administrative zones with well-defined jurisdiction, thereby leading to better organisation and supervision of field work.

(ii) Stratification by natural characteristics helps in improving the sampling design. For example, in area and yield surveys, there may be different types of sampling problems in plains, deserts and hilly areas which may need different approaches. Hence, it would be advantageous to constitute a separate stratum for each such areas.

(iii) Stratification is particularly more effective when there are extreme values in the population which can be segregated into separate strata, thereby reducing the variability within strata. Separate estimates obtained for individual stratum can be combined into a precise estimate for the whole population.

(iv) Stratification makes it possible to use different sampling designs in different strata. In many practical situations, the information for stratification is not uniformly available for all units in the population. In these cases, the whole population is sub-divided into strata according to the nature of the information available and some suitable sampling scheme of selection of units within these strata is adopted.

(v) Stratification ensures adequate representation to various groups of the population, which may be of some interest or importance.

(vi) Stratification also ensures selection of a better cross-section of the population than that under unstratified population.

(vii) Stratification brings a gain in the precision in estimation of a characteristic of a population. To achieve it, a heterogeneous population is sub-divided into sub-populations, each of which is homogeneous within itself. If each stratum is homogeneous such that measurement, within it vary little from one unit to another, a more precise estimate can be obtained by taking a relatively smaller sample.

3.4 NOTATIONS

Let i denote for the stratum and j for the sampling unit within the stratum. The following symbols refer to stratum i:

N_i = total number of units

n_i = number of units in sample

$W_i = N_i/N$ stratum weight

$f_i = n_i/N_i$ sampling fraction in the stratum

Let y_{ij} be the value of the jth unit in the stratum.

$$\hat{\bar{Y}}_i = \sum_j^{N_i} y_{ij}/N_i \qquad \text{stratum mean}$$

$$\bar{y}_i = \sum_j^{n_i} y_{ij}/n_i \qquad \text{sample mean}$$

$$S_i^2 = \sum_j^{N_i} (y_{ij} - \bar{Y}_i)^2/(N_i - 1) \qquad \text{stratum variance}$$

3.5 ESTIMATION OF THE POPULATION MEAN AND ITS VARIANCE

Suppose that a population of N units is divided into k strata. The population mean per unit can be written as

$$\bar{Y}=\sum_i^k \sum_j^{N_i} y_{ij}/N=\sum_i^k N_i \bar{Y}_i/N=\sum_i^k W_i \bar{Y}_i$$

An estimator $\bar{y}_{st}$ (st for *stratified*) for the population mean $\bar{Y}$ can be written as

$$\bar{y}_{st}=\sum_i^k N_i \bar{y}_i/N \qquad (3.5.1)$$

which is different from the overall sample mean

$$\bar{y}=\sum_i^k n_i\bar{y}_i/n \qquad (3.5.2)$$

THEOREM 3.5.1 If in every stratum the sample estimator $\bar{y}_i$ is unbiased and samples are drawn independently in different strata, then $\bar{y}_{st}$ is an unbiased estimator of the population mean and its sampling variance is given by

$$V(\bar{y}_{st})=\sum_i^k N_i^2 \, V(\bar{y}_i)/N^2=\sum_i^k W_i^2 \, V(\bar{y}_i) \qquad (3.5.3)$$

Proof we have that

$$E(\bar{y}_{st})=E\{\sum_i^k (N_i\bar{y}_i/N)\}$$

$$=\sum_i^k N_i \, E(\bar{y}_i)/N=\sum_i^k N_i \, \bar{Y}_i/N=\bar{Y} \qquad (3.5.4)$$

This shows that $\bar{y}_{st}$ is an unbiased estimator.

To obtain the sampling variance, we note that sampling is done independently in each stratum and, therefore,

$$V(\bar{y}_{st})=V(\sum_i^k N_i \, \bar{y}_i/N)=\sum_i^k W_i^2 \, V(\bar{y}_i)$$

THEOREM 3.5.2 For stratified random sampling, wor, the sample estimator $\bar{y}_{st}$ is unbiased and its sampling variance is given by

$$V(\bar{y}_{st})=\sum_i^k N_i(N_i-n_i) \, S_i^2/N^2 n_i=\sum_i^k (1-f_i)W_i^2 S_i^2/n_i \qquad (3.5.6)$$

Proof Since, in each stratum, a simple random sample is taken, $\bar{y}_i$ is an unbiased estimator of $\bar{Y}_i$ and hence, by Theorem 3.5.1, it can be shown that $\bar{y}_{st}$ is an unbiased estimator of $\bar{Y}$.

Further, by Theorem 2.3.1 applied to an individual stratum, we have

$$V(\bar{y}_i) = (N_i - n_i)\, S_i^2 / N_i\, n_i$$

Substituting the value in the result of Theorem 3.5.1, we get

$$V(\bar{y}_{st}) = \sum_i^k N_i^2\, V(\bar{y}_i)/N^2 = \sum_i^k N_i\,(N_i - n_i)\, S_i^2/N^2 n_i$$

$$= \sum_i^k (1 - f_i)\, W_i^2\, S_i^2/n_i$$

COROLLARY 1   If $\hat{Y}_{st} = N\,\bar{y}_{st}$ is the estimator of the population total Y, then $\hat{Y}_{st}$ is an unbiased estimator and its sampling variance is given by

$$\left.\begin{aligned} V(\hat{Y}_{st}) &= \sum_i^k N_i\,(N_i - n_i)\, S_i^2/n_i \\[2mm] &= \sum_i^k N^2\,(1 - f_i)\, W_i^2\, S_i^2/n_i \end{aligned}\right\} \tag{3.5.7}$$

COROLLARY 2   If in every stratum $(n_i/n) = (N_i/N)$, the variance of $\bar{y}_{st}$ reduces to

$$V(\bar{y}_{st}) = \frac{(N-n)}{N} \sum_i^k N_i\, S_i^2/nN = \frac{(1-f)}{n} \sum_i^k W_i\, S_i^2 \tag{3.5.8}$$

COROLLARY 3   If in every stratum $n_i/n = N_i/N$, and the variance in all strata have the same value S_w^2, the result reduces to

$$V(\bar{y}_{st}) = (N-n)\, S_w^2/nN = (1-f)S_w^2/n \tag{3.5.9}$$

COROLLARY 4   For stratified random sampling, wr, $\bar{y}_{st}$ is an unbiased estimator and its sampling variance is given by

$$V(\bar{y}_{st}) = \sum_i^k N_i^2\, S_i^2/N^2 n_i = \sum_i^k W_i^2\, S_i^2/n_i \tag{3.5.10}$$

THEOREM 3.5.3   In stratified random sampling, wor, with sample size $n_i[i = 1, 2, \ldots, k; \sum_i^k n_i = n]$, an unbiased estimator of the population proportion P is given by

$$\hat{P}_{st} = \sum_{i}^{k} N_i p_i / N = \sum_{i}^{k} W_i p_i \qquad (3.5.11)$$

with its variance

$$V(\hat{P}_{st}) = \sum_{i}^{k} (1-f_i) W_i^2 \, N_i P_i (1-P_i)/(N_i-1) \, n_i \qquad (3.5.12)$$

where p_i is the sample estimate of proportion P_i in the ith stratum. The proof is obvious.

COROLLARY 1 $N_i/(N_i-1)$ can be taken as unity, we have

$$V(\hat{P}_{st}) \cong \sum_{i}^{k} (1-f_i) \, W_i^2 \, P_i (1-P_i)/n_i \qquad (3.5.13)$$

COROLLARY 2   If stratified random sampling is with replacement, the sampling variance is given by

$$V(\hat{P}_{st}) = \sum_{i}^{k} W_i^2 P_i (1-P_i)/n_i \qquad (3.5.14)$$

COROLLARY 3   If in every stratum $n_i/n = N_i/N$, the sampling variance reduces to

$$V(\hat{P}_{st}) = \sum_{i}^{k} \frac{(1-f)N}{n} \frac{W_i^2 P_i(1-P_i)}{(N_i-1)} \qquad (3.5.15)$$

$$\cong \sum_{k}^{k} \frac{(1-f)}{n} W_i P_i (1-P_i) \qquad (3.5.16)$$

The variance of $\hat{P}_{st}$ depends on the product of P_i and $(1-P_i)$ within strata. The product is small if P_i is near zero or unity. Thus, the efficiency can be increased if the strata are formed such that units belonging to a given class are allocated in the same stratum.

3.6   ESTIMATE OF VARIANCE

If a simple random sample is taken within each stratum, an unbiased estimator of S_i^2 is given by

$$s_i^2 = \sum_{j}^{n_i} (y_{ij} - \bar{y}_i)^2/(n_i - 1) \qquad (3.6.1)$$

Then, an unbiased estimator of sampling variance of $\bar{y}_{st}$ can be obtained as follows:

THEOREM 3.6 With stratified random sampling, wor, an unbiased estimator of the variance of $\bar{y}_{st}$ is given by

$$v(\bar{y}_{st}) = \sum_i^k N_i (N_i - n_i) s_i^2 / N^2 n_i$$

$$= \sum_i^k W_i^2 \, s_i^2 / n_i - \sum_i^k W_i \, s_i^2 / N \qquad (3.6.2)$$

The proof is obvious.

COROLLARY 1 With stratified random sampling, wor, an unbiased estimator of $V(\bar{y}_{st})$ reduces to

$$v(\bar{y}_{st}) = \sum_i^k W_i^2 S_i^2 / n_i \qquad (3.6.3)$$

COROLLARY 2 With stratified random sampling, wor, an unbiased estimator of $V(\hat{P}_{st})$ is given.

$$v(\hat{P}_{st}) = \sum_i^k (1 - f_i) \, W_i^2 \, p_i (1 - p_i) / (n_i - 1) \qquad (3.6.4)$$

3.7 ALLOCATION OF SAMPLE SIZE IN DIFFERENT STRATA

In stratified sampling, the allocation of the sample to different strata is done by the consideration of three factors, viz.,

(i) the total number of units in the stratum, i.e. stratum size
(ii) the variability within the stratum, and
(iii) the cost in taking observations per sampling unit in the stratum

A good allocation is one where maximum precision is obtained with minimum resources, or in other words, the criterion for allocation is to minimize the budget for a given variance or minimize the variance for a fixed budget, thus making the most effective use of the available resources.

There are four methods of allocation of sample sizes to different strata in a stratified sampling procedure. These are:

(i) Equal allocation
(ii) Proportional allocation
(iii) Neyman allocation
(iv) Optimum allocation

3.7.1 Equal Samples From Each Stratum

This is a situation of considerable practical interest for reasons of administrative or field work convenience. In this method, the total sample size n is divided equally among all the strata, i.e. for the ith stratum

$$n_i = n/k \qquad (3.7.1)$$

3.7.2 Proportional Allocation

This allocation, generally known as proportional allocation, was originally proposed by Bowley (1926). This procedure of allocation is very common in practice because of its simplicity. When no other information except N_i, the total number of units in the ith stratum, is available, the allocation of a given sample of size n to different strata is done in proportion to their sizes, i.e. in the ith stratum,

$$n_i = nN_i/N \text{ (or } f_i = f) \qquad (3.7.2)$$

This means that the sampling fraction is the same in all strata. It gives a *self-weighting* sample by which numerous estimates can be made with greater speed and a higher degree of precision.

3.7.3 Neyman Allocation

This allocation of the total sample size to strata is called *minimum-variance allocation* and is due to Neyman (1934). This result appears to have been first discovered by Tschuprow (1923) but remained unknown until it was rediscovered independently by Neyman. The allocation of samples among different strata is based on a joint consideration of the stratum size and the stratum variation. In this allocation, it is assumed that the sampling cost per unit among different strata is the same and the size of the sample is fixed. Sample sizes are allocated by

$$n_i = n \frac{W_i S_i}{\sum\limits_i^k W_i S_i} = n \frac{N_i S_i}{\sum\limits_i^k N_i S_i} \qquad (3.7.3)$$

A formula for the minimum variance with fixed n is obtained by substituting the value of n_i in Eq. (3.5.3) when we get

$$V_{\min}(\bar{y}_{st}) = (\sum\limits_i^k W_i S_i)^2/n - \sum\limits_i^k W_i S_i^2/N \qquad (3.7.4)$$

There may be difficulty in using this method as the value of S_i will usually be unknown. However, the stratum variances may be obtained from previous surveys or from a specially planned pilot survey. The other alternative is to conduct the main survey in a phased manner and utilize the data collected in the first phase for ensuring better allocation in the second phase.

3.7.4 Optimum Allocation

In this method of allocation the sample sizes n_i in the respective strata are determined with a view to minimize $V(\bar{y}_{st})$ for a specified cost of conducting the sample survey or to minimize the cost for a specified value of $V(\bar{y}_{st})$. The simplest cost function in stratified sampling that can be taken is

$$C = a + \sum_i^k n_i c_i \qquad (3.7.5)$$

where the overhead cost a is constant and c_i is the average cost of surveying one unit in the ith stratum, which may depend upon the nature and size of the units in the stratum.

To determine the optimum value of n_i, we consider the function

$$\psi = V(\bar{y}_{st}) + \lambda C \qquad (3.7.6)$$

where λ is some unknown constant.

Using the calculus method of Lagrange multipliers, we select n_i and the constant λ to minimize ψ.

Differentiating with respect to n_i, we have

$$- W_i^2 S_i^2 / n_i^2 + \lambda c_i = 0 \quad (i = 1, 2, \ldots, k)$$

or

$$n_i = W_i S_i / \sqrt{\lambda c_i} \qquad (3.7.7)$$

Summing over all strata, we get

$$n = \sum_i^k (W_i S_i / \sqrt{\lambda c_i}) \qquad (3.7.8)$$

From relations (3.7.7) and (3.7.8) we can obtain

$$n_i = n \frac{(W_i S_i / \sqrt{c_i})}{\sum_i^k (W_i S_i / \sqrt{c_i})} = n \frac{N_i S_i / \sqrt{c_i}}{\sum_i^k (N_i S_i / \sqrt{c_i})} \qquad (3.7.9)$$

Thus the relation (3.7.9) leads to the following important conclusions that, in a given stratum, we have to take a larger sample if

(i) the stratum size is larger;

(ii) the stratum has larger variability; and

(iii) the cost per unit is cheaper in the stratum.

If c_i's are the same from stratum to stratum, relation (3.7.9) will lead to the Neyman allocation. Similarly, if c_i's and S_i's do not vary from stratum to stratum, relation (3.7.9) will lead to proportional allocation.

The total sample size n required for estimating the population with a specified cost C is given by

$$n = \frac{(C - a) \sum_i^k (W_i S_i / \sqrt{c_i})}{\sum_i^k (W_i S_i \sqrt{c_i})} \qquad (3.7.10)$$

If V is fixed, we find

$$n = \frac{(\sum_i^k W_i S_i / \sqrt{c_i}) \; \sum_i^k (W_i S_i \sqrt{c_i})}{V + \sum_i^k W_i S_i^2 / N} \qquad (3.7.11)$$

The values of the stratum variances can be obtained from earlier surveys or from a knowledge of the measurements within each stratum.

An alternative proof of the allocation results due to Stuart (1954) may be summarized as follows:

Let V' and C' be the parts of V and C that depend on n_i or in a simple form, $V' = \sum A_i / n_i$ and $C' = \sum c_i n_i$.

Minimizing V for fixed C or vice versa are both equivalent to minimizing the product

$$V'C' = (\sum A_i / n_i) (\sum c_i n_i) \qquad (3.7.12)$$

Now, by Cauchy–Schwartz's inequality,

$$(\sum x_i^2)(\sum y_i^2) \geqslant (\sum x_i y_i)^2,$$

we have

$$(\sum A_i / n_i) (\sum c_i n_i) \geqslant (\sum \sqrt{A_i c_i})^2$$

The equality being if and only if y_i is proportional to x_i, i.e.

$$\frac{c_i n_i}{A_i / n_i} = \text{constant}$$

which yields $\qquad\qquad n_i \propto \sqrt{A_i / c_i} \qquad\qquad (3.7.13)$

(i) In case $A_i = W_i^2 S_i^2$, we get from relation (3.7.13) the optimum sample sizes within strata as given in relation (3.7.9).

(ii) In case $c_i = c$ and $A_i = W_i^2 S_i^2$, the optimum sample sizes for Neyman allocation are given by the above relation.

(iii) In case $A_i = W_i$ and $c_i = c$, proportional allocation is also obtained by the above relation.

3.8 RELATIVE PRECISION OF STRATIFIED RANDOM SAMPLING WITH SIMPLE RANDOM SAMPLING

In this section, we shall make a comparative study of the usual estimators under simple random sampling, without stratification and stratified random sampling employing various schemes of allocation, i.e. proportional and optimum allocations. The variances of these estimators of mean are denoted by V_{SR}, V_{prop} and V_{opt}, respectively.

THEOREM 3.8.1   If fpc is ignored, show that

$$V_{opt} \leqslant V_{prop} \leqslant V_{SR} \tag{3.8.1}$$

Proof   When fpc is ignored, we have

$$V_{SR} = S^2/n$$

$$V_{prop} = \sum_i^k N_i S_i^2 / Nn = \sum_i^k W_i S_i^2 / n$$

$$V_{opt} = (\sum_i^k N_i S_i)^2 / N^2 n = (\sum_i^k W_i S_i)^2 / n$$

Since terms $1/N_i$ and $1/N$ are negligible, we can have

$$NS^2 = \sum_i^k N_i S_i^2 + \sum_i^k N_i (\overline{Y}_i - \overline{Y})^2$$

Hence, we have

$$V_{SR} = S^2/n = \frac{1}{nN} [\sum_i^k N_i S_i^2 + \sum_i^k N_i (\overline{Y}_i - \overline{Y})^2]$$

$$= V_{prop} + \sum_i^k N_i (\overline{Y}_i - \overline{Y})^2 / nN \tag{3.8.2}$$

or $V_{SR} - V_{prop} =$ a positive quantity.

Thus, it proves that $V_{prop} \leqslant V_{SR}$

Hence, it may be concluded that the larger the difference in the stratum means, the greater is the gain in precision with proportional allocation over simple random sampling.

Similarly, $$V_{prop} - V_{opt} = \frac{1}{nN} [\sum_i^k N_i S_i^2 - (\sum_i^k N_i S_i)^2 / N]$$

$$= \frac{1}{nN} (\sum_i^k N_i (S_i - \overline{S})^2 \tag{3.8.3}$$

where $\qquad \overline{S} = \sum_{i}^{k} N_i S_i/N$

$\qquad\qquad\qquad = $ a positive quantity

Also $\qquad V_{\text{SR}} = V_{\text{opt}} + [\sum_{i}^{k} N_i (S_i - \overline{S})^2$

$$+ \sum_{i}^{k} N_i (\overline{Y}_i - \overline{Y})^2]/nN \qquad (3.8.4)$$

which proves that $V_{\text{opt}} \leqslant V_{\text{prop}}$

Hence, $\qquad\qquad V_{\text{opt}} \leqslant V_{\text{prop}} \leqslant V_{\text{SR}}$

It concludes that Neyman's optimum allocation gives better results than proportional allocation. We observe from Eq. (3.8.4) that as we change from simple random sampling to stratified random sampling with Neyman's allocation, the gain in precision of the estimates results from two factors, viz.

(i) the differences between stratum means and

(ii) the difference between the stratum standard deviations.

COROLLARY If fpc cannot be neglected, show that

$$V_{\text{opt}} \leqslant V_{\text{prop}} \leqslant V_{\text{SR}}$$

EXAMPLE 3.1 2000 cultivators' holdings in Uttar Pradesh (India) were stratified according to their sizes. The number of holdings (N_i), mean area under wheat per holdings ($\overline{Y}_i$) and s.d. of area under wheat per holding (S_i) are given below for each stratum:

Stratum number	Number of holdings (N_i)	Mean area under wheat per-holding ($\overline{Y}_i$)	s.d. of area under wheat per-holding (S_i)
1	394	5.4	8.3
2	461	16.3	13.3
3	381	24.3	15.1
4	334	34.5	19.8
5	169	42.1	24.5
6	113	50.1	26.0
7	148	63.8	35.2

For a sample of 200 farms, compute the sample size in each stratum under proportional and optimum allocations. Calculate the sampling variance of the estimated area under wheat from the sample (i) if the farms are selected under proportional allocation by with and without replacement methods, (ii) if the farms are selected under Neyman's allocation by with and without replacement methods. Also compute the gain in efficiency from these procedures as compared to simple random sampling.

The relevant calculations have been shown in Table 3.1 below:

Table 3.1 Calculations of allocations, means and sampling variances

Stratum number	N_i	$\overline{Y}_i$	S_i	W_i	nW_i	W_iS_i	$\dfrac{nW_iS_i}{\Sigma W_iS_i}$	$W_i\overline{Y}_i$	$W_i\overline{Y}_i^2$	$W_iS_i^2$
(1)	(2)	(3)	(4)	(5)	(6)	(7)	(8)	(9)	(10)	(11)
1	394	5.4	8.3	0.1970	40	1.04	19	1.06	5.72	13.612
2	461	16.3	13.3	0.2305	46	3.07	36	3.76	61.29	40.831
3	381	24.3	15.1	0.1905	38	2.88	34	4.63	112.51	43.488
4	334	34.5	19.8	0.1670	33	3.31	39	5.76	198.72	65.538
5	169	42.1	24.5	0.0845	17	2.07	24	3.56	149.88	50.715
6	113	50.1	26.0	0.0565	11	1.47	17	2.83	141.78	38.220
7	148	63.8	35.2	0.0740	15	2.61	31	4.72	301.14	91.872
Total	2000			1.0000	200	17.05	200	26.32	971.04	344.276

Proportional Allocation $n_i \propto N_i$ or $n_i = nW_i$

The numbers of holdings to be selected from strata are given in column (6) of the table and are 40, 46, 38, 33, 17, 11 and 15, respectively.

$$V(\bar{y}_{\text{prop}}) \text{ wor} = [(N - n)/nN] \sum W_iS_i^2$$

$$= 1.5492$$

If fpc is ignored, we have

$$V(\bar{y}_{\text{prop}}) \text{ wr} = 1/n \sum W_iS_i^2 = 1.7214$$

Optimum Allocation

$$n_i \propto N_i S_i \text{ or } n_i = n \frac{W_i S_i}{\sum W_i S_i}$$

The numbers of holdings to be selected from strata are given in column (8) of the table and are 19, 36, 34, 39, 24, 17 and 31, respectively.

$$V(\bar{y}_{opt}) \text{ wor} = (\sum W_i S_i)^2/n - \sum W_i S_i^2/N = 1.2813$$

If fpc is ignored, we get

$$V(\bar{y}_{opt}) \text{ wr} = (\sum W_i S_i)^2/n = 1.4535$$

Simple Random Sampling The variance of the estimate of mean is given by

$$V(\bar{y}_{SR}) \text{ wor} = \left(\frac{1}{n} - \frac{1}{N}\right) S^2$$

$$= V_{opt} + \frac{(N-n)}{n(N-1)} [\sum W_i \bar{Y}_i^2 - (\sum W_i \bar{Y}_i)^2$$

$$+ \sum W_i S_i^2 - (\sum W_i S_i)^2] = 3.0420$$

If fpc is ignored, we get

$$V(\bar{y}_{SR}) \text{ wr} = V_{prop} + \sum W_i (\bar{Y}_i - \bar{Y})^2/n = 3.1129$$

(i) The relative precision of proportional allocation is given by
   (a) without replacement method $V_{SR}/V_{prop} = 1.9636$
   (b) with replacement methods $V_{SR}/V_{prop} = 1.8084$
(ii) The relative precision of optimum allocation is given by
   (a) without replacement method $V_{SR}/V_{opt} = 2.3742$
   (b) with replacement method $V_{SR}/V_{opt} = 2.1416$

3.9 ESTIMATION OF GAIN IN PRECISION DUE TO STRATIFICATION

In comparing the precision of stratified with unstratified random sampling, it was assumed that the population values of stratum means and variances were known.

It is sometimes of interest to examine, from a survey, how useful the mode of stratification has been. What is available is only a stratified sample and the problem is to estimate the gain in precision due to stratification. An estimate of the variance of the estimate, in case of unstratified sampling, is obtained from a stratified sample and a comparison can be made with a situation in which no stratification is done.

Let $n_1, n_2, \ldots, n_k$ represent the stratified sample. Also, stratum means and variances are estimated from the sample. Let us denote the estimated stratum means by $\bar{y}_1, \bar{y}_2, \ldots, \bar{y}_k$ and the estimated variances by $s_1^2, s_2^2, \ldots, s_k^2$. The problem is, therefore, to get an unbiased estimate of $V(\bar{y}_{SR})$ based on the given stratified sample. Here, it can be shown that an estimate $s^2 = \sum_i \sum_j (y_{ij} - \bar{y})^2/(n-1)$ will not be an unbiased estimate of S^2.

An unbiased estimator of $V(\bar{y}_{SR})$ based on stratified sample is given by

$$v(\bar{y}_{SR}) = \left(\frac{1}{n} - \frac{1}{N}\right)\sum_i^k W_i s_i^2$$

$$+ \frac{(N-n)}{(N-1)N}[\sum_i^k W_i (\bar{y}_i - \bar{y}_{st})^2$$

$$- \sum_i^k W_i (1 - W_i)\ s_i^2/n_i] \qquad (3.9.1)$$

The estimate of the relative gain in precision due to stratification is thus obtained by

$$\frac{v(\bar{y}_{SR}) - v(\bar{y}_{st})}{v(\bar{y}_{st})} \qquad (3.9.2)$$

If the sample allocation is large enough in each stratum, i.e., $n_i > 50$ the relation (3.9.1) reduces to

$$v(\bar{y}_{SR}) = \frac{N-n}{nN}\ [\sum W_i s_i^2 + \sum W_i \bar{y}_i^2 - (\sum W_i \bar{y}_i)^2] \qquad (3.9.3)$$

In actual practice, it is found that sampling variance is not much sensitive to small or even moderate deviations in the allocations. If allocation is proportional, i.e. $n_i = nW_i$; we can write s^2, by applying the technique of analysis of variance, as

$$s^2 = \sum W_i s_i^2 + \sum W_i \bar{y}_i^2 - (\sum W_i \bar{y}_i)^2$$

giving that s^2 is an unbiased estimator in proportional allocation.

EXAMPLE 3.3 The number of pepper standards for selected villages in each of the three strata of Trivandrum zone are as follows:

Stratum	Total number of villages in the stratum	Number of villages selected from the stratum	Number of pepper standards in each of the selected villages
1	441	11	41, 116, 19, 15, 144, 159, 212, 57, 28, 119, 76.
2	405	12	39, 70, 38, 37, 161, 38, 27, 119, 36, 128, 30, 208.
3	103	7	252, 385, 192, 296, 115, 159, 120.

Estimate the total number of pepper standards along with its standard error in Trivandrum zone. Also, estimate the gain in precision due to stratification.

The relevant calculations have been done in Table 3.2.

An estimate of the total number of pepper standards is given by

$$\hat{Y}_{st} = \sum_i^k N_i \, \bar{y}_i = 87376.54$$

An estimate of variance of $\tilde{Y}_{st}$ is

$$v\,(\hat{Y}_{st}) = \sum_i^k N_i^2 \left(\frac{1}{n_i} - \frac{1}{N_i} \right) s_i^2 = 116{,}432{,}940.29$$

or standard error of $\hat{Y}_{st} = \sqrt{116{,}432{,}940.29}$
$$= 10{,}790.41$$

Estimate of the variance of the total number of pepper standards, when simple random sampling is assumed, is given by

$$v\,(\hat{Y}_{SR}) = \left(\frac{1}{n} - \frac{1}{N} \right) \left[N \left\{ \sum_i^k N_i \, s_i^2 - \sum_i^k \frac{N_i \, s_i^2}{n_i} + \sum_i^k N_i \, \bar{y}_i^2 \right\} \right.$$
$$\left. + \sum_i^k \frac{N_i \, s_i^2}{n_i} - (\sum_i^k N_i \, \bar{y}_i)^2 \right]$$
$$= 167{,}269{,}559.37$$

Thus, the percentage gain in precision due to stratification

$$= \frac{v\,(\hat{Y}_{SR}) - v\,(\hat{Y}_{st})}{v\,(\hat{Y}_{st})} \times 100$$
$$= \frac{167{,}269{,}559.37 - 116{,}432{,}940.29}{116{,}432{,}940.29} \times 100$$
$$= 43.65$$

Table 3.2 Calculations of estimates of means and variances

Stratum	N_i	n_i	$\sum_j y_{ij}$	$\bar{y}_i$	$N_i \bar{y}_i$	$\sum_j y^2_{ij}$	$\sum_j y^2_{ij}/n_i$	s_i^2	$\left(\dfrac{1}{n_i} - \dfrac{1}{N_i}\right)$
(1)	(2)	(3)	(4)	(5)	(6)	(7)	(8)	(9)	(10)
1	441	11	986	89.63	39,529.81	130,654	88,381.45	4227.2	0.0886
2	405	12	743	91.92	25,481.22	70,469	46,004.08	2224.08	0.0809
3	103	7	1520	217.71	22,365.71	389,885	330,057.14	9971.44	0.1331
Total	949	30			87,376.54				

	$N_i^2\left(\dfrac{1}{n_i} - \dfrac{1}{N_i}\right) s_i^2$	$N_i s_i^2$	$\dfrac{N_i s_i^2}{n_i}$	$N_i \bar{y}_i^2$	$\dfrac{N_i^2 s_i^2}{a_i}$
	(11)	(12)	(13)	(14)	(15)
	72,839,892.71	1,864,219.28	169,474.48	3,543,275.94	74,738,245.68
	29,512,745.97	900,753.74	75,062.81	1,577,720.88	30,400,438.62
	14,080,301.61	1,027,062.04	146,723.15	4,856,556.93	15,112,484.24
Total	116,432,940.29	3,792,035.06	391,260.44	9,977,523.75	120,251,168.54

3.10 FORMATION OF STRATA

The problem of formation of strata has been discussed by Hagood and Bernert (1945), Dalenius (1950, 1952), Dalenius and Gurney (1951), Mahalanobis (1952) and Dalenius and Hodges (1959). Ekman (1959) suggested approximations to the theoretical solutions. Cochran (1961) has examined the application of these approximations through empirical studies. Sethi (1963) derived the solutions for optimum points of stratification in case of certain populations and Hess, Sethi and Balakrishnan (1966) have applied these solutions to some empirical studies. In this section, we shall discuss in brief the outlines for the construction of strata on these lines, although the results derived are necessarily approximate and of limited importance from the point of view of application.

In order to form k strata, the range of main variates under study is to be subdivided at the points $y_1, y_2, \ldots, y_{k-1}$ such that $a \leqslant y_1 < y_2 < \ldots < y_{k-1} \leqslant b$. It has been seen earlier in Eq. (3.5.10) that the variance of $(\bar{y}_{st})$ (with proportional allocation) is given by

$$V(\bar{y}_{st}) = \frac{1}{n} \sum_i^k W_i S_i^2$$

$$= \frac{1}{n} \left[\sum_i^k \sum_j^{N_i} Y_{ij}^2 / N - \sum_i^k W_i \bar{Y}_i^2 \right] \qquad (3.10.1)$$

As mentioned in the beginning, the main purpose of stratification is to increase precision, therefore $V(\bar{y}_{st})$ is to be minimized. Since the term $\sum_i^k \sum_j^{N_i} Y_{ij}^2$ in relation (3.10.1) is a constant and independent of stratification, the problem is to obtain a value of y_i which maximizes the term $\sum_i^k W_i \bar{Y}_i^2$. Assuming that all points except y_i are fixed, it can be proved that the value of y_i which maximizes the term $\sum_i^k W_i \bar{Y}_i^2$ is given by iterative procedures as

$$y_i = \tfrac{1}{2} (\bar{Y}_i + \bar{Y}_{i+1}) \qquad (3.10.2)$$

This shows that the best y_i is the average of the two strata means which it separates. Hence, all the points y_i $(i = 1, \ldots, k - 1)$ can be obtained and the desired strata formed.

The problem of strata construction in equal and optimum allocations can also be treated similarly. The sampling variances in equal allocation and optimum allocation are

$$\frac{k}{n} \sum_{i}^{k} W_i^2 S_i^2 \text{ and } \frac{1}{n} (\sum_{i}^{k} W_i S_i)^2, \text{ respectively.}$$

Proceeding as before, the best points of stratification in the two cases are given by

$$W_i [S_i^2 + (y_i - \overline{Y}_i)^2] \cong W_{i+1} [S_{i+1}^2 + (y_i - \overline{Y}_{i+1})^2] \quad (3.10.3)$$

and

$$\frac{S_i^2 + (y_i - \overline{Y}_i)^2}{S_i} \cong \frac{S_{i+1}^2 + (y_i - \overline{Y}_{i+1})^2}{S_{i+1}} \quad (3.10.4)$$

for $i = 1, \ldots, k - 1$, respectively.

Since it is difficult to apply them in practical situations, due to the heavy computational work, some approximate solutions and simpler methods are summarized below:

(i) *Equalization of $W_i S_i$* Dalenius and Gurney (1951) suggested that the construction of strata on the basis of equalization of $W_i S_i$, and giving equal allocation to the strata, would lead to optimum stratification. This method is not convenient in practice since it involves the calculation of S_i for different stratification points.

(ii) *Equalization of strata totals* Another rule widely used was proposed by Mahalanobis (1952). He suggested making strata which have the same aggregate size $W_i Y_i$ with equal allocation. This would lead to efficient stratification if the strata coefficients of variation are the same. This procedure was found to give poorer results when applied to normal, gamma and beta distributions (Sethi, 1963).

(iii) *Equalization of $W_i R_i$* Among the other approximate rules suggested is the one by Ayoma (1954), in which he suggested the formation of strata on the basis of equalization of strata ranges and with equal allocation. Ekman (1959) suggested also the equalization of $W_i R_i$ for forming strata with equal allocation, where R_i is the range of the variate in the ith stratum. This method is quite simple and can be useful in practice. A good performance of Ekman's method has also been reported in all these cases.

(iv) *Equalization of cumulatives of $\sqrt{f}(y)$* Dalenius and Hodges (1959) suggested construction of strata by equalization of the cumulative of $\sqrt{f}(y)$, where $f(y)$ is the frequency function. The assumptions of the rule are that the distribution is bounded and that the number of strata is large. Sethi (1963) has shown that this rule provides optimum stratification even when the number of strata is as small as 2 or 3.

(v) *Equalization of* $\frac{1}{2}[f(y) + l(y)]$ Durbin (1959) suggested the equalization of the cumulative frequencies of a distribution, $\phi(y)$, which is in between the original distribution $f(y)$ and rectangular distribution $l(y)$ over the given range of y, $l(y)$ is given by $F(y)/R$, where $F(y)$ is the distribution function of y and R is the range of the study variate. Thus, the best points of stratification are obtained by equalization of the cumulatives of the function

$$\phi(\bar{y}) = \frac{1}{2}[f(y) + l(y)]$$

Rules for obtaining optimum points of stratification with different types of estimators are given by Singh (1967), in which optimum properties have also been studied.

3.11 DETERMINATION OF NUMBER OF STRATA

In context with the decision about the number of strata, the points to be discussed are

(i) how $V(\bar{y}_{st})$ decreases as k increases?

(ii) how the cost of survey is affected by a change in k?

Dalenius and Gurney (1951), Dalenius (1953), Cochran (1961), Sethi (1963), and Hess, Sethi and Balakrishnan (1966) have discussed the first question and some models have been proposed for the determination of the number of strata. For the second aspect regarding the cost with a change in k, Sethi (1963) suggested that an increase in k beyond 6 would seldom be profitable. Some results which have practical significance in surveys have been discussed in this section in the light of the above mentioned works.

It can be shown easily that stratification increases efficiency if the number of strata is increased by further sub-division of the strata. The ultimate limit of stratification is the sample size so as to arrive at the point of selecting only one unit from each stratum. But it can also be shown that multiplication of strata beyond a reasonable number is not profitable. In large-scale surveys, there may not be much advantage by increasing the number of strata up to the maximum possible extent because it may adversely affect the cost of the survey. It has been conjectured by Dalenius (1953) that the ratio of the variance of an estimator of the population mean based on k optimum strata to that based on $(k-1)$ optimum strata is $(k-1)^2/k^2$ approximately, i.e.

$$V_k\,(\bar{y}_{\text{st}}\,/V_{k-1}\,(\bar{y}_{\text{st}}) = (k-1)^2/k^2 \qquad (3.11.1)$$

This relationship has been verified by Cochran (1961) for $k = 2, 3$ and 4, for eight distributions relating to different characteristics and found to hold good in all cases.

For determitnation of the optimum number of strata, the cost function may be defined as

$$C = a + kc_1 + nc_2 \qquad (3.11.2)$$

where a is the overhead cost and c_1 and c_2 are the costs per stratum and per unit, respectively.

Sethi (1963) has shown that for optimum stratification with proportional and equal allocations, in case of gamma distribution, the relationship between sampling variance and number of strata is of the form

$$V_k\,(\bar{y}_{\text{st}}) = \frac{S^2}{n}\,(bk^2 + ck + d)^{-1} \qquad (3.11.3)$$

where b, c, d are constants to be determined by considering the values of the variance ratio for $k = 1, 2$ and 3.

The optimum value of k for a given cost can be determined with the value which minimises Eq. (3.11.3) along with restrictions given by Eq. (3.11.2). It can be shown easily that the optimum values of k and n are given by

$$k = 2\,(C - a)/3c_1$$

and
$$n = (C - a)/3c_2 \qquad (3.11.4)$$

For ensuring a specified efficiency with minimum cost, a similar procedure can be followed. Generally, this is being done by graphing the variance or cost against the number of strata for different values of n. The optimum number of strata is then to be located in them carefully.

3.12 METHOD OF COLLAPSED STRATA

Sometimes the population is so heterogeneous that, to achieve a better representation of the population, stratification is carried to the point that only one unit is selected from each stratum. In such cases, it will not be possible to estimate the variance of the estimated mean An approximate estimate of the variance of the estimated mean may be obtained by grouping the strata in pairs. The method of grouping together pairs of strata whose means do not differ much is called the *technique of collapsed strata.*

Let the units drawn in a typical pair be y_{j_1} and y_{j_2} where j takes values from 1 to $k/2$, k being an even number. Then, averaging over all samples from this pair,

$$E(y_{j_1} - y_{j_2})^2 = E(y_{j_1}^2) + E(y_{j_2}^2) - 2E(y_{j_1} y_{j_2})$$

$$= \overline{Y}_{j_1}^2 + \frac{(N_j - 1)}{N_j} S_{j_1}^2 + \overline{Y}_{j_2}^2$$

$$+ \frac{(N_j - 1)}{N_j} S_{j_2}^2 - 2\bar{y}_{j_1} \bar{y}_{j_2}$$

$$= (\overline{Y}_{j_1} - \overline{Y}_{j_2})^2 + \frac{(N_j - 1)}{N_j}(S_{j_1}^2 + S_{j_2}^2) \quad (3.12.1)$$

where the paired strata have the same size N_j. Hence, it is easy to see that

$$E\sum_{j}^{k/2} (y_{j_1} - y_{j_2})^2 = V(\bar{y}_{st}) + \sum_{j}^{k/2} (\overline{Y}_{j_1} - \overline{Y}_{j_2})^2 \quad (3.12.2)$$

This shows that the quantity $\sum_{j}^{k/2} (y_{j_1} - y_{j_2})^2$, used as an estimate of the variance, will over estimate $V(\bar{y}_{st})$. The second term will be negliglible if the strata are so grouped in pairs that their sample values differ very little from each other. In other words, the pairing should be so arranged that the strata forming the pair are about equal in size in total of y.

Hartley, Rao and Keifer (1969) have suggested a method where collapsing of strata is not involved. They have used one or more auxiliary variates on which the stratum means are supposed to have a linear regression. This method may lead to smaller bias in variance estimation, in many situations, than the method discussed above. Fuller (1970) has also developed a method of strata construction that provides an unbiased estimate of $V(\bar{y}_{st})$ with one unit per stratum. The designs are developed for equal and unequal probability sampling.

3.13 POST-STRATIFICATION (STRATIFICATION AFTER SELECTION OF SAMPLE)

In stratified sampling, it is presumed that the stratum sizes and the sampling frame in each stratum are available. However, there are instances where the latter is not always available, e.g. from the voter list of a given locality, the age of an individual voter is available although the lists of voters belonging to different age groups are not. With some characteristic which is suitable for stratification, it is not possible to know in advance to which stratum a sampling unit belongs until the sample is selected. So the technique of *post-stratification* consists in classfiying the population and selected sample into a given number of strata after selection of the sample. The problem of post-stratification has been discussed by Hansen, Hurwitz and Madow (1953). A simple procedure for getting approximations to the variance of a post-stratified estimator has been given by Williams (1962). We shall examine here the gain in precision due to such post-stratification.

Let us assume that N_i and W_i $(i = 1, \ldots, k)$ are known. If the sample is considered as a stratified sample, then instead of sample mean $\bar{y}$ we use the weighted mean $\bar{y}_w$ $(= \sum_{i}^{k} W_i \bar{y}_i)$ which would be an unbiased estimator of the population mean $\bar{Y}$. Since, for each i

$$E(\bar{y}_i) = E_1 E_2 (\bar{y}_i | n_i) = E_1 (\bar{Y}_i) = \bar{Y}_i \tag{3.13.1}$$

Hence

$$E(\bar{y}_w) = \sum_{i}^{k} W_i E(\bar{y}_i) = \sum_{i}^{k} W_i \bar{Y}_i = \bar{Y} \tag{3.13.2}$$

Let us further suppose that the sample size is so large that none of the n_i's $(i = 1, \ldots, k)$ is zero. If n_i $(i = 1, \ldots, k)$ are fixed, then we have

$$V(\bar{y}_w | n_1, \ldots, n_k) = \sum_{i}^{k} (1/n_i - 1/N_i) W_i^2 S_i^2 \tag{3.13.3}$$

Using the conditional variance formula given in relation 1.3.10 we have

$$V(\bar{y}_w) = E_1 V_2 (\bar{y}_w | n_1, n_2, \ldots, n_k)$$
$$+ V_1 E_2 (\bar{y}_w | n_1, n_2, \ldots, n_k)$$
$$= E_1 V_2 (\bar{y}_w | n_1, n_2, \ldots, n_k)$$

since the second term is independent of n_i.

$$= E_1 \left[\sum_{i}^{k} (1/n_i - 1/N_i) W_i^2 S_i^2 \right]$$

$$= \sum_{i}^{k} [E(1/n_i) - 1/N_i] W_i^2 S_i^2 \tag{3.13.4}$$

An exact expression for relation (3.13.4) cannot be derived. However, for large n and N, Stephan (1945) has shown that

$$E(1/n_i) \cong 1/nW_i + (1 - W_i)/n^2W_i^2 \tag{3.13.5}$$

Substituting the value in relation (3.13.4), we get

$$V(\bar{y}_w) = (1 - f)/n \sum_i^k W_iS_i^2 + 1/n^2 \sum_i^k (1 - W_i) S_i^2 \tag{3.13.6}$$

The first term in relation (3.13.6) is the value of $V(\bar{y}_{st})$ with proportional allocation. The second term represents the adjustment in the variance due to post-stratification, which will be small in comparison to the first term if n is large. Hence post-stratification is almost as precise as proportional stratified sampling for large samples.

3.14   DEEP STRATIFICATION

Suppose there are two alternative criteria of stratification. The question then arises as to which of the two criteria of stratification is to be preferred. An obvious choice is to opt for two-way stratification which yields k rows and k' columns on the basis of those two stratification variables. There shall be kk' strata and thus a sample of size $n = kk'$ is required to estimate the population mean. If it is required to estimate the variance of the estimated mean, at least two observations should be taken from each stratum so the minimum sample size must be $2kk'$. A problem arises when $n < kk'$ and it is also desired to give proportional allocation to each criterion of stratification. Bryant, Hartley and Jessen (1960) have given an interesting and simple solution to this problem. The steps involved in this procedure are as follows:

(i) construct a square of n^2 cells with n rows and n columns,

(ii) select n of the cells with equal probability such that no two sclected cells belong to the same column or row,

(iii) amalgamate the n rows to form k strata such that the ith stratum has an allocation of n_i units, and

(iv) amalgamate the n columns to form k' strata such that the jth stratum has an allocation of n_j units.

Let n_{ij} be the number of cells from the deep ijth stratum selected in the sample. An unbiased estimator of the population mean is given by

$$\bar{y}_u = 1/n \sum_i \sum_j G_{ij} \, n_{ij}\bar{y}_{ij} \tag{3.14.1}$$

where $\bar{y}_{ij}$ is the sample mean in the ijth stratum, G_{ij} is the weighting factor defined by

$$G_{ij} = \frac{n^2 W_{ij}}{n_i n_j}$$

and

$$W_{ij} = N_{ij}/N$$

In case of proportional allocation when $W_{ij} = W_i . W_{.j}$, the variance of the estimator in relation (3.14.1) can be written as

$$V(\bar{y}_u) \cong \frac{1}{n} \sum_i \sum_j W_{ij} S_{ij}^2 - \frac{1}{(n-1)} \left[\sum_i \sum_j W_{ij} (\bar{y}_{ij} - \bar{Y}_i)^2 \right.$$

$$\left. - \sum_j W_j (\bar{Y}_j - \bar{Y})^2 \right] \qquad (3.14.2)$$

The estimation of variance in the case of two-way stratification is quite cumbersome and will not be discussed here as limitations of space do not permit giving an account of this aspect. For this purpose, the readers are referred to Bryant (1955), where some methods for variance estimate are discussed under different conditions.

3.15 CONTROLLED SELECTION

Goodman and Kish (1950) suggested a procedure of selection, called *controlled selection*, in those situations in which it is considered desirable that certain combinations of units or preferred combinations be given greater probabilities of selection, and less probabilities of selection for some or all non-preferred combinations. Suppose, in a block growing wheat, it is proposed to select a simple random sample of 2 fields to estimate the average yield per hectare. In the sample, there are three possibilities (i) both the fields are irrigated, (ii) both the fields are unirrigated, or (iii) one field is irrigated and the other unirrigated. Undoubtedly, all samples will give an unbiased estimate of the average yield but it is likely that the first two samples will be over estimating or under estimating, while in the last case it is likely to be much closer to the parameter. Hence, it is desirable to group the fields into two strata, one with irrigated fields and the other with unirrigated, and then to select a simple random sample of one field from each stratum. Such samples are called *preferred* samples and others may be called *non-preferred* samples. At this stage, it may be mentioned that the probabilities of some or all preferred combinations of units are increased and those of some or all non-preferred combinations are reduced. Thus, stratified sampling can be considered a method of *controlled selection.* However, in this section the term controlled selection is used in the sense as having additional controls in stratification for reducing variance of the estimate of the parameter under study. Hess, Riedel and Fitzpatrick (1961) illustrated the method and derived approximate formulae for sampling variance of the estimator and its estimated variance. Another approach, using balanced incomplete block designs, has been discussed by Avadhani and Sukhatme (1973). In this section, we shall illustrate the method of examining how probabilities of preferred samples can be increased.

Let A, B, C, D be the four irrigated fields forming stratum I and let a, b, c, d, e be the five unirrigated fields forming stratum II. Suppose it is further known that fields A, B, a and b are unmanured, having code 0, and that fields C, D, c, d and e are manured, having code 1. Representation of each of two types of manure application is also desirable, but only one unit (field) is to be drawn from each stratum. In numbering the units within strata, a subscript indicates the type of manure application. In controlled selection, an attempt is made to increase the probability of preferred samples. One way of achieving it is to revise the order of units in stratum II so that the preferred samples come first. All possible samples in stratified and controlled sampling, with their probabilities are shown in Table 3.3.

Table 3.3 All possible samples in stratified and controlled sampling with their probabilities of selection

Stratum	Original order	Stratified sampling				Revised order	Controlled selection	Prob.
		Sample	Prob.	Sample	Prob.			
I	A_0	$A_0 a_0$	0.05	$C_1 a_0$	0.05	A_0	$A_0 c_1$	0.20
	B_0	$A_0 b_0$	0.05	$C_1 b_0$	0.05	B_0	$A_0 d_1$	0.05
	C_1	$A_0 c_1$	0.05	$C_1 c_1$	0.05	C_1	$B_0 d_1$	0.15
	D_1	$A_0 d_1$	0.05	$C_1 d_1$	0.05	D_1	$B_0 c_1$	0.10
II	a_0	$A_0 e_1$	0.05	$C_1 e_1$	0.05	c_1	$C_1 e_1$	0.10
	b_0	$B_0 a_1$	0.05	$D_1 a_0$	0.05	d_1	$C_1 a_0$	0.15
	c_1	$B_0 b_1$	0.05	$D_1 b_0$	0.05	e_1	$D_1 a_0$	0.05
	d_1	$B_0 c_1$	0.05	$D_1 c_1$	0.05	a_0	$D_1 b_0$	0.20
	e_1	$B_0 d_1$	0.05	$D_1 d_1$	0.05	b_0		
		$B_0 e_1$	0.05	$D_1 e_1$	0.05			

In stratified sampling, if unit A_0 or B_0 is drawn from stratum I, we would like to draw units c_1, d_1 or e_1 from stratum II so that both types of stratification are present with sample size 2. Similarly, C_1 or D_1 (stratum I) is desired with a_0 or b_0 (stratum II). The probability of a desired combination is $0.5 \times 0.6 + 0.5 \times 0.4 = 0.5$. Controlled selection makes the probability of these preferred samples as high as possible. Here we find that the only non-preferred sample is C_1e_1 and the total probability of preferred sample is 0.90.

It can be seen from the above example that we have conceptually attempted to select 2 units from 4 deep strata formed on the basis of two stratification systems. It should also be noted that the selection is not made independently, but in a dependent manner. Yet the original probabilities of selection of units are preserved. If the probabilities are p_i and p_j, then an unbiased estimator of the total is given by

$$\hat{Y} = \frac{y_i}{p_i} + \frac{y_j}{p_j} \tag{3.15.1}$$

The sampling variance of $\hat{Y}$ is given by

$$V(\hat{Y}) = \sum \frac{y_i^2}{p_i} + \sum \frac{y_j^2}{p_j} + \sum p_{ij} \frac{y_i y_j}{p_i p_j} \tag{3.15.2}$$

which is different from that given by independent selection.

SET OF PROBLEMS

3.1 A population is divided into 2 strata of sizes N_1 and N_2 units. Let ϕ denote $[n_1/n_2]/[n_1'/n_2']$, where $n_1\ n_2$ is a general allocation of the total sample size n, and n_1', n_2' is the optimum allocation for the estimation of mean in stratified random sampling.

(i) Show that the relative precision of a general allocation to the optimum allocation is given by

$$RP = \phi(N_1 U + N_2)^2/(\phi N_1 U + N_2)(N_1 U + \phi N_2)$$

where $\qquad U = S_1/S_2$

(ii) Show that the relative precision is never less than $4\phi(1 + \phi)^{-2}$.

3.2 Prove that

$$V_{SR} = V_{prop} + \frac{(N - n)}{nN(N - 1)} \{ \overset{k}{\underset{i}{\Sigma}} N_i (\bar{Y}_i - \bar{Y})^2$$

$$- \frac{1}{N} \overset{k}{\underset{i}{\Sigma}} (N - N_i) S_i^2 \}$$

3.3 The variate x_i which is non-negative follows the exponential distribution $e^{-x_i} dx_i$. The population is divided into 2 strata at a specified point x_0, and a stratified random sample of size n is taken with proportional allocation. Derive $V(\bar{x}_{st})$ as a function of x_0. Also find the optimum value of x_0 which will minimize the variance.

3.4 The variate y has rectangular distribution in the interval $(a, a + d)$. The interval is divided into k equal sub-intervals which form k strata of equal size. From each stratum, a simple random sample of size n/k units is drawn. Let V_1 and V_2 be the variances for stratified and unstratified samples of size n, respectively, prove that $V_1/V_2 = k^{-2}$.

3.5 For the normal population $\mathcal{N}(0, 1)$, $f(y)$ denotes the ordinate at y and w the weight between y_1 and y_2. If the distribution is truncated at y_1 and y_2, show that the mean M and variances S^2 of the truncated distribution are given by

$$M = \{f(y_1) - f(y_2)\}/w$$

and

$$S^2 - 1 = \frac{y_1 f(y_1) - y_2 f(y_2)}{w} - \frac{[f(y_1) - f(y_2)]^2}{w^2}$$

3.6 A population is divided into 2 strata of sizes N_1 and N_2 and sd's S_1 and S_2. The cost of the survey is fixed and given by $C = c_1 n_1 + c_2 n_2$. Assuming that S_1 and S_2 are nearly equal and fpc can be ignored, show that

$$V_{prop}/V_{opt} = N(N_1 c_1 + N_2 c_2)/(N_1 \sqrt{c_1} + N_2 \sqrt{c_2})^2.$$

If $N_1 = N_2$, compute the relative increases in precision by using proportional allocation when $c_2/c_1 = 1$ and 2.

3.7 An investigator desires to take a stratified random sample with the following assumptions:

Stratum	N_i	S_i	C_i (in Rs.)
1	400	10	4
2	600	20	9

(i) Estimate the values of n_1/n and n_2/n which minimize the total field cost $C = c_1 n_1 + c_2 n_2$ for a given value of $V(\bar{y}_{st})$.

(ii) Estimate the total sample size required, under the scheme of optimum allocation, to make $V(\bar{y}_{st}) = 1$, when fpc is ignored.

(iii) Also estimate the cost of the survey.

3.8 After the sample in Excercise 3.7 was taken, it was found that the field costs were actually Rs. 2 per unit in stratum 1 and Rs 10 in stratum 2.

(i) How much has the field cost changed from the anticipated value?

(ii) If the exact field cost is known in advance, how could the value $V(\bar{y}_{st}) = 1$ be obtained for the original estimated cost given in Exercise 3.7.

3.9 For estimating the average catch of fish on a certain part of the Indian coast, the coastal strip was divided into two geographical strata of equal number of fish-landing centres. From among the large number of centres in each stratum, a simple random sample of 5 centres was selected for observation and from each such centre, out of the large number of operating units, 3 units were selected by simple random sampling for recording the weight of eatch. The data obtained are given below:

S. no. of centres		Catches of units selected (kg)		
		1	2	3
Stratum I	1	610	754	688
	2	297	411	515
	3	1187	92	487
	4	1297	533	1130
	5	860	357	656
Stratum II	1	1085	956	980
	2	817	736	926
	3	920	616	109
	4	511	328	412
	5	906	990	736

Obtain an estimate of the average catch per unit, for the coast, with its standard error. For a given number of centres to be selected, what is the optimum break up between the two strata when the number of units selected at a centre are 3 and 4. Calculate the number of centres and the optimum break up necessary to estimate the average catch with 5 per cent standard error.

3.10 The following data show the wheat acreage of a stratified random sample of 1 in 20 taken from a certain district:

Size group	3	4	5	6
Acreage	(21–50)	(51–50)	(151–300)	(301–)
No. of farms	18	26	20	13
	8, 0, 0	49, 0, 13	20, 0, 71	72, 84, 158
	0, 8, 0	10, 27, 0	24, 36, 48	92, 0, 62
	0, 5, 9	27, 10, 16	30, 0, 70	69, 102, 0
	0, 0, 0	23, 24, 28	59, 56, 62	78, 13, 51
	0, 5, 0	4, 19, 5	17, 18, 0	92
	0, 0, 0	30, 14, 23	76, 17, 0	873
	35	0, 4, 22	80, 32	
		0, 0, 3	716	
		13, 12		
		386		

Size group 1:   (1–5) acreage: 22 farms, no wheat
Size group 2:   (6–20) acreage: 25 farms, 7 acreage of wheat on one farm.

Estimate the total wheat acreage of the district and the mean acreage of wheat per farm. Estimate also the number of farms growing wheat. What are the standard errors of the estimates?

3.11 A sample survey for estimating the number of orchards of apple was conducted in Mahasu district of Himachal Pradesh (India) during a given year. Four strata A, B, C and D of villages, according to the acreage of temperate fruit trees as obtained from the revenue records, were formed. The sizes of strata (in acres) were 0–3, 3–6, 6–15 and 15 and above, respectively. A simple random sample of villages in each stratum was selected and the number of apple orchards was noted in the selected villages. The numbers of apple orchards for various strata are given below:

Stratum	Total number	Number of villages selected	Number of orchards in the selected villages
A	275	15	2, 5, 1, 9, 6, 7, 0, 4, 7, 0, 5, 0, 0, 3, 0.
B	146	10	21, 11, 7, 5, 6, 19, 5, 24, 30, 24.
C	93	12	3, 10, 4, 11, 38, 11, 4, 46, 4, 18, 1, 19.
D	62	11	30, 42, 20, 38, 29, 22, 31, 28, 66, 41, 15.

Estimate the number of orchards in the district. Calculate whether there is any gain due to stratification, over simple random sampling. Using the value of s_i^2, the mean square within the ith stratum as the variance, give an optimum allocation of 48 villages.

3.12 In a pilot sample survey conducted in the district of Wardha (India), for estimation of total cattle population in the district, a two way stratification of the district by tehsils and sizes of the villages, as judged by the number of households, was adopted. A total sample of 125 villages was distributed equally among all 9 strata. Villages were sampled with equal probability and without replacement within each stratum. All households in a village were enumerated for total cattles. The total cattle and mean square between villages within each stratum are given on page 78.

Calculate the sampling variance of the estimated total cattle population in the district, for the sample of 60 villages, if the villages were selected by the method of

(i) simple random sampling without stratification, and

(ii) simple random sampling within each stratum and allocated in proportion to the product $N_i S_i$

Tehsils	Size according to number of households	Stratum	Total no. of villages N_i	No. of villages sampled (n_i)	Total cattle in sampled villages	Mean square between villages within strata s_i^2
Arvi	0–50	1	127	13	1486	$(78.9279)^2$
	51–125	2	109	14	5085	$(126.2198)^2$
	126 and above	3	81	14	10141	$(339.9322)^2$
Wardha	0–50	4	86	14	1224	$(73.0564)^2$
	51–125	5	121	14	4282	$(55.2853)^2$
	126 and above	6	115	14	11334	$(432.2364)^2$
Huigaughat	0–50	7	123	14	2909	$(128.2665)^2$
	51–125	8	103	14	5753	$(101.3872)^2$
	126 and above	9	63	14	10002	$(298.5456)^2$
			928	125	52228	

3.13 For a socio-economic survey, all the villages in a region, including the un-inhabited ones, were grouped into four strata on the basis of their altitude above sea level and population density. From each stratum, 10 villages were selected with srs wr. The data on the number of households in each of the sample villages are given below:

Stratum S. no.	Total no. of villages	Total number of households in sample villages									
		1	2	3	4	5	6	7	8	9	10
1	1411	43	84	98	0	10	44	0	124	13	0
2	4705	50	147	62	87	84	158	170	104	56	160
3	2558	228	262	110	232	139	178	334	0	63	220
4	14,997	17	34	25	34	36	0	25	7	15	31

(i) Obtain an estimate of the total number of households and its standard errror.

(ii) Estimate the gain due to use of stratification as compared to unstratified srs wr.

(iii) Compare the efficiency of the present allocation with that of the optimum allocation, keeping the total sample size fixed.

3.14 Using the data given below and considering the size classes as strata, compare the efficiencies of the following alternative allocations of a sample of

3000 factories for estimating the total output. The sample is to be selected with sr wor within each stratum:
  (i) Proportional allocation
  (ii) Allocation proportional to total output
  (iii) Optimum allocation

S. no.	Size class no. of workers	No. of factories	Output per factory (in '000 Rs)	Standard deviation (in '000 Rs
1	1–49	18260	100	80
2	50–99	4315	250	200
3	100–249	2233	500	600
4	250–999	1057	1760	1900
5	1000 and above	567	2250	2500

3.15 A survey is to be conducted for estimating the total number of literates in a town inhabited by three communities, some particulars of which are given bleow on the basis of the results of a pilot survey:

Community	Total number of persons	Percentage of literates
1	60000	40
2	10000	80
3	30000	60

  (i) Treating the community as strata and assuming srs wr in each stratum, allocate a total sample size of 2000 persons to the strata in an optimum manner for estimating the overall proportion of literates in the town.
  (ii) Estimate the efficiency of stratification as compared to unstratified sampling.

REFERENCES

Avadhani, M.S. and B.V. Sukhatme, "Controlled sampling with equal probabilities and without replacement", *Int. Statist. Rev*, **41**, 175–183, (1973).

Ayoma, H., "A study of the stratified random sampling". *Ann. Inst. Statist. Math.*, **6**, 1–36, (1954).

Bowley, A.L., "Measurement of precision attained in sampling," *Bull. Inter. Statist. Inst.*, **22**, 1–62, (1926).

Bryant, E.C., "An analysis of some two-way stratifications" *Ph. D. Thesis*, Iowa State University, Iowa (1955).

Bryant, E.C., H.O. Hartley and R.J. Jessen, "Design and estimation in two-way stratification," *J. Amer. Statist. Assoc.*, **55**, 105–124, (1960).

Cochran, W.G., "Comparison of methods for determining stratum boundaries," *Bull. Int. Statist. Inst.*, **38**, 345-358, (1961).

Dalenius, T., "The problem of optimum stratification," *I. Skand. Akt.*, **33**, 203-213, (1950).

"The problem of optimum stratification in a special type of design," *Skand. Akt.*, **35**, 61-70, (1952).

"Multivariate sampling problem," *Skand. Akt.*, **36**, 92-122, (1953).

and M. Gurney, "The problem of optimum stratification II," *Skand. Akt.* **34**, 133-148, (1951).

and J.L. Hodges, "The choice of stratification points," *Skand. Akt.*, **40**, 198-203, (1957).

"Minimum variance stratification," *J. Amer. Statist. Assoc.*, **54**, 88-101, (1959).

Durbin, J., "Review of the book sampling in Sweden," *J.R. Statist. Soc.* **122**, 246-248, (1959).

Ekman, G., "An approximation useful in univariate stratification," *Ann. Math. Statist.* **30**, 219-229, (1959).

Fuller, W.A., "Sampling with random stratum boundaries," *J.R. Statist. Soc.*, **B 32**, 209-206, (1970).

Goodman, R. and I. Kish, "Controlled selection—a technique in probability sampling," *J. Amer. Statist. Assoc.*, **45**, 350-372, (1950).

Hagood, M.J. and E.H. Bernert, "Component indexes as a basis for stratification in sampling," *J. Amer. Statist Assoc.*, **40**, 330-341, (1945).

Hansen, M.H., W.N. Hurwitz and W.G. Madow, *Sample Surveys: Methods and Theory*, John Wiley and Sons, New York, (1953).

Hartley, H.O., J.N.K. Rao and G. Kiefer, "Variance estimation with one unit per stratum," *J. Amer. Statist. Assoc.*, **64**, 841-851, (1969).

Hess, I., V.K. Sethi and T.R. Balakrishnan, "Stratification—a practical investigation," *J. Amer. Statist. Assoc.*, **61**, 74-90, (1966).

Hess, I., D.C. Riedel and T.B. Fitzpatrick, *Probability Sampling of Hospitals and Patients*, University of Michigan, Ann. Arbor, (1961).

Mahalanobis, P.C., "Some aspects of the design of sample surveys," *Sankhya*, **12**, 1-7, (1952).

Neyman, J., "On the two different aspects of the representative method," *J.R. Statist. Soc.*, **97**, 558-606, (1934).

Sethi, V.K., "A note on optimum stratification for estimating the population means," *Aust. J. Statist.*, **5**, 20-33, (1963).

Singh, Ravindra, *Some contributions to the theory of construction of strata*, Ph.D. Thesis, I.A.S.R.I., New Delhi, (1967).

Stephan, F.F., "The expected value and variance of the reciprocal and other negative powers of a positive Bernoulli variate," *Ann. Math. Statist.*, **16**, 50-61, (1945).

Stuart, A., "A simple presentation of optimum sampling results," *J.R. Statist. Soc.*, **B 16**, 239-241, (1959).

Tschuprow, A A., "On the mathematical expectation of the moment of frequency distribution in the case of correlated observations," *Metron*, **2**, 461-493, 646-683, (1923).

Williams, W.H., "On the variance of an estimator with post-stratification," *J. Amer. Statist. Assoc.*, **57**, 622-627, (1962).

4

Systematic Random Sampling

> *At 6 p.m. the well marked 1/2 inch of water, at nightfall 3/4 and at daybreak 7/8 of an inch. By noon of the next day there was 15/16 inch and on the next night 31/32 of an inch of water in the hold. The situation was desperate. At this rate of increase few, if any, could tell where it would rise to in a few days.*
>
> Stephen Leacock

4.1 INTRODUCTION

In the previous chapters we have discussed those sampling techniques where sampling units were selected randomly. Now we shall discuss a sampling technique which has a nice feature of selecting the whole sample with just one random start. A sampling technique in which only the first unit is selected with the help of random numbers and the rest get selected automatically according to some pre-designed pattern is known as *systematic random sampling*. Systematic random sampling is also referred to briefly as systematic sampling. Suppose N units of the population are numbered from 1 to N in some order. Let $N = nk$, where n is the sample size and k is an integer, and a random number less than or equal to k be selected and every kth unit thereafter. The resultant sample is called *every kth* systematic sample and such a procedure termed *linear systematic sampling*. If $N \neq nk$, and every kth unit be included in a circular manner till the whole list is exhausted, it will be called *circular systematic* sampling.

Systematic sampling is simple and fool proof. Apart from its simplicity, this procedure, in many situations, provides estimates more efficient than simple random sampling and is widely used in various types of surveys. Systematic sampling has been discussed at length by Madow and Madow (1944), Madow (1949, 1953) and others. Its applications in different fields has been demonstrated by Finney (1948) and Sukhatme *et al* (1958). Singh *et al* (1968) have suggested a modified systematic sampling procedure and its suitability has been illustrated by its application to a survey for estimating milk yield.

4.2   SAMPLE SELECTION PROCEDURES

Systematic sampling is a commonly used technique if a complete and up-to-date sampling frame is made available. We shall first discuss these sample selection procedures before we evaluate their advantages and disadvantages.

4.2.1   Linear Systematic Sampling

As mentioned above, a common procedure is the linear systematic sampling scheme. We suppose that the population is linearly ordered in some way such that units can be referred to by number, without ambiguity. Further, let N be expressible in the form $N = nk$ and let the selected random number be $i \, (\leqslant k)$, k being called the *sampling interval*. In this procedure, the sample comprises the units, $i, i + k, i + 2k, \ldots, i + (n - 1) k$. The technique will generate k systematic samples with equal probability, which may be shown in the schematic diagram given below:

Table 4.1   Schematic Diagram Showing k Systematic Samples in the Population

		Sample Number		
1	2	... $\quad$ i	...	k
y_1	y_2	... $\quad$ y_i	...	y_k
y_{1+k}	y_{2+k}	... $\quad$ y_{i+k}	...	y_{2k}
y_{1+2k}	y_{2+2k}	... $\quad$ y_{i+2k}	...	y_{3k}
.	.	... $\quad$.	...	.
.	.	... $\quad$.	...	.
$y_{1+(j-1)k}$	$y_{2+(j-1)k}$	... $\quad$ $y_{i+(j-1)k}$	...	y_{jk}
.	.	... $\quad$.	...	.
$y_{1+(n-1)k}$	$y_{2+(n-1)k}$	... $\quad$ $y_{i+(n-1)k}$	...	y_{nk}
Mean $\bar{y}_1$.	$\bar{y}_2$.	... $\quad$ $\bar{y}_i$.	...	$\bar{y}_k$.

Another practical situation is that N is not expressible in the form $N = nk$. In this case, the present sampling scheme will give rise to samples of unequal size. k is taken as an integer nearest to N/n. Then a random number is chosen from 1 to k and every kth unit is drawn in the sample. Under this condition, the sample size is not necessarily n and in some cases it may be $(n-1)$. For example, if $N=11$, $n=4$, then the value of k is 3 and possible samples are $(1, 4, 7, 10)$; $(2, 5, 8, 11)$ and $(3, 6, 9)$, which are not of the same size. We shall discuss an improvement over it in the next section.

4.2.2 Circular Systematic Sampling

To overcome the difficulty of varying sample size under the situation $N \neq nk$, the procedure is modified slightly by which a sample of constant size is always obtained. The procedure consists in selecting a unit, by a random start, from 1 to N and then thereafter selecting every kth unit, k being an integer nearest to N/n, in a circular manner, until a sample of n units is obtained. This technique is generally known as *circular systematic sampling*. Suppose that a unit with random number i is selected. The sample will then consist of the units corresponding to the serial numbers

$$i + jk; \quad \text{if } i + jk \leqslant N$$
$$\text{for } j = 0, 1, \ldots, (n-1)$$
$$i + jk - N; \quad \text{if } i + jk > N$$

It can be easily verified that every unit has got an equal probability of selection $(1/N)$ in this method.

As an illustration, let $N = 11$, and $n = 4$. Then $k = 3$.

The possible samples are, therefore,

$(1, 4, 7, 10)$; $(2, 5, 8, 11)$; $(3, 6, 9, 1)$; $(4, 7, 10, 2)$; $(5, 8, 11, 3)$; $(6, 9, 1, 4)$; $(7, 10, 2, 5)$; $(8, 11, 3, 6)$; $(9, 1, 4, 7)$; $(10, 2, 5, 8)$; and $(11, 3, 6, 9)$.

Further, it will be seen that systematic sampling has the drawback of not yielding an unbiased estimate of sampling variance with single sample. To overcome this difficulty, a new systematic sampling procedure has been suggested by Singh and Singh (1977), which we shall explain in Section 4.9 of this chapter.

4.3 ADVANTAGES AND DISADVANTAGES

The main advantage of systematic sampling is its simplicity of selection, operational convenience and even spread of the sample over the population. It has, therefore, been found very useful in forest surveys for estimating the volume of timber, in fisheries for estimating the total catch of fish, in milk yield surveys for estimation of the lactation yield, etc. Another advantage is that, except for populations with periodicities, systematic sampling provides an efficient estimate as compared to alternative designs. Sometimes systematic sampling variances are much smaller than the variances for random selection of units within strata.

In case of periodicity in the population, systematic sampling has to be used with considerable care. If, for periodic population, sampling interval is an odd multiple of half the period of the cycle, systematic sampling provides zero variance. When the sampling interval is a simple multiple of the period of the cycle, systematic sampling is no better than selecting one unit at random. A serious disadvantage of systematic sampling lies in its use with populations having unforeseen periodicity which may substantially contribute bias to the estimate of the population mean value. Another disadvantage concerns the drawback of estimating the sampling variance of estimators with single sample.

4.4 ESTIMATION OF MEAN AND ITS SAMPLING VARIANCE

Let y_{ij} denote the jth member of the ith systematic sample, ($j = 1, 2, \ldots, n; i = 1, 2, \ldots, k$). The mean of the ith sample may be denoted by $\bar{y}_i.$. We have to consider the problem of estimating the population mean under two different situations: (i) when $N = nk$ and (ii) when $N \neq nk$. First we shall discuss the situation when $N = nk$.

THEOREM 4.4.1 In systematic sampling with interval k, sample mean $\bar{y}_{sy}$ is an unbiased estimator of the population mean $\bar{Y}$. Its sampling variance is given by

$$V(\bar{y}_{sy}) = \frac{(k-1)}{k} S_c^2 \qquad (4.4.1)$$

where S_c^2 denotes the mean square between the column means in the population (c stands for column).

Proof   Since the probability of selection of the ith systematic sample is $1/k$, we have

$$E(\bar{y}_{sy}) = \frac{1}{k} \sum_{i}^{k} \bar{y}_{i\cdot} = \sum_{i}^{k} \sum_{j}^{n} y_{ij}/nk = \bar{Y}$$

Hence, $\bar{y}_{sy}$ is an unbiased estimator.

If the ith sample is considered the ith column, the total sum of squares due to column means is given by

$$\sum_{i}^{k} (\bar{y}_{i\cdot} - \bar{Y})^2 = (k - 1) S_c$$

Hence, the variance of the sample means $\bar{y}_{sy}$ is given by

$$V(\bar{y}_{sy}) = E(\bar{y}_{i\cdot} - \bar{Y})^2$$
$$= \sum_{i}^{k} (\bar{y}_{i\cdot} - \bar{Y})^2/k = \frac{(k - 1)}{k} S_c^2$$

It should not be inferred from the obove formula that the variance of systematic sample mean will decrease if the sample size is increased. This makes the point clear that systematic sampling is a delicate device and should be used carefully.

THEOREM 4.4.2   The sampling variance of sample mean $\bar{y}_{sy}$ is given by

$$V(\bar{y}_{sy}) = (N - 1) S^2/N - (n - 1) S_w^2/n = \sigma^2 - \sigma_w^2 \quad (4.4.2)$$

where

$$S_w^2 = \sum_{i}^{k} \sum_{j}^{n} (y_{ij} - \bar{y}_{i\cdot})^2/k(n - 1)$$

with $(n - 1) S_w^2 = n\sigma_w^2$, and σ^2 has its usual meaning.

Proof   We know that the total sum of squares is given by

$$\text{TSS} = (N - 1) S^2 = \sum_{i} \sum_{j} (y_{ij} - \bar{Y})^2$$
$$= \sum_{i} \sum (\bar{y}_{ij} - \bar{y}_{i\cdot})^2 + n \sum_{i} (\bar{y}_{i\cdot} - \bar{Y})^2$$

Also, we know that the variance of $\bar{y}_{sy}$ is

$$V(\bar{y}_{sy}) = \sum_{i}^{k} (\bar{y}_{i\cdot} - \bar{Y})^2/k$$

and

$$(N - 1) S^2 = k(n - 1) S_w^2 + nk V(\bar{y}_{sy})$$

Hence,

$$V(\bar{y}_{sy}) = (N - 1) S^2/N - (n - 1)S^2/n$$
$$= \sigma^2 - \sigma_w^2$$

Since σ^2 is fixed for a given population, it is obvious from this result that in order to reduce $V(\bar{y}_{sy})$ it is necessary to increase the within-sample variance. This is possible by making an arrangement of units in such a way that the units within each systematic sample are heterogeneous to the maximum extent. This shows that one has to be very careful in using systematic sampling and should always ensure that units similar to one another with respect to the study variate are put together in one zone so that the units included in the sample from different zones have the maximum possible variation.

COROLLARY The systematic sample mean $\bar{y}_{sy}$ is more efficient than the simple random sample mean $\bar{y}$ if $S_w^2 > S^2$.

This result states that systematic sampling is more precise than simple random sampling if the variance within the systematic samples is more than the total variation in the population. In other words, systematic sampling will be more precise only if units within the sample are heterogeneous.

We shall now briefly consider the case when $N \neq nk$. Let us suppose that $N = nk + l$, where $l < k$. In this situation the sample size will vary, being either n or $(n + 1)$, depending upon the random start. Suppose y_i. be the sample total, then the sample mean is

$$\bar{y}_i. = y_i./(n + 1) = \sum_{j}^{n+1} y_{ij}/(n + 1) \qquad \text{if } i \leqslant l \qquad (4.4.3)$$

$$= y_i./n = \sum_{j}^{n} y_{ij}/n \qquad \text{if } i > l$$

Hence, $$E(\bar{y}_i.) = 1/k \sum_{i}^{k} \bar{y}_i. \neq \bar{Y} \qquad (4.4.4)$$

Thus, $\bar{y}_i.$ is a biased estimator of the population mean. However, an unbiased estimator of mean can be obtained by

$$\bar{y}'_{sy} = k\,\bar{y}_i./N \qquad (4.4.5)$$

It can be shown that $(\bar{y}'_{sy})$ is unbiased, for

$$E(\bar{y}'_{sy}) = E(ky_i./N) = \sum_{i}^{k} \frac{k}{N} y_i./k$$

$$= \sum_{i} \sum_{j} y_{ij}/N = \bar{Y} \qquad (4.4.6)$$

The difference between $\bar{y}'_{sy}$ and $\bar{y}_{sy}$, viz.,

$$\frac{k}{N} \sum_{i} y_i. - \bar{y}_{sy} = \frac{n'k - 1}{N} \bar{y}_{sy},$$

where n' is the number of units expected in the sample, is likely to be negligible if N is sufficiently large. Hence, the bias involved in using the sample mean as an estimator of the population mean will be negligible in case of sample selected from a large population.

Also,
$$V(\bar{y}'_{sy}) = V(ky_i./N) = \frac{k^2}{N^2} V(y_i.)$$

$$= \frac{k^2}{N^2} \cdot \frac{(k-1)}{k} S'^2_c$$

$$= k(k-1)S'^2_c/N^2 \qquad (4.4.7)$$

where
$$S'^2_c = \sum_i^k (\bar{y}_i. - \bar{Y})^2/(k-1)$$

As we have already mentioned, the difficulty of varying sample size can be avoided by the use of circular systematic sampling. This method gives rise to N possible samples, each with probability of selection of $1/N$. It can also be shown that in case of circular systematic sampling, the sample mean is an unbiased estimator of the population mean. The variance of the estimator is given by

$$V(\bar{y}_{csy}) = E(\bar{y}_i. - \bar{Y})^2 = \sum_i^N (\bar{y}_i. - \bar{Y})^2/N \qquad (4.4.8)$$

For the sake of convenience and simplicity, we shall assume $N=nk$ for further discussion in this chapter, unless mentioned otherwise.

4.5 COMPARISON OF SYSTEMATIC WITH RANDOM SAMPLING

If $\bar{y}$ is the mean of a simple random sample of size n from a population of size N then its sampling variance is given by

$$V(\bar{y}) = (N-n)S^2/nN$$

where S^2 is the mean square between units in the population.

Also, we can write

$$V(\bar{y}_{sy}) = \sum_i^k (\bar{y}_i. - \bar{Y})^2/k = \frac{1}{k} \sum_i^k \{\sum_i^n y_{ij}/n - \bar{Y}\}^2$$

$$= \frac{1}{n^2 k} \sum_i^k \{\sum_j^n (y_{ij} - \bar{Y})\}^2$$

$$= \frac{1}{n^2 k} [\sum_i^k \sum_j^n (y_{ij} - \bar{Y})^2$$

$$+ \sum_i^k \sum_{j \neq j'}^n (y_{ij} - \bar{Y})(y_{ij'} - \bar{Y})]$$

$$= \frac{1}{n^2 k} [(nk - 1) S^2 + \rho (nk - 1)(n - 1) S^2]$$

$$= \frac{(nk - 1) S^2}{n^2 k} [1 + (n - 1) \rho] \tag{4.5.1}$$

where ρ is the intra-class correlation between the units of the same systematic sample and is defined by

$$\rho = E (y_{ij} - \overline{Y}) (y_{ij'} - \overline{Y})/E (y_{ij} - \overline{Y})^2$$

$$= \frac{\sum_i \sum_{j \neq j'} (y_{ij} - \overline{Y}) (y_{ij'} - \overline{Y})/nk (n - 1)}{(nk - 1) S^2/nk} \tag{4.5.2}$$

From relation (4.5.1) it can be seen that a positive correlation between units in the same sample inflates the sampling variance of the estimate. The increase is quite significant even for small positive correlation because of the term $(n - 1)$. Since S^2 is fixed for a given population, it is clear from the above relation that an arrangement with the largest possible negative value of ρ should be preferred. This again leads to the same conclusion that the units within the same systematic sample should be as heterogeneous as possible with respect to the characteristic under study. Since V_{sy} is never less than zero, ρ cannot be less than $-1/(n - 1)$. Thus the minimum value of ρ is $- 1/(n - 1)$ and in this case $V_{\text{sy}} = 0$.

The relative precision of systematic sample mean with sample random mean is given by

$$V_{\text{sy}}/V_{\text{SR}} = \frac{(N - 1)}{(N - n)} [1 + \rho (n - 1)] \tag{4.5.3}$$

It can be seen that the relative precision depends on the value of ρ. For $\rho = -1/(N - 1)$, the two methods give estimates of equal precision ; for $\rho < -1/(N - 1)$, the estimate based on systematic sample is more precise; while if $\rho > - 1/(N - 1)$, it reverses. The maximum value which ρ can attain is 1, when the relative precision of systematic sampling is given by $(k - 1)/(N - 1)$.

4.6 COMPARISON OF SYSTEMATIC WITH STRATIFIED RANDOM SAMPLING

Let us suppose that the population of $N = nk$ units is divided into n strata corresponding to n rows of the schematic diagram in Table 4.1

and that one unit is drawn randomly from each stratum, thus giving a stratified sample of size n. Then the mean of the jth stratum is

$$\bar{y}._j = \sum_i^k y_{ij}/k, (j = 1, 2, \ldots, n) \qquad (4.6.1)$$

and the population mean is

$$\overline{Y} = \sum_i^k \sum_j^n y_{ij}/nk = \sum_i^k \bar{y}_{i.}/k = \sum_j^n \bar{y}._j/n$$

Similarly, the stratum mean square of the units within the jth stratum (row) is defined by

$$S_j^2 = \sum_i^k (y_{ij} - \bar{y}._j)^2/(k - 1), (j = 1, 2, \ldots, n) \qquad (4.6.2)$$

Hence, the pooled mean square between units within stratum is defined by

$$S_{wst}^2 = \sum_j^n \sum_i^k (y_{ij} - \bar{y}._j)^2/n(k - 1)$$

$$= \sum_j^n S_j^2/n \qquad (4.6.3)$$

Clearly, the variance of the mean of this stratified sample will be

$$V_{st}(\bar{y}) = \frac{1}{n^2} \sum_j^n \frac{(1 - 1)}{k} S_j^2$$

$$= \frac{(k - 1)}{nk} S_{wst}^2 \qquad (4.6.4)$$

Now we shall express the variance of the systematic sample in a suitable form for a comparative study. Equation (4.4.1) can be written as

$$V_{sy}(\bar{y}) = \frac{1}{k} \sum_i^k \left[\frac{1}{n} \sum_j^n y_{ij} - \frac{1}{n} \sum_j^n \bar{y}._j \right]^2$$

$$= \frac{1}{n^2 k} \sum_i^k [\sum_j^n (y_{ij} - \bar{y}._j)]^2$$

$$= \frac{1}{n^2 k} [\sum_i^k \sum_j^n (y_{ij} - \bar{y}._j)^2 + \sum_i^k \sum_{i \neq j'}^n (y_{ij} - \bar{y}._j)(y_{ij'} - \bar{y}._{j'})]$$

$$= \frac{1}{n^2 k} [n(k-1) S_{wst}^2 + n(n - 1)(k - 1) \rho_{wst} S_{wst}^2]$$

$$= \frac{(k - 1)}{nk} S_{wst}^2 [1 + (n - 1) \rho_{wst}] \qquad (4.6.5)$$

where ρ_{wst} is a non-circular serial correlation coefficient defined by

$$\rho_{wst} = E(y_{ij} - \bar{y}_{.j})(y_{ij'} - \bar{y}_{.j'})/E(y_{ij} - \bar{y}_{.j})^2 \qquad (4.6.6)$$

$$= \sum_{i}^{k} \sum_{j \neq j'}^{n} (y_{ij} - \bar{y}_{.j})(y_{ij} - \bar{y}_{.j'})/(n-1)n(k-1)S_{wst}^2$$

Comparing relations (4.6.4) and (4.6.5), we get

$$\text{R.P.} = V_{st}/V_{sy} = [1 + (n-1)\rho_{wst}]^{-1} \qquad (4.6.7)$$

Thus we see that the relative precision of systematic sampling over stratified random sampling depends upon the values of ρ_{wst}. If ρ_{wst} is positive then stratified random sampling will provide a better estimate of $\bar{Y}$. However, if $\rho_{wst} = 0$, it may be concluded that both systematic sampling and stratified random sampling are equally good.

EXAMPLE 4.1   Given below are the daily milk yield (in litres) records of the first lactation of a specified cow belonging to the Tharparkar herd maintained at the Government Cattle Farm, Patna. The milk yields of the first five days were not recorded, being the colostrum period.

Day	1	2	3	4	5	6	7	8	9	10
Milk yield	10	11	14	10	14	9	10	8	11	10
Day	11	12	13	14	15	16	17	18	19	20
Milk yield	6	9	8	7	9	10	11	11	13	12
Day	21	22	23	24	25	26	27	28	29	30
Milk yield	12	10	11	11	14	15	12	17	18	16
Day	31	32	33	34	35	36	37	38	39	40
Milk yield	13	14	14	15	16	16	16	13	16	17
Day	41	42	43	44	45	46	47	48	49	50
Milk yield	14	16	15	14	14	15	17	15	16	17
Day	51	52	53	54	55	56	57	58	59	60
Milk yield	25	22	23	19	18	16	22	21	21	23
Day	61	62	63	64	65	66	67	68	69	70
Milk yield	21	19	19	19	19	19	19	19	19	19

Day	71	72	73	74	75	76	77	78	79	80
Milk yield	18	19	21	20	17	16	18	18	18	22

Day	81	82	83	84	85	86	87	88	89	90
Milk yield	22	22	20	20	20	18	20	21	21	20

Day	91	92	93	94	95	96	97	98	99	100
Milk yield	18	21	22	22	20	21	21	21	21	21

Day	101	102	103	104	105	106	107	108	109	110
Milk yield	19	20	21	20	21	20	21	20	21	20

Day	111	112	113	114	115	116	117	118	119	120
Milk yield	19	21	18	21	20	22	21	21	21	16

Day	121	122	123	124	125	126	127	128	129	130
Milk yield	19	15	15	16	19	12	16	14	15	17

Day	131	132	133	134	135	136	137	138	139	140
Milk yield	16	20	15	19	16	16	20	20	18	21

Day	141	142	143	144	145	146	147	148	149	150
Milk yield	22	22	21	22	21	21	21	18	20	17

Day	151	152	153	154	155	156	157	158	159	160
Milk yield	20	20	21	21	21	20	20	16	16	15

Day	161	162	163	164	165	166	167	168	169	170
Milk yield	18	19	18	20	19	18	16	14	14	13

Day	171	172	173	174	175	176	177	178	179	180
Milk yield	16	16	16	18	16	15	16	18	18	15

Day	181	182	183	184	185	186	187	188	189	190
Milk yield	18	16	17	18	17	16	13	14	13	12

Day	191	192	193	194	195	196	197	198	199	200
Milk yield	16	10	13	8	8	6	8	9	4	5

Day	201	202	203
Milk yield	6	6	4

Find the efficiency of systematic sampling at 7 and 14 days' interval of recording, with respect to corresponding simple random sampling, in estimating the lactation yield of the cow.

(i) For $N = 203$, $k = 7$, the value of n is $\dfrac{N}{k} = 29$.

The data may be arranged in a $n \times k$ table as given below:

	1	2	3	4	5	6	7
1	10	11	14	10	14	9	10
2	8	11	10	6	9	8	7
3	9	10	11	11	13	12	12
4	10	11	11	14	15	12	17
5	18	16	13	14	14	15	16
6	16	16	13	16	17	14	16
7	15	14	14	15	17	15	16
8	17	25	22	23	19	18	16
9	22	21	21	23	21	19	19
10	19	19	19	19	19	19	19
11	18	19	21	20	17	16	18
12	18	18	22	22	22	20	20
13	20	18	20	21	21	20	18
14	21	22	22	20	21	21	21
15	21	21	19	20	21	20	21
16	20	21	20	21	20	19	21
17	18	21	20	22	21	21	21
18	16	19	15	15	16	19	12
19	16	14	15	17	16	20	15
20	19	16	16	20	20	18	21
21	22	22	21	22	21	21	21
22	18	20	17	20	20	21	21
23	21	20	20	16	16	15	18
24	19	18	20	19	18	16	14
25	14	13	16	16	16	18	16
26	15	16	18	18	15	18	16
27	17	18	17	16	13	14	13
28	12	16	10	13	8	8	6
29	8	9	4	5	6	6	4
Total	477	495	481	494	486	472	465
Mean ($\bar{y}_{i.}$)	16.45	17.07	16.59	17.03	16.76	16.28	16.03

Hence $$Y = \sum_{i=1}^{7} y_{i.} = 3370$$

Taking random start $i = 5$, we get the 5th sample.

In this case, the unbiased estimate of the total milk yield is given by

$$\hat{Y}_{sy} = 7 \times 486 = 3402$$

The variance of systematic sampling for the population total estimate is given by

$$V(\hat{Y}_{sy}) = \frac{N^2}{k}\left[\sum_{i=1}^{k} \bar{y}_{i.}^2 - \frac{Y^2}{k}\right]$$

$$= \frac{203^2}{7}\left[16.45^2 + 17.07^2 + \ldots + 16.03^2 - \frac{3370^2}{7}\right] = 5192$$

The variance of simple random sample estimate of total milk yield is given by

$$V(\hat{Y}_{SR}) = N^2\left(\frac{1}{n} - \frac{1}{N}\right)S^2$$

where $$S^2 = \frac{1}{N-1}\left[\sum_{i=1}^{k}\sum_{j=1}^{n} y_{ij}^2 - \frac{Y^2}{N}\right]$$

Thus, $$V(\hat{Y}_{SR}) = (203)^2\left(\frac{1}{29} - \frac{1}{203}\right)\frac{3865.7228}{202}$$

$$= 23194.33$$

Therefore, the relative precision of systematic sampling over simple random sampling will be

$$\frac{23194.33}{5192} \times 100 = 446.73\%$$

(ii) For $N = 203$ and $k = 14$, we have $n = 15$ or 14

Taking a random start $i = 10$, we get the 10th sample. In this case, the unbiased estimate of total milk yield is given by

$$\hat{Y}_{sy} = 14 \times 246 = 3444$$

Its variance is given by

$$V(\hat{Y}_{sy}) = k\sum_{i} y_i^2 - Y^2$$

$$= 14 \times 812028 - (2270)^2 = 11492$$

The data can be arranged in a $k \times n$ table as given below:

$j \backslash i$	1	2	3	4	5	6	7	8	9	10	11	12	13	14
1	10	11	14	10	14	9	10	8	11	10	6	9	8	7
2	9	10	11	11	13	12	12	10	11	11	14	15	12	17
3	18	16	13	14	14	15	16	16	16	13	16	17	14	16
4	15	14	14	15	17	15	16	17	25	22	23	19	18	16
5	22	21	21	23	21	19	19	19	19	19	19	19	19	19
6	18	19	21	20	17	16	18	18	18	22	22	22	20	20
7	20	18	20	21	21	20	18	21	22	22	20	21	21	21
8	21	21	19	20	21	20	21	20	21	20	21	20	19	21
9	18	21	20	22	21	21	21	16	19	15	15	16	19	12
10	16	14	15	17	16	20	17	19	16	16	20	20	18	21
11	22	22	21	22	21	21	21	18	20	17	20	20	21	21
12	21	20	20	16	16	15	18	19	18	20	19	18	16	14
13	14	13	16	16	16	18	16	15	16	18	18	15	18	16
14	17	18	17	16	13	14	13	12	16	10	13	8	8	6
15	8	9	4	5	6	6	4	—	—	—	—	—	—	—
Total ($y_{i.}$)	249	247	246	248	247	241	238	228	248	235	246	239	231	227
Mean ($\bar{y}_{i.}$)	16.60	16.47	16.40	16.53	16.47	16.06	15.87	16.28	17.71	16.78	17.57	17.07	16.50	16.21

The variance of a simple random sample estimate of total milk yield will be

$$V(\hat{Y}_{\text{sR}}) = N^2 \left(\frac{1}{n} - \frac{1}{N} \right) S^2$$

$$= (203)^2(1/14 - 1/203) \times 3865.7228/202$$

$$= 52187.2578$$

Therefore, the relative precision of systematic estimate over simple random sample estimate is

$$52187.2578 \times 100/11492 = 454.12\%$$

4.7 ESTIMATION OF VARIANCE

An unbiased estimate of the variance is not available for a systematic sample with one random start because a systematic sample is regarded as a random sample of one unit (cluster). However, some biased estimators are possible on the basis of a systematic sample. If two or more systematic samples are available, an unbiased estimate of the variance of the estimated mean can be made.

Suppose m independent systematic samples are available, each of size n. Then, an unbiased estimate of the estimated mean is given by

$$v(\bar{y}_{\text{sy}}) = \sum_{i=1}^{m} [\bar{y}_{i\cdot} - (\bar{y}_{\text{sy}})]^2/m \, (m - 1) \qquad (4.7.1)$$

where $\qquad (\bar{y}_{\text{sy}}) = \sum_{i=1}^{m} \bar{y}_i/m$

An approximate and biased estimate of the variance of a systematic sample mean is given by

$$v(\bar{y}_{\text{sy}}) = \left(\frac{1}{n} - \frac{1}{N} \right) s_{\text{wc}}^2 \qquad (4.7.2)$$

where $\qquad s_{\text{wc}}^2 = (\sum_{j}^{n} y_{ij}^2 - n\bar{y}_{i\cdot}^2)/(n - 1)$

Another approximate and biased estimate of variance of a systematic sample mean is given by

$$v(\bar{y}_{\text{sy}}) = \left(\frac{1}{n} - \frac{1}{N} \right) \sum_{j=1}^{n-1} (y_{ij} - y_{ij+1})^2/2(n - 1) \qquad (4.7.3)$$

These estimators, given by relations (4.7.2) and (4.7.3), are biased and should be used with caution otherwise they may provide misleading results in practice. It is, therefore, desirable to get an idea of the bias in a specific case by comparing it with the sampling variance obtained by using the entire data. These estimators are based on the results given by Cochran (1946) and Yates (1948).

EXAMPLE 4.2 In an experimental agricultural census carried out by I.A.S.R.I., New Delhi, in the Loni block of Meerut District in U.P. (India) during 1967-68, two villages of the block were selected randomly. Out of 225 holdings in two villages, namely Panch-lok and Agrola, 45 holdings were selected by systematic sampling (with 5 as the sampling interval). The total arable land (in kacha bigha)* for the 45 selected holdings are given below:

S. No.	1	2	3	4	5	6	7
Total arable land	60	50	14	10	1	0	0
S. No.	8	9	10	11	12	13	14
Total arable land	0	0	0	150	150	100	20
S. No.	15	16	17	18	19	20	21
Total arable land	0	25	192	25	0	13	0
S. No.	22	23	24	25	26	27	28
Total arable land	0	50	0	10	0	0	0
S. No.	29	30	31	32	33	34	35
Total arable land	85	30	30	70	30	35	0
S No.	36	37	38	39	40	41	42
Total arable land	30	0	0	10	0	20	70
S. No.	43	44	45				
Total arable land	16	15	35				

*(5 kacha bigha = 1 acre)

Estimate the total arable land in the two villages and also the approximate standard error of the estimate.

An estimate of the total arable land is given by

$$\hat{Y} = \frac{N}{n} \sum_{i}^{n} y_{i.} = \frac{225}{45} \times 1348 = 6740$$

An estimate of the variance of $\hat{Y}$ is given by

$$v(\hat{Y}) = \frac{N(k-1)}{2(n-1)} \left[\sum_i^{n-1} (y_{i+1} - y_i)^2 \right]$$

$$= \frac{225 \times (5-1)}{2 \times (45-1)} \times 116845$$

$$= 11,950,031.25$$

$\therefore$ Estimated standard error of

$$\hat{Y} = \sqrt{11,950,031.25} = 3456.88.$$

4.8 INTERPENETRATING SYSTEMATIC SAMPLING

If it is essential to have a rigorous estimate of the sampling variance, it can be done by taking systematic sub-samples with independent random starts, each containing n/m units to keep the total sample size the same. Let $\bar{y}_i, \bar{y}_2, \ldots, \bar{y}_m$ be estimates based on m independent systematic sub-samples. An unbiased estimator of the population mean is the pooled mean given by

$$\bar{y} = \sum_i^m \bar{y}_i/m \qquad (4.8.1)$$

Then, an unbiased estimate of the variance of $\bar{y}$ is

$$v(\bar{y}) = \sum_i^m (\bar{y}_i - \bar{Y})^2/m(m-1) \qquad (4.8.2)$$

In case of m systematic sub-samples of $n/m(=n')$ units, each with random starts from 1 to $mk(=k')$, the variance and estimated variance of the pooled mean $\bar{y}$ are

$$V(\bar{y}) = \frac{(k'-m)}{k'm(k'-1)} \sum_i^{k'} (\bar{y}_i - \bar{Y})^2 \qquad (4.8.3)$$

where $\bar{y}_i$. is the ith sample mean $(i = 1, \ldots, k')$, and

$$v(\bar{y}) = \frac{(k'-m)}{k'm(m-1)} \sum_i^m (\bar{y}_i. - \bar{y})^2 \qquad (4.8.4)$$

Though these variance estimators are unbiased, they are less precise if m is small. Similarly, if m is increased by decreasing the sub-sample size, the combined estimator of the population mean is likely to be less efficient. Hence, one has to arrive at a decision between the desire of getting a good estimate of variance or a good estimate of the population mean.

4.9 NEW SYSTEMATIC SAMPLING*

As pointed out in the previous section, systematic sampling suffers from the drawback of not being able to provide an unbiased estimator of sampling variance on the basis of a single sample. One possible way of getting an unbiased variance estimator is to use interpenetrating systematic sampling, but this results in loss in precision and simplicity. Singh and Singh (1977) have suggested a new systematic sampling procedure which provides an unbiased variance estimator. In this section, we shall discuss briefly the outlines of the main results.

Suppose a population consists of N distinct units and a sample of size n is to be drawn from it. Let $u \, (\leqslant n)$ and d be two predetermined integers which are chosen in such a way that (i) every sample contains distinct units, and (ii) the inclusion probability for each pair units is non-zero. Starting with a random number $r \, (\leqslant N)$, we select u units continuously, and thereafter the remaining $n - u (=v)$ units with interval d such that $d \leqslant u$ and $u + vd \leqslant N$. With these restrictions, a sample of desired sample size n can be drawn in two or more phases by this sampling scheme. Here the term 'phase' is used in the sense that the ultimate sample of the given size is selected in stages. The number of phases p required for selecting a sample of n units from a population of size N is given by

$$p \geqslant \{\log \log (N/2) - \log \log (n/2)\}/\log 2 \qquad (4.9.1)$$

The sample space corresponding to the selection procedure contains N possible samples, and the sample point corresponding to the random number r is given by $s_r = (s_r', s_r'')$ where s_r' consists of the unit indices $r + m \, (m = 0, 1, \ldots, u - 1)$ and s_r'' consists of the unit indices $r + u - 1 + m'd \, (m' = 1, 2, \ldots, v)$.

The probability of selecting each sample is $1/N$, and thus, the probability measure associated with the selection procedure is given by

$$P = \{P(s_r)\} : (P(s_r) = 1/N, \quad r = 1, 2, \ldots, N) \qquad (4.9.2)$$

Thus, the inclusion probabilities for individual units and pairwise units are given by π_i and π_{ij}, which are known under the new systematic sampling procedure. Hence, the Horvitz–Thompson estimator of the population mean simplifies to

$$\bar{y}_{n \text{ sy}} = \frac{1}{N} \sum_i^N y_i/\pi_i \qquad (4.9.3)$$

*May be omitted at the first reading.

The sampling variance of the estimator $\bar{y}_{nsy}$ in Yates–Grundy form reduces to

$$V(\bar{y}_{nsy}) = \frac{1}{N^2} \sum_i^N \sum_{j>i}^N \left(1 - \pi_{ij} - \frac{N^2}{n^2}\right)(y_i - y_j)^2 \qquad (4.9.4)$$

which can be written as

$$V(\bar{y}_{nsy}) = \left\{\left(\frac{N-1}{N}\right)S^2 - \left(\frac{n-1}{n}\right)S^2_{wnsy}\right\} \qquad (4.9.5)$$

where $S^2_{wnsy} = \sum_r^N \sum_i^n (y_{ri} - \bar{y}_r)^2 / N(n-1)$.

An unbiased estimator of $V(\bar{y}_{nsy})$ can be obtained by

$$v(\bar{y}_{nsy}) = \frac{1}{N^2} \sum_i^n \sum_{j>i}^n (1/\pi_{ij} - N^2/n^2)(y_i - y_j)^2 \qquad (4.9.6)$$

In a comparative study, the efficiency of the new systematic sampling over the usual systematic sampling and simple random sampling has been examined by considering different types of populations. It was concluded, in this study, that for many natural populations, new systematic sampling provides better results than the usual systematic sampling.

4.10 COMPARISON OF SYSTEMATIC WITH SIMPLE AND STRATIFIED RANDOM SAMPLES FOR SOME SPECIFIED POPULATIONS

The performance of systematic sampling in relation to stratified or simple random sampling depends very much on the nature of the population in which the characteristic under study has some simple trend in terms of units alone. Some main trends are discussed here as under.

4.10.1 Populations with Linear Trend

Suppose the values of the population units increase in accordance with a linear model such that

$$y_i = a + bi$$

where a and b are constants and i goes from 1 to N.

Hence

$$\bar{Y} = \frac{1}{N} \sum_i^N (a + bi) = a + b(N+1)/2$$

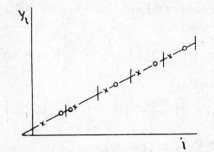

Fig. 4.1 Systematic sampling in a population with linear trend

and
$$S^2 = \sum_{i}^{N} (y_i - \bar{Y})^2/(N - 1)$$

$$= \sum_{i}^{N} \frac{b^2}{(N - 1)} \left(i - \frac{N + 1}{2} \right)^2$$

$$= \frac{N(N + 1)\, b^2}{12} = \frac{nk(nk + 1)\, b^2}{12}$$

Therefore
$$V_{sr} = \frac{(N - n)}{Nn} \cdot S^2$$

$$= \frac{k - 1}{nk} \cdot \frac{nk(nk + 1)}{12} \cdot b^2$$

$$= (k - 1) \left(\frac{nk + 1}{12} \right) b^2 \tag{4.10.1}$$

Similarly, to find variance within strata, we have to put k for N, and we get

$$V_{st} = \frac{N - n}{Nn} S_w^2 = \frac{(k - 1)}{nk} \cdot \frac{k(k+1)}{12} \cdot b^2$$

$$= (k - 1) \frac{(k + 1)}{12n} b^2 \tag{4.10.2}$$

For systematic sampling, the values can be put in Eq. (4.3.2)

$$V_{sy} = \frac{1}{k} \sum (\bar{y}_i - \bar{Y})^2 = \frac{1}{k} \cdot \frac{k(k + 1)(k - 1)}{12} \cdot b^2$$

$$= (k - 1) \frac{(k + 1)}{12} b^2 \tag{4.10.3}$$

which proves

$$V_{\text{st}} : V_{\text{sy}} : V_{\text{sr}} = (k+1)/n : k + 1 : nk + 1 \left.\begin{array}{c} \\ \\ \end{array}\right\}$$
$$\simeq 1/n : 1 : n = 1 : n : n^2$$

(4.10.4)

It is seen that the variance of a stratified sample is only $1/n$th of the variance of a systematic sample in a population with linear trend. Similarly, the variance of a systematic sample is approximately $1/n$th of the variance of a random sample. Hence, stratified sampling is the most efficient of all the methods for eliminating the effect of linear trend. The performance of systematic sampling in the presence of a linear trend can be improved by applying the end corrections given by Yates (1948) or the modified method of Singh *et al* (1968), which chooses pairs of units equidistant from the ends of the population.

4.10.2 Populations with Periodic Variations

Suppose the population consists of a periodic trend which is represented by

$$y = \sin\left(\frac{\pi i}{n} + \alpha\right)$$

where i varies from 0 to an integral multiple of $2n$. Assuming $2n = 20$, successive sampling units will repeat themselves after every 20th value.

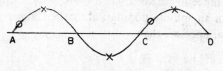

Fig. 4.2 A population with regular
periodic variation

A five per cent systematic sample from such a population will consist of sampling units drawn from the same position of each cycle. An estimate from such a sample will be as good as a single value. On the other hand, a five per cent random sample will contain units from different parts of the population and estimates from such samples will be more precise for the effect of a periodic trend. If we select a 10 per cent systematic sample, then the estimate will be as good as the population value, thus making systematic sampling the most precise of all sampling methods. Thus, the precision of systematic sampling from populations having a periodic trend depends upon the periodicity. The most favourable case occurs when i is an odd multiple of the half period.

4.10.3 Natural Populations

Systematic sampling is both operationally convenient and efficient in sampling some natural populations like forest areas for estimating the volume of timber, hardwood seedlings, areas under different types of covers, etc. Osborne (1942), Yates (1948) and Finney (1948, 1950) have examined the relative efficiency of systematic sampling in certain natural populations and observed that performance of systematic sampling had been comparatively better.

4.10.4 Auto-correlated Populations

It has already been observed that the efficiency of systematic sampling in comparison with sample random sampling or stratified sampling depends upon the nature of the population. In various natural populations, units close to each other are more alike than units far apart. For the study of such populations, called auto-correlated populations, we assume that the observations y_i and y_j $(i, j = 1, 2, \ldots, N)$ are positively correlated and the serial correlation coefficient ρ_d is a function of the distance in between, i.e. $d = y_i - y_j$. Suppose the units y_i are drawn from an infinite population (called *super population*) with mean μ and variance σ^2, then

$$\left. E(y_i) = \mu, \qquad E(y_i - \mu)^2 = \sigma^2 \atop \text{and} \qquad E(y_i - \mu)(y_j - \mu) = \rho_d \sigma^2 \right\} \qquad (4.10.5)$$

for $i = 1, 2, \ldots, N$ and $d = 1, 2, \ldots, (N-1)$

where

$$\rho_d \geqslant \rho_{d'} \geqslant 0 \text{ whenever } d < d'$$

The graph of ρ_d as a function of d is called a correlogram. It has been observed that ρ_d increases as d decreases. Madow and Madow (1944) have shown that if the units in the population are ordered, then it is enough to adopt systematic sampling instead of simple random sampling. Cochran (1946) has shown that if, besides the assumptions taken in Eq. (4.10.5), it is assumed that

$$\delta_d^2 = \rho_{d-1} + \rho_{d+1} - 2\rho_d \geqslant 0 \qquad (4.10.6)$$

for all $d = 2, 3, \ldots, (nk - 2)$,

then $$E(V_{sy}) \leqslant E(V_{st}) \qquad (4.10.7)$$

The condition, $\delta_d^2 \geqslant 0$, about the second difference of serial correlation coefficients, implies that the correlogram is concave upwards, and for such populations systematic sampling is more efficient than stratified sampling.

Singh *et al* (1968) have discussed different cases, viz.

(i) population with linear trend plus random element,

(ii) population with linear trend and periodic variation, and

(iii) population with parabolic trend,

and they have shown that the performance of modified systematic sampling over stratified or simple random sampling is quite satisfactory.

4.11 TWO-DIMENSIONAL SYSTEMATIC SAMPLING

So far we have discussed one-dimensional systematic sampling in which the population units are supposed to be serially ordered on a line. There are many situations where the population units are naturally arranged on an area or a plane instead of a line. A systematic sampling procedure for such a situation is known as *plane systematic* or *two-dimensional systematic* sampling. The simplest extension of a linear systematic sample to the two-dimensional plane is popularly known as a *squared grid* or *aligned* sample. Quenouille (1949) and Das (1950) have discussed the problem of two-dimensional systematic sampling at length. In this section, we describe briefly a procedure for selecting a two-dimensional systematic sample.

Let us assume that the population consists of N square (or rectangular) grid areas of equal size and a sample of n grid areas is to be taken. Also assume that the grids are arranged in $l \times m$ ($= nk = N$) form of cells and each cell is arranged in the form of r rows and s columns. The simplest way is to select a pair of random numbers (i, j) such that $i \leqslant r, j \leqslant s$. Thus, the random location of a grid in the cell is done uniquely. A systematic sample selected by this procedure is an aligned or square-grid sample and the selected grids are shown by the dots in Fig. 4.3 (a).

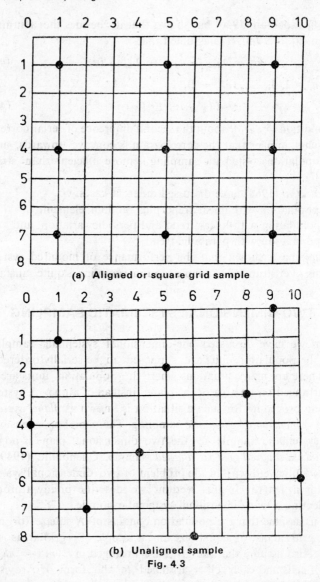

(a) **Aligned or square grid sample**

(b) **Unaligned sample**

Fig. 4.3

Another procedure is as follows: the grids are arranged in the form of $r \cdot l$ rows and $m \cdot s$ columns and it is required to select a systematic sample of size $n = rs$ grids.

Select r independent random numbers $i_1, i_2, \ldots, i_r$ each less than or equal to l; and s independent random numbers $j_1, j_2, \ldots, j_s$ each less than or equal to m. The grids included in the sample are $(i_{x+1} + xl, j_{x+1} + ym)$ for $x = 0, 1, \ldots, (r - 1)$ and $y = 0, 1, \ldots, (s - 1)$.

A sample selected by this procedure is known as an unaligned sample and the selected grid points are shown by the dots in Fig. 4.3(b).

Investigations given by Quenouille (1949) and Das (1950) indicated that an unaligned pattern will be superior both to a square grid and a stratified random sample. Milne (1959) has proposed a central square grid technique in which the sample point lies at the centre of the square. A two- and three-dimensional lattice sampling technique has been discussed by Yates (1960).

SET OF PROBLEMS

4.1 What is systematic sampling? Give the circumstances under which it is to be preferred to simple random sampling. Explain how you will estimate the variance of a systematic sample with a random start.

4.2 Describe 'linear' and 'circular' systematic sampling procedures and discuss briefly their advantages and disadvantages. In a finite population of size N, show that systematic sampling will be more efficient than random sampling with equal probability, wor, if the intra-class correlation coefficient

$$\rho < - 1/(N - 1).$$

4.3 Discuss the situations under which systematic samples are preferred to other types of samples in censuses and surveys. Show that a systematic sample mean is a more efficient estimator of the population mean than a simple random mean, but less efficient than a stratified random sample mean in a population with linear trend.

4.4 Describe precisely the procedure of drawing a systematic sample of n units from a population of N units, where N is not necessarily a multiple of n. (i) How would you estimate the mean of a characteristic in the population and examine its unbiasedness? (ii) How would you combine the estimates from three systematic samples of size n into a single-pooled estimate which is unbiased and estimate its standard error?

4.5 In a two-dimensional population with $N^2 = n^2k^2$ units, the linear trend is given by $y_{ij} = i + j$ $(i, j = 1, 2, \ldots, nk)$, where y_{ij} is the item value in the ith row and jth column.

A systematic square grid sample of n^2 units is taken by taking two independent random starts (i_1, j_1) each between 0 and k and $(i_1 + xk, j_1 + yk)$ for $x, y = 0, 1, \ldots, (n - 1)$. Show that the mean of this sample has the same precision as the sample mean of size n^2. (Cochran, 1977)

4.6 In a two-dimensional population with NM units, the linear trend is given by $y_{ij} = i + j$ for $i = 1, 2, \ldots, N$, and $j = 1, 2, \ldots, M$, where y_{ij} is the item value on the ith row and jth column.

A systematic two-dimensional sample of nm units, where $n = N/k$ and $m = M/k'$, k and k' being integers, is taken by taking two independent random starts i_1, j_1 each between 0 and k and k', respectively. The units included in the sample are

$$(i_1 + xk, j_1 + yk')$$

for $\qquad x = 1, 2, \ldots, (n - 1)$

and $\qquad y = 1, 2, \ldots, (m - 1)$

Show that the sample mean is an unbiased estimate for the population mean. Derive its sampling variance and compare it with a simple random sample mean of nm units from the population.

4.7 In a population with quadratic trend $y_i = i^2$ ($i = 1, \ldots, 25$), compare the values of $E(\bar{y}_{sy} - \bar{Y})^2$ given by every kth systematic sample of size 5 by (i) Yates method and (ii) Singh *et al* method.

4.8 Data on the number of seedlings in every individual foot of a sown bed which is 80 feet in length, are given below:

1–10	11–20	21–30	31–40	41–50	51–60	61–70	71–80
26	16	27	37	4	36	20	21
28	9	20	14	5	20	21	26
11	22	25	14	11	43	15	16
16	26	39	24	9	27	14	18
7	17	24	18	25	20	13	11
22	39	25	17	16	21	9	19
44	21	18	14	13	18	25	27
26	14	44	38	22	19	17	29
31	40	55	36	18	27	7	31
26	30	39	29	9	30	30	29

(i) Find the standard error of the estimate of the total number of seedlings based on a systematic sample consisting of every 10th foot of the sown bed.

(ii) Also find the relative efficiency of systematic sampling when compared to (a) simple random sampling, wor, with the sample size as 8 one-foot bed lengths (b) a stratified sample of size 8 with 2 units per stratum.

4.9 For investigating the possibility of estimating the catch of marine fish, a pilot survey was conducted in a sample of fishing centres on the Malabar Coast of India. At each landing centre in the sample, a count was made of the number of boats landing every hour from 6 am to 6 pm. Of the boats landing each hour, the first one was selected for observation on the weight of fish, product of this with the number of boats giving an estimate of the fish during the hour.

Data on the number of boats landing and the catch of fish at a particular centre on a particular day are given below:

Hours	1	2	3	4	5	6	7	8	9	10	11	12
No. of boats (x)	42	52	19	5	23	56	36	59	14	14	2	6
Catch of fish (y) (quintals)	563	887	223	88	352	1295	934	1265	466	433	98	0

Calculate the relative efficiencies of linear systematic sampling as compared to srs wor for estimating the population totals of x and y when the sample sizes are 2, 3, and 6, taking each hour as the sampling unit.

4.10 Table given below furnishes complete enumeration data on the length of strip (x) and volume of timber (y) for each strip in three blocks of the Black Mountain Forest, California.

	Block I			Block II			Block III	
Strip no.	x	y	Strip no.	x	y	Strip no.	x	y
1	12	762	1	9	471	1	6	165
2	12	651	2	9	426	2	6	224
3	12	461	3	9	448	3	6	192
4	12	521	4	9	402	4	6	161
5	12	653	5	9	372	5	6	104
6	12	544	6	9	372	6	5	94
7	12	542	7	9	411	7	5	102
8	12	590	8	9	323	8	5	115
9	11	533	9	9	381	9	4	110
10	11	517	10	9	430	10	4	109
11	11	520	11	9	434	11	4	83
12	11	539	12	9	324	12	4	36
13	10	509	13	9	543	13	4	61
14	10	449	14	9	607	14	4	92
15	10	492	15	8	416	15	4	75
16	10	498	16	8	326	16	4	64

(i) Examine the behaviour of the sampling variance of estimates of volume of timber based on systematic samples of sizes 2, 3, 4, 6, 8 and 12.

(ii) Compare the efficiency of systematic sampling with those of simple random sampling with and without replacement for the sample sizes considered in (i).

(iii) Study the efficiency of sampling the strips with probability proportional to the length of the strips with replacement.

4.11 Given below are data for 10 systematic samples of size 4 from a population of 40 units.

				Systematic sample numbers					
1	2	3	4	5	6	7	8	9	10
0	1	2	1	4	5	6	7	7	9
7	8	9	10	12	13	15	6	16	17
18	18	19	20	21	20	24	13	28	29
29	30	31	31	33	32	35	37	38	63

Work out the relative efficiency of systematic sampling over random sampling.

4.12 Table given below shows a list of 70 villages in a tehsil of India along with their population in 1981 and cultivated area in the same year. Making use of the population in 1981 as preliminary information, rearrange the villages in linear order for estimating the total cultivated area from a systematic sample.

S. no.	Population	Cultivated area (acres)	S. no.	Population	Cultivated area (acres)	S. no.	Population	Cultivated area (acres)
1	226	678	26	1007	680	51	441	622
2	670	663	27	1567	970	52	555	342
3	4505	1290	28	5271	1850	53	827	387
4	1732	1170	29	659	340	54	2867	322
5	2874	1390	30	3209	2450	55	726	636
6	2282	1110	31	2902	1760	56	633	410
7	793	760	32	2955	2120	57	680	427
8	895	730	33	1746	1220	58	587	496
9	1157	950	34	1045	860	59	1901	936
10	3201	1700	35	666	620	60	2419	1226
11	1117	909	36	904	760	61	1258	836
12	1236	1169	37	773	602	62	1225	634
13	5201	1840	38	1040	532	63	1447	978
14	848	660	39	760	438	64	1314	724
15	1238	1140	40	2084	633	65	1298	422
16	1917	1360	41	828	277	66	728	493
17	1800	1509	42	4877	1640	67	851	396
18	2335	1810	43	911	424	68	786	732
19	4396	2240	44	1205	822	69	663	422
20	1607	1225	45	1139	555	70	740	370
21	2071	1250	46	4064	347			
22	2155	1690	47	1114	744			
23	7780	3200	48	547	372			
24	2746	1744	49	1178	644			
25	2549	2400	50	1159	732			

(i) Draw five circular systematic samples of size 7 each, from rearranged frame.

(ii) From each of the five samples, estimate the total cultivated area in the tehsil using the figures for cultivated area for the selected villages as given in Table.

(iii) Obtain a single combined estimate from the five sample estimates. Also, calculate the standard error of this combined estimate.

REFERENCES

Cochran, W.G., "Sampling theory when the sampling units are of unequal sizes," *J. Amer. Statist. Assoc.*, **37**, 199–212, (1942).

——"Relative accuracy of systematic and stratified random samples for a certain class of populations," *Ann. Math. Statist.*, **17**, 164–177, (1946).

——*Sampling Techniques*, John Wiley & Sons, New York, (1977).

Das, A.C., "Two dimensional systematic sampling," *Sankhya*, **10**, 95–108, (1950).

Finney, D.J., "Random and systematic sampling in timber surveys," *Forestry*, **22**, 1–36, (1948).

——"An example of periodic variation in forest sampling," *Forestry*, **23**, 96–111, (1950).

Madow, W.G., "On the theory of systematic sampling II," *Ann. Math. Statist.*, **20**, 333–354, (1949).

——"On theory of systematic sampling III," *Ann. Math. Statist.*, **24**, 101–106, (1953).

——and L.H. Madow, "On the theory of systematic sampling I," *Ann. Math. Statist.*, **15**, 1–24, (1944).

Milne, A., "The centric systematic area sample treated as a random sample," *Biometrics*, **15**, 270–297, (1959).

Osborne, J.G., "Sampling errors of systematic and random surveys of cover-type areas," *J. Amer. Statist. Assoc.*, **37**, 256-264, (1942).

Quenouille, M.H., "Problems in plane sampling," *Ann. Math. Statist.*, **20**, 355-375, (1949).

Singh, D., K.K. Jindal and J.N. Garg, "On modified systematic sampling," *Biometrika*, **55**, 541-546 (1968).

Singh, D. and P. Singh. "New systematic sampling," *J. Statist. Plan and Inference*, **1**, 163-177, (1977).

Sukhatme, P.V., V.G. Panse and K.V.R. Sastri, "Sampling techniques for estimating the catch of sea fish in India," *Biometrics*, **14**, 78–96, (1958).

Yates, F., "Systematic sampling," *Phil. Trans. R. Soc.*, A **241**, 345–377, (1948).

——*Sampling Methods for Censuses and Surveys*, Charles Griffin & Co., London, (1960).

Varying Probability Sampling

In a minute there is time
For decisions and revision which a minute will reverse.

T.S. Eliot

5.1 INTRODUCTION

In Chapter 2 we have discussed simple random sampling in which the
selection probabilities were equal for all units of the population. Whenever the units vary in size, simple random sampling is not an appropriate procedure as no importance is given to the size of the unit.
Such ancillary information about the size of the units can be utilized
in selecting the sample so as to set more efficient estimators of the
population parameters. One such method is to assign unequal probabilities of selection to different units in the population depending on
their sizes. Let us consider again Example 2.2 (Chapter 2). With
orchards having varying numbers of fruit trees, it may be desirable to
provide a sampling scheme in which orchards are selected with probabilities proportional to the number of trees in the orchards. When
units vary in their sizes and the variate under study is highly correlated
with the size of the unit, the probability of selection may be assigned
in proportion to the size of the unit. This type of sampling procedure
where the probability of selection is proportional to the size of the
unit is known as *probability proportional to size* sampling, abbreviat-
ed as *pps sampling*.

There is a basic difference between simple random sampling and pps sampling procedures. In simple random sampling, the probability of drawing any specified unit at any given draw is the same, while in pps sampling, it differs from draw to draw. The theory of pps sampling is consequently more complex than that of simple random sampling. We shall first discuss the theory appropriate to *pps sampling with replacement* (pps wr) and then present the theory of *pps sampling without replacement* (pps wor).

5.2 PROCEDURES OF SELECTING A SAMPLE

The procedure of selecting a sample consists in associating with each unit a number or set of numbers equal to its size. The selection of units is done corresponding to a number chosen at random from the totality of numbers associated. There are two methods of selection:

(i) Cumulative total method

(ii) Lahiri's method

5.2.1 Cumulative Total Method

Let the size of the ith unit be X_i $(i = 1, 2, \ldots, N)$, the total being $X = \sum_{i}^{N} X_i$. We associate the numbers 1 to X_1 with the first unit, the numbers $(X_1 + 1)$ to $(X_1 + X_2)$ with the second unit, and so on. A number k is chosen at random from 1 to X and the unit with which this number is associated is selected. Clearly, the ith unit in the population is being selected with a probability proportional to X_i. If a sample of size n is required, the procedure is repeated n times with replacement of the units selected. This procedure of selection is known as the cumulative total method for the method needs cumulation of the unit sizes.

The main difficulty in this procedure is the compulsion to complete successive cumulative totals, which becomes time consuming and costly when the population size is large.

EXAMPLE 5.1 A village has 10 holdings consisting of 50, 30, 45, 25, 40, 26, 24, 35, 28 and 27 fields, respectively. Select a sample of four holdings with the replacement method and with probability proportional to the number of fields in the holding.

The first step in the selection of holdings is to form cumulative totals as shown on next page.

S. no. of holdings	Size (X_i)	Cumulative size	Numbers associated
1	50	50	1– 50
2	30	80	51– 80
3	45	125	81–125
4	25	150	126–150
5	40	190	151–190
6	26	216	191–216
7	44	260	217–260
8	35	295	261–295
9	28	323	296–323
10	27	350	324–350

To select a holding, a random number not exceeding 350 is drawn with the help of a random number table. Suppose the random number thus selected is 272. It can be seen from the cumulative totals that the number is associated with the group 261–295, i.e. the 8th holding is selected corresponding to the random number 272. Similarly, we select three more random numbers. Suppose these numbers are 346, 165 and 094. Then the holdings selected corresponding to these random numbers are the 10th, 5th and 3rd, respectively. Hence, a sample of 4 holdings selected with probability proportional to size will contain the 8th, 10th, 5th and 3rd holdings.

5.2.2 Lahiri's Method

Lahiri (1951) suggested an alternative procedure in which cumulations are avoided completely. It consists in selecting a number at random between 1 and N and noting down the unit with the corresponding serial number, provisionally. Another random number is then chosen between 1 and M, where M is the maximum size of the N units of the population.

If the second random number in smaller than the size of the unit provisionally selected, the unit is selected into the sample. If not, the entire procedure is repeated until a unit is finally selected. For selecting a sample of n units, the procedure is to be repeated until n units are selected.

EXAMPLE 5.2 Select a sample of four holdings in the previous example by Lahiri's method of pps, wr.

In this case, $N = 10$, $M = 50$. First we have to select a random number which is not greater than 10 and a second random number which is not greater than 50. Referring to the random number table, the pair is $(10, 13)$. Hence, the 10th unit is selected in the sample. Similarly, choosing other pairs, we can have $(4, 26)$, $(5, 35)$, $(7, 26)$. The pair $(4, 26)$ is rejected as 26 is greater than the size value (25) and so another pair is drawn which turns out to be $(8, 16)$. Hence, the sample will consist of the holdings with serial numbers 10, 5, 7 and 8.

5.3 ESTIMATION IN pps SAMPLING WITH REPLACEMENT: TOTAL AND ITS SAMPLING VARIANCE

Consider a population of N units and let y_i be the value of the characteristic under study for the unit U_i of the population $(i = 1, \ldots, N)$. Suppose further that $p_i = X_i/X$ be the probability that the unit U_i is selected in a sample of one such that $\sum_i^N p_i = 1$. Let n independent selections be made with the replacement method and the value of y_i for each selected unit be observed. Further, let (y_i, p_i) be the value and probability of selection of the ith unit of the sample. It can be seen that the random variates y_i/p_i $(i = 1, 2, \ldots, n)$ are independently and identically distributed. If $p_i = 1/N$, it gives rise to a simple random sample. This shows that simple random sampling is a particular case of pps sampling. In this section, we shall describe briefly a procedure of estimating the population total by pps sampling, wr.

THEOREM 5.3.1 In pps sampling, wr, an unbiased estimator of the population total Y is given by

$$\hat{Y}_{\text{pps}} = \frac{1}{n} \sum_i^n (y_i/p_i) \qquad (5.3.1)$$

with its sampling variance

$$V(\hat{Y}_{\text{pps}}) = \frac{1}{n} \sum_i^N p_i (y_i/p_i - Y)^2 \qquad (5.3.2)$$

Proof Let us define random variates $z_i = y_i/p_i$ $(i = 1, \ldots, n)$ which are independently and identically distributed. Hence,

$$E(z_i) = \sum_i^N p_i (y_i/p_i) = Y$$

Now let us consider $\bar{z} = \dfrac{1}{n} \sum\limits_{i}^{n} z_i$

Since, $E(\hat{Y}_{\text{pps}}) = E(\bar{z}) = \sum\limits_{i}^{n} \dfrac{1}{n} E(z_i) = Y,$

$\hat{Y}_{\text{pps}}$ is an unbiased estimator.

To obtain the sampling variance of $\hat{Y}_{\text{pps}}$, we have

$$V(\hat{Y}_{\text{pps}}) = V(\bar{z}) = \dfrac{1}{n^2} \sum\limits_{i}^{n} V(z_i) = \dfrac{1}{n} \sum\limits_{i}^{N} p_i \left(\dfrac{y_i}{p_i} - Y \right)^2$$

the covariance terms $\text{Cov}(y_i/p_i, y_j/p_j)$, $i \neq j$ being zero.

This shows that the variance of the estimator is inversely proportional to the sample size n as in simple random sampling, wr.

COROLLARY An unbiased estimator of the population mean, $\bar{Y}$ is given by

$$\hat{\bar{Y}}_{\text{pps}} = \dfrac{1}{nN} \sum\limits_{i}^{N} (y_i/p_i) \tag{5.3.3}$$

with its sampling variance

$$V(\hat{\bar{Y}}_{\text{pps}}) = \dfrac{1}{n} \sum\limits_{i}^{N} p_i (y_i/Np_i - \bar{Y})^2 \tag{5.3.4}$$

THEOREM 5.3.2 In pps sampling, wr, an unbiased estimator of $V(\hat{Y}_{\text{pps}})$ is given by

$$v(\hat{Y}_{\text{pps}}) = \dfrac{1}{n(n-1)} \sum\limits_{i}^{n} (y_i/p_i - \hat{Y}_{\text{pps}})^2 \tag{5.3.5}$$

$$= \dfrac{1}{n(n-1)} \left[\sum\limits_{i}^{n} (y_i/p_i)^2 - n\hat{Y}_{\text{pps}}^2 \right]$$

Proof In pps sampling, wr, $z_i (= y_i/p_i)$, $i = 1, 2, \ldots, n$, are n independent unbiased estimators of Y having the same variance. Like Theorem 2.4.2, let us define

$$s^2 = \sum\limits_{i}^{n} (z_i - \bar{z})^2/(n-1)$$

It can be shown easily that s^2 is an unbiased estimator of S^2 (with the usual meaning) and hence the theorem can be proved.

COROLLARY   An unbiased estimator of $V(\hat{\overline{Y}}_{pps})$ is given by

$$v(\hat{\overline{Y}}_{pps}) = \frac{1}{n(n-1)} \sum_i^n (y_i/Np_i - \hat{\overline{Y}}_{pps})^2 \qquad (5.3.6)$$

$$= \frac{1}{n(n-1)N^2} [\sum_i^n (y_i/p_i)^2 - n\hat{Y}_{pps}^2] \qquad (5.3.7)$$

Equation (5.3.7) is mainly used for computational purposes.

5.4   GAIN DUE TO pps SAMPLING WITH REPLACEMENT

One may be interested to know whether it is possible to estimate the gain due to pps sampling, as compared to simple random sampling, from the pps sample itself. In simple random sample, wr, the variance is given by

$$V(\hat{Y}_{SR}) = \frac{N^2}{n} S^2 = \frac{N}{n} [\sum_i^N y_i^2 - N\overline{Y}^2] \qquad (5.3.8)$$

An unbiased estimator of $\sum_i^N y_i$ is given by $\frac{1}{n} \sum_i^n y_i^2/p_i$ and that of $N\overline{Y}^2$ is obtained from

$$[\hat{Y}_{pps}^2 - v(\hat{Y}_{pps})]$$

After substituting these values, we have an unbiased estimator of the variance of the estimate based on simple random sample on the basis of pps sample estimate,

$$v_{pps}(\hat{Y}_{SR}) = \frac{N}{n^2} \sum_i^n y_i^2/p_i - \frac{1}{n} \{\hat{Y}_{pps}^2 - v(\hat{Y}_{pps})\}$$

$$= \frac{1}{n^2} [N \sum_i^n \frac{y_i^2}{p_i} - n\hat{Y}_{pps}^2] + \frac{1}{n} v(\hat{Y}_{pps}) \quad (5.3.9)$$

Hence, the gain due to pps sampling can be estimated by

$$\text{Gain} = [v_{pps}(\hat{Y}_{SR}) - v(\hat{Y}_{pps})]/v(\hat{Y}_{pps})$$

$$= \left[\frac{1}{n^2} \sum_i^n \frac{y_i^2}{p_i}\left(N - \frac{1}{p_i}\right)\right]/v(\overline{Y}_{pps}) \qquad (5.3.10)$$

EXAMPLE 5.3   A sample survey was conducted to study the yield of wheat in Haryana. A sample of 20 farms from a total of 100 was taken, with probability proportional to the area under wheat crop, with replacement method. The total area under wheat crop (X) was

484.5 hectares. The area under crop (x) and yield (y) were noted in hectares and quintals per hectare, respectively.

The sample selected by the cumulative method was

Area under crop	5.2	5.9	3.9	4.2	4.7	4.8	4.9
Yield of crop	28	29	30	22	22	25	28
Area under crop	6.8	4.7	5.7	5.2	5.2	4.9	4.0
Yield of crop	37	26	32	25	38	31	16
Area under crop	1.3	7.4	7.4	4.8	6.2	6.2	
Yield of crop	6	61	61	29	47	47	

The sample selected by Lahiri's method was

Area under crop	4.8	4.1	1.3	5.2	6.9	6.0	2.0	6.3	5.2	4.2
Yield of crop	22	19	6	25	54	43	4	40	28	29
Area under crop	4.8	5.9	5.8	5.8	5.1	4.7	5.6	5.2	4.0	4.6
Yield of crop	22	39	39	45	30	27	34	31	18	31

(i) Estimate the average yield per farm along with its standard error for these two samples.

(ii) Estimate the gain in efficiency due to pps sampling compared to simple random sampling with replacement.

(i) *Cumulative Total Method*

(a) The estimate of average yield/farm is given by

$$\hat{\bar{Y}}_{pps} = \frac{X}{nN} \sum_i^n \frac{y_i}{x_i}$$

Here $N = 100$, $n = 20$, $X = 484.5$

Therefore,

$$\hat{\bar{Y}}_{pps} = 484.5 \times 1205930/100 \times 20 = 29,21$$

and $\quad v(\hat{\bar{Y}}_{pps}) = \dfrac{1}{20 \times 19} [(484.5)^2 \times 728.1481 - 20 \, (19.21)^2] = 0.06$

or standard error of $\hat{\bar{Y}}_{pps} = \sqrt{0.06} = 0.24$

(b) The estimate of variance of the estimate based on simple random sample on assumption of the pps sample is given by

$$v'\,(\hat{\bar{Y}}_{\text{pps}}) = \frac{100}{20}\left[\frac{484.5\times4156.05}{20\times(100)^2} - \frac{1}{100}\times(29.21-0.06)\right]$$

$$= 8.31$$

Hence, the percentage gain in precision due to pps sampling as compared to simple random sampling, wr, is given by

$$\text{Percentage gain in precision} = \frac{v'\,(\hat{\bar{Y}}_{\text{SR}}) - v\,(\hat{\bar{Y}}_{\text{pps}})}{v\,(\hat{Y}_{\text{pps}})} \times 100$$

$$= \frac{8.31-0.06}{0.06}\times100 = 13750$$

(ii) *Lahiri's Method*

(a) The estimate of average yield/farm is given by

$$\hat{\bar{Y}}_{\text{pps}} = \frac{484.5\times115.71}{100\times20} = 28.03$$

and $\quad v(\hat{\bar{Y}}_{\text{pps}}) = \dfrac{1}{20\times19}\,[23.4740\times708.6056 - 20\times(28.03)^2] = 2.43$

Standard error of $\hat{\bar{Y}}_{\text{pps}} = 1.56$

(b) The estimate of variance of the estimate based on simple random sample, on assumption of pps sample, is given by

$$v\,(\hat{\bar{Y}}_{\text{SR}}) = \frac{100\times80}{20\times99}\left[\frac{484.5\times3734.74}{20\times(100)^2} - \frac{1}{100}(28.03-2.43)\right]$$

$$= 35.52$$

Therefore, the percentage gain in efficiency due to pps sampling over simple random sampling, wr, is

$$\frac{(35.52-2.43)}{2.43}\times100 = 1362$$

5.5 pps SAMPLING WITHOUT REPLACEMENT

Since the effective sample size in case of without replacement sampling is expected to be larger compared to sampling with replacement, the sampling without replacement can provide a more efficient estimator than sampling with replacement. A lot of work in the field of sampling with varying probabilities, wor, has been done, but most procedures are more complex and not easily applicable in large-scale surveys. If the sampling fraction is small, in large-scale surveys, the efficiencies of sampling with or without replacement will differ insignificantly. However, if the sampling fraction is larger or moderately large, it is expected that the gain in efficiency due to pps sampling, wor, will be substantial. In the remaining part of this chapter, we shall discuss some results which have become most significant these days.

5.6 PROCEDURES OF SELECTION OF A pps SAMPLE WITHOUT REPLACEMENT

There are several procedures for selecting samples with unequal probabilities, wor. A brief discussion of some important procedures is given below.

5.6.1 General Selection Procedure

A generalization of the sampling scheme, wr, as given in Section 5.2, would be to select a pps sample of size unity and remove the selected unit from the population. From the remaining units, another pps sample of size one is taken as before and the selected unit removed from the population. This process is repeated until n selections are made.

Suppose n units are selected one by one, with probability proportional to size measure x, at each draw, without replacing the units selected in the previous draws. The probability of selection at the first draw for the jth unit is given by

$$p_j = X_j/X, \qquad j = 1, \ldots, N$$

where $X = \sum_{j}^{N} X_j$.

Similarly, the probability that the ith unit is selected at the second draw, when the jth unit has been selected at the first draw, is given by

$$p_{i/j} = p_i/(1 - p_j) \quad i \neq j, \text{ and so on}$$

This set up of sampling comprises an ordered set of sample values $(y_1, y_2, \ldots, y_n)$ with probabilities $(p_1, p_2, \ldots, p_n)$.

EXAMPLE 5.4 In a village, there are 8 orchards with 50, 30, 25, 40, 26, 44, 20 and 35 trees, respectively.

Select a sample of 2 orchards with probability proportional to the number of trees in the orchard and without replacement.

(i) For selecting a unit by probability proportional to the number of trees, using Lahiri's method of selection, consider the following arrangement:

Orchard number	1	2	3	4	5	6	7	8
Number of trees	50	30	25	40	26	44	20	35

Selecting a pair of random numbers $(i, j)/(i \leqslant 8, j \leqslant 50)$, using the random number table, we get the pair (5, 17). Since the number of trees X_i for orchard number 5 is greater than the second number (17) of the selected random pair, the 5th orchard is selected in the sample.

(ii) For selecting the second unit by probability proportional to the number of trees in the orchard, we prepare the following arrangement once again after deleting the 5th orchard:

Orchard number	1	2	3	4	5	6	7
Number of trees	50	30	25	40	44	20	35

As in (i), a pair of random numbers (i, j) $(i \leqslant 7, j \leqslant 50)$ has to be selected, using the random number table. Referring to the table of random numbers, the pair selected is (6, 18). As the size of the 6th unit in the above arrangement is greater than the second number of the random pair selected, orchard 6 is selected into the sample. Thus, the sample selected consists of the units at serial numbers 5 and 7 of the original list with the number of trees being 26 and 20, respectively.

5.6.2 Narain's Scheme of Sample Selection

In this section, we shall discuss a scheme of sample selection due to Narain (1951). The scheme consists of constructing revised probabilities of selection p_i' $(i = 1, 2, \ldots, N)$ such that the inclusion probabilities π_i are proportional to the original probabilities of selection p_i $(i = 1, 2, \ldots, N)$ and sampling is done without replacement.

Here the inclusion probabilities π_i are given by

$$\pi_i = np_i, \qquad i = 1, 2, \ldots, N$$

The probabilities of selection and inclusion at the second and subsequent draws are proportional to revised probabilities p_i' on lines similar to those at the first draw. As an illustration, let us consider the simple case $N = 4$ and $n = 2$. The problem is to evaluate π_{ij} given p_i $(i, j = 1, 2, 3, 4)$. The relationship $\sum_{j \neq i} \pi_{ij} = \pi_i$ provides a system of four linear equations with six unknowns. Thus, the problem is to choose the values of two arbitrary parameters with the restriction that all the above values are positive. The computations become tedious for n greater than 2. The methods of solving the set of equations have been discussed by several workers. Brewer and Undy (1962) have given a faster iterative method of solution. Further improvements in the sampling scheme have been introduced by Singh (1954) and shown some situations in which sampling without replacement will be less efficient as compared to that with replacement.

5.6.3 Sen-Midzuno Method

A simple procedure of selecting a sample was suggested by Midzuno (1952) and independently by Sen (1952), which consists in selecting the first unit with pps and the remaining $(n - 1)$ units from $(N - 1)$ units of the population by simple random sampling, wor.

For this selection procedure, the inclusion probabilities for individual and pairwise units are given by

$$\pi_i = p_i + (1 - p_i)\, \frac{n - 1}{N - 1} \qquad \text{for } i = 1, 2, \ldots, N$$

$$= \frac{N - n}{N - 1}\, p_i + \frac{n - 1}{N - 1}$$

and

$$\pi_{ij} = p_i\, \frac{n - 1}{N - 1} + p_j\, \frac{n - 1}{N - 1}$$

$$+ (1 - p_i - p_j)\, \frac{(n - 1)\,(n - 2)}{(N - 1)\,(N - 2)}$$

$$\text{for } i \neq j = 1, 2, \ldots, N$$

$$= \frac{(n - 1)}{(N - 2)}\left[\frac{(N - n)}{(N - 2)}\,(p_i + p_j) + \frac{(n - 2)}{(N - 2)}\right]$$

By extension of the above argument, we can have $y_i, y_j, \ldots, y_q$, a sample of n units. The probability of including these n units in the sample is given by

$$\pi_{ij \ldots q} = \frac{1}{\binom{N-1}{n-1}}\,(p_i + p_j + \ldots + p_q)$$

5.7 ESTIMATION IN pps SAMPLING WITHOUT REPLACEMENT: TOTAL AND ITS SAMPLING VARIANCE

As mentioned earlier, sampling without replacement is more efficient, than sampling with replacement. This rule also applies to pps sampling. Thus, we find that there are enumerable sampling procedures for selecting a sample with varying probabilities without replacement. A tremendous development of theory of varying probability sampling has taken place during the last thirty years. A good number of interesting results can be quoted, but limitations of space do not permit us a detailed discussion. We shall, therefore, confine ourselves to the results which appear to be worthwhile, with a substantial gain in efficiency. Since the probability of inclusion changes by draws or the order of selected units, we shall examine both ordered and unordered estimators in the following sections.

5.8 ORDERED ESTIMATORS

Let us consider those estimators which are based on the order of units selected in the sample and do not require calculations of inclusion probabilities. Das (1951) and Des Raj (1956) have proposed such estimators which make use of conditional probabilities without calculating π_i and π_{ij} which are generally difficult to compute for many sampling schemes. Here, we shall consider the estimator proposed by Des Raj, first for the case when $n = 2$, and then generalize the result.

Des Raj's Ordered Estimator

(i) *When $n = 2$* Let us suppose that the initial probabilities of selection of units U_i are p_i, $i = 1, 2, \ldots, N$, where $p_i = X_i/X$. The first draw is made with probability p_i. The second draw is taken with conditional probability $p_j/(1 - p_i)$. Suppose y_1 and y_2 are the values of the units drawn at the first and second draws respectively, and p_1 and p_2 are their corresponding initial probabilities. Now let us define the estimators

$$z_1 = y_1/p_1$$

and

$$z_2 = y_1 + y_2 (1 - p_1)/p_2$$

Then,

$$\hat{Y}_D = (z_1 + z_2)/2$$
$$= [y_1 (1 + p_1)/p_1 + y_2 (1 - p_1)/p_2] \quad (5.8.1)$$

THEOREM 5.8.1   In pps sampling, wor, the estimator $\hat{Y}_D$ is an unbiased estimator and its sampling variance is given by

$$V(\hat{Y}_D) = \left(1 - \frac{1}{2} \sum_i^N p_i^2\right)\left\{\frac{1}{2} \sum_i^N (y_i/p_i - Y)^2 p_i\right\}$$

$$- \frac{1}{4} \sum_i^N (y_i/p_i - Y)^2 p_i^2 \qquad (5.8.2)$$

Proof   We observe that the expected value of z_1 is clearly Y, i.e.
$E(z_1) = \sum (y_i/p_i) p_i = \sum y_i = Y$.

Similarly, for given y_1 drawn at the first draw, we have

$$E_2 [y_2 (1 - p_1)/p_2|y_1] = \sum y_j \frac{(1 - p_i)}{p_j} \frac{p_j}{(1 - p_i)}$$

where the summation is taken over all the values of y except y_1.

Hence
$$E_2 \left[y_2 \frac{(1 - p_1)}{p_2}\bigg| y_1\right] = Y - y_1$$

Therefore, $E(z_2) = E_1 E_2 (z_2|y_1) = y_1 + Y - y_1 = Y$

Thus, it can be shown that $E(\hat{Y}_D) = Y$

Again, to derive the variance, we have
$$V(z_1) = \sum_{i > j} \sum p_i p_j (y_i/p_i - y_j/p_j)^2$$

Also,
$$V(z_2) = E_1 V_2 (z_2) + V_1 E_2 (z_2)$$

Since
$$E_2 (z_2) = Y, \qquad V_1 E_2 (z_2) = 0$$

Hence, we have
$$V(z_2) = \sum_{i > j} \sum p_i p_j (y_i/p_i - y_j/p_j)^2 (1 - p_i - p_j)$$

Thus, the variance $\hat{Y}_D$ is given by

$$V(\hat{Y}_D) = \frac{1}{4} \sum_{i > j} \sum p_i p_j \left(\frac{y_i}{p_i} - \frac{y_j}{p_j}\right)^2 (2 - p_i - p_j) \qquad (5.8.3)$$

After algebraic simplification, we obtain the results in the given form.

COROLLARY   An unbiased estimator of $V(\hat{Y}_D)$ is obtained by

$$v(\hat{Y}_D) = \frac{1}{4} (z_1 - z_2)^2 = \frac{1}{4}(1 - p_1)^2 (y_1/p_1 - y_2/p_2)^2 \qquad (5.8.4)$$

(ii) *The General Case* Let $y_1, y_2, \ldots, y_n; p_1, p_2, \ldots, p_n$ be the values of the units in the ordered sequence of selection and their initial probabilities of selection. Define $z_1 = y_1/p_1$

and
$$z_i = \sum_{k}^{i-1} y_k + (y_i/p_i)(1 - \sum_{k}^{i-1} p_k)$$

for
$$i = 2, \ldots, n$$

We have an estimator

$$\hat{Y}_D = \frac{1}{n} \sum_{i}^{n} z_i = \bar{z} \quad \text{(say)} \tag{5.8.5}$$

THEOREM 5.8.2 In pps sampling, wor, the estimator $\hat{Y}_D$ is an unbiased estimator of the population total Y and its sampling variance is given by

$$V(\hat{Y}_D) = \frac{1}{n^2} \sum_{i > j} \sum p_i p_j (y_i/p_i - y_j/p_j)^2 \{1 + r_{ij}(1) + \ldots$$
$$+ r_{ij}(k) + \ldots + r_{ij}(n-1)\} \tag{5.8.6}$$

where $r_{ij}(k)$ is the probability that y_i and y_j are not included in the sequence.

Proof It can be seen as before that $E(z_1) = Y$

and
$$E(z_i | y_1, y_2, \ldots, y_{i-1}) = Y, \quad i = 2, \ldots, n$$

Hence
$$E(z_i) = Y \quad \text{for } i = 2, \ldots, n$$

It follows that $\hat{Y}_D = \bar{z} = \sum_{i}^{n} z_i/n$ is an unbiased estimator. Further, to derive the sampling variance one can see that $E(z_i z_j) = Y^2$, which shows that z_i and z_j are uncorrelated. Hence, with similar treatment, one can derive the result. For detailed discussion, the reader is referred to Des Raj (1966). Although the expression for $V(\hat{Y}_D)$ is somewhat complex, it can be modified to a simpler form as

$$V(\hat{Y}_D) = V(\sum_{i}^{n} \frac{z_i}{n}) = \frac{1}{n^2} \sum_{i}^{n} V(z_i) \tag{5.8.7}$$

COROLLARY An unbiased estimator of $V(\hat{Y}_D)$ can be obtained by

$$v(\hat{Y}_D) = \sum_{i}^{n} (z_i - \bar{z})^2/n(n-1) \tag{5.8.8}$$

An improvement in Des Raj's estimator has been suggested by Pathak and Shukla (1966). They have shown that the variance estimator is always positive.

5.9 UNORDERED ESTIMATORS

In the previous section, we have considered the estimators which depend on the order of the units drawn in the sample. Now we shall discuss such estimators which do not depend on the order in which the units are drawn within the sample. These estimators are generally called unordered estimators. Horvitz and Thompson (1952), Murthy (1957) and Basu (1958) have shown that these estimators are more efficient than ordered estimators. A comparative study will be made after their brief discussion.

5.9.1 Horvitz-Thompson Estimator

Horvitz and Thompson (1952) suggested an estimator which is an unbiased estimator of the population total. Let us suppose that the initial probability of selection of the unit U_i is p_i, where $p_i = X_i/X$ for $i = 1, 2, \ldots, N$. The probability that the unit U_i is included in the sample would be given by

$$\pi_i = p_i + \sum_{j \neq i} p_j p_i/(1 - p_j) = p_i \left[1 + \sum_{j \neq i} p_j/(1 - p_j) \right]$$

Further, the probability that both the units U_i and U_j are included in the sample is

$$\pi_{ij} = p_i p_j/(1 - p_i) + p_j p_i/(1 - p_j)$$
$$= p_i p_j \left[1/(1 - p_i) + 1/(1 - p_j) \right]$$

Suppose that y_i be the value of the ith unit with π_i the probability of inclusion in the sample. The H-T estimator is defined by

$$\hat{Y}_{\text{HT}} = \sum_i^n y_i/\pi_i \qquad (5.9.1)$$

THEOREM 5.9.1 In pps sampling, wor, $\hat{Y}_{\text{HT}}$ is unbiased and its sampling variance is given by

$$V_{\text{HT}}(\hat{Y}_{\text{HT}}) = \sum_i^N \frac{(1 - \pi_i) y_i^2}{\pi_i} + \sum_i^N \sum_{j \neq i}^N \frac{(\pi_{ij} - \pi_i \pi_i)}{\pi_i \pi_j} y_i y_j$$

$$(5.9.2)$$

where π_{ij} is the probability of inclusion of both the ith and jth units in the sample.

(The suffix HT with variance stands for the expression due to Horvitz-Thompson).

Proof A most general form of linear estimator can be written as

$$\hat{Y} = \sum_i^N a_i c_i y_i$$

where a_i is a random variate that takes the value 1 if the ith unit is drawn and zero otherwise; c_i are the constants attached to units U_i $(i = 1, 2, \ldots, N)$.

Obviously, a_i follows a binomial distribution for a sample of size 1 with probability p_i. Hence, $E(a_i) = \pi_i$, $V(a_i) = \pi_i (1 - \pi_i)$. Since $a_i a_j$ is 1 only if both units are distinct and appear in the sample,

$$\text{Cov}(a_i, a_j) = \pi_{ij} - \pi_i \pi_j$$

Now,

$$E(\hat{Y}) = \sum_i^N \pi_i c_i y_i = Y \qquad \text{if it is unbiased}$$

Therefore,

$$c_i = 1/\pi_i$$

Hence $\hat{Y}_{\text{HT}} = \sum_i^n y_i/\pi_i$ is an unbiased estimator. Further, the sampling variance is given by

$$V_{\text{HT}}(\hat{Y}_{\text{HT}}) = \sum_i^N \frac{y_i^2}{\pi_i^2} V(a_i) + \sum_i^N \sum_{j \neq i}^N \frac{y_i y_j}{\pi_i \pi_j} \text{Cov}(a_i, a_j)$$

$$= \sum_i^N \pi_i (1 - \pi_i) \frac{y_i^2}{\pi_i^2} + \sum_i^N \sum_{j \neq i}^N \frac{(\pi_{ij} - \pi_i \pi_j)}{\pi_i \pi_j} y_i y_j$$

The variance of $\hat{Y}_{\text{HT}}$ depends on quantities π_{ij} and π_i which are calculated from the sampling procedure. This is the reason that various sampling schemes have been proposed to evaluate these probabilities. If we take $\pi_i = n y_i/Y$, $V(\hat{Y})$ becomes zero. Thus, if there is some way of choosing values of π_i so that they are close to ny_i/Y, $\hat{Y}$ will have a very small variance.

COROLLARY 1   An unbiased sample estimator of $V(\hat{Y}_{HT})$ is given by

$$v_{HT}(\hat{Y}_{HT}) = \sum_i^n \frac{(1-\pi_i)\,y_i^2}{\pi_i^2} + 2\sum_i \sum_{j>i} \frac{(\pi_{ij} - \pi_i\,\pi_j)}{\pi_{ij}}\frac{y_i\,y_j}{\pi_i\,\pi_j} \qquad (5.9.3)$$

provided that none of the π_{ij} in the population is zero.

It may be noted that the Horvitz-Thompson estimator can be used for any sampling design when the estimator has only distinct units in the sample. The main drawback of this variance estimator is that the variance term does not reduce to zero even when all the values are equal, a case for which the variance is necessarily zero. Another drawback is that the variance term may assume negative values for some samples.

For $n = 2$, the H–T estimator of the total is

$$\hat{Y}_2 = \left(\frac{y_i}{\pi_1} + \frac{y_2}{\pi_2}\right)$$

and its estimated variance is

$$v(\hat{Y}_2) = \frac{(\pi_1\,\pi_2 - \pi_{12})}{\pi_{12}}\left(\frac{y_1}{\pi_1} - \frac{y_2}{\pi_2}\right)^2$$

where the values of π's are calculated as shown in the above scheme.

COROLLARY 2   Another expression for the variance of the H–T estimator derived by Yates and Grundy (1953) is given by

$$V_{YG}(\hat{Y}_{HT}) = \sum_i^N \sum_{j>i}^N (\pi_i\,\pi_j - \pi_{ij})\left(\frac{y_i}{\pi_i} - \frac{y_j}{\pi_j}\right)^2 \qquad (5.9.4)$$

and its unbiased estimator is

$$v_{YG}(\hat{Y}_{HT}) = \sum_i^n \sum_{j>i}^n \frac{(\pi_i\,\pi_j - \pi_{ij})}{\pi_{ij}}\left(\frac{y_i}{\pi_i} - \frac{y_j}{\pi_j}\right)^2 \qquad (5.9.5)$$

The estimator in relation (5.9.5) is not perfectly satisfactory as it may also assume negative value in some cases.

From these variance expressions, it can be seen that the variance does not reduce to zero when the variate y and its size measure x are the same, unless the inclusion probabilities π_i are made proportional to X_i. Further, it can be seen that the Sen-Midzuno scheme ensures the required probabilities of inclusion of the units in the sample and thus the Yates-Grundy estimate of the variance of the H-T estimator is never negative. Of course, some restrictions over the initial probabilities of selection confine the use of the system in practice. Even then, it has been observed that the H-T estimator under the Sen-Midzuno scheme of sampling is more efficient than the other estimators with pps and replacement.

For the case when $n=2$, the H-T estimator and its estimated variance can be calculated according to the Sen–Midzuno sampling scheme, or any other desired choice, and a comparative study can be easily made.

5.9.2 Murthy's Unordered Estimator

Murthy (1957) suggested that an unordered estimator can be obtained by weighting all possible ordered estimators with their respective probabilities. In sampling n units without replacement from a finite population of N units, there will be $\binom{N}{n}$ unordered sample (s). Each unordered sample (s) of size n can be ordered in M ($= n!$) ways, i.e. an unordered sample corresponds to M ordered samples. Consider a scheme of selection in which the probability of selecting the sample s_i is p (s_i). Then, the probability of getting the unordered sample (s) is the sum of the probabilities of getting the ordered samples corresponding to (s), i.e. $p_i = \sum_{i}^{M} p_{si}$. Let y_{si} be an estimator of a population parameter θ based on the ordered sample (s_i). An unordered estimator of θ is given by

$$\hat{y}_M = \sum_{i}^{M} y_{si} \, p'_{si} \tag{5.9.6}$$

where $p'_{si} = p_{si}/p_s$.

THEOREM 5.9.2 In pps sampling, wor, $\hat{y}_M$ is an unbiased estimator of θ and its sampling variance is given by

$$V(\hat{y}_M) = \sum_{s} p_s \{\sum_{i} y_{si} \, p'_{si}\}^2 - (\sum_{s} \sum_{i} y_{si} \, p_{si})^2 \tag{5.9.7}$$

The proof is obvious.

COROLLARY 1 In pps sampling, wor, the ordered estimator y_{si} is an unbiased estimator of θ and its sampling variance is given by

$$V(y_{s_i}) = \sum_{s} \sum_{i} p_{si} y_{si}^2 - (\sum_{s} \sum_{i} y_{s_i} \, p_{si})^2 \tag{5.9.8}$$

Hence, or otherwise, show that

$$V(y_{si}) - V(\hat{y}_M) = \sum_{s} \sum_{i} \{y_{si} - \sum_{i} y_{si} \, p'_{si}\}^2 \, p_{si} \tag{5.9.9}$$

This proves that the variance of the unordered estimator $\hat{y}_M$ is less than or equal to that of the ordered estimator.

COROLLARY 2 An unbiased estimator of the sampling variance of $\hat{y}_M$ is obtained by

$$v(\hat{y}_M) = \frac{1}{[p_s]^2} \sum_i^n \sum_{j>i}^n [p_s \, p(s_{i,j}) - p_{s_i} \, p_{s_j}] \, p_i \, p_j \left(\frac{y_i}{p_i} - \frac{y_j}{p_j} \right)^2 \quad (5.9.10)$$

where $p(s_{i,j})$ is the conditional probability of getting the sth sample given that the units i and j are selected in the first two draws.

For a comparative study, let us discuss the case when $n=2$. Suppose that y_i and y_j are the values of the units selected with varying probability and without replacement. Since two units can be ordered in two ways only, $M=2$. If y_i and y_j have p_i and p_j as their initial probabilities of selection, then the ordered estimator corresponding to order (y_i, y_j) is given by

$$y_{s_i} = \tfrac{1}{2} [(1 + p_i) \, y_i/p_i + (1 - p_i) \, y_j/p_j] \quad (5.9.11)$$

Also, the probability of obtaining the values y_i and y_j at the first and second draw, respectively is given by

$$p_{s_1} = p_i \, p_j/(1 - p_i) \quad (5.9.12)$$

Similarly, if y_j and y_i were the values of the units drawn at the first and second draws, respectively, the ordered estimator corresponding to the order (y_j, y_i) is given by

$$y_{s^2} = \tfrac{1}{2} [(1 + p_j) \, y_j/p_j + (1 - p_j) \, y_i/p_i] \quad (5.9.13)$$

and the probability of obtaining the ordered sample is

$$p_{s_2} = p_i \, p_j/(1 - p_j) \quad (5.9.14)$$

Thus, the unordered estimator is given by

$$\hat{y}_M = \sum_i^2 y_{s_i} \, p_{s_i}/\sum_i^2 p_{s_i}$$

$$= [(1 - p_j) \, y_i/p_i + (1 - p_i) \, y_j/p_j]/(2 - p_1 - p_j) \quad (5.9.15)$$

Further, substituting these results in Eq. (5.9.7) and simplifying, it can be shown that the sampling variance of $\hat{y}_M$ is given by

$$V(\hat{y}_M) = \sum_s^N p_i \, p_j \frac{(1 - p_i - p_j)}{(2 - p_i - p_j)} \left(\frac{y_i - y_j}{p_i \, p_j} \right)^2 \quad (5.9.16)$$

An unbiased estimator of variance is obtained as

$$v(\hat{y}_M) = \frac{(1 - p_i)(1 - p_i)(1 - p_i - p_j)}{(2 - p_i - p_j)^2} \left(\frac{y_i}{p_i} - \frac{y_j}{p_j} \right)^2 \quad (5.9.17)$$

which is always non-negative and less than the ordered estimator's variance.

5.10 pps SYSTEMATIC SAMPLING

We shall now describe another method of sampling, based on systematic selection of units, which is always better than pps sampling, wr. Madow (1949) suggested that generalization can be made about systematic sampling with equal probabilities. The method consists of arranging the units at random and forming cumulative totals $T_i = \sum_j^i X_j$ ($i = 1, 2, ..., N$), where X_j is the size of the jth unit. In selecting n units by pps systematic sampling, a random number R is chosen from 1 to k, where $k = \dfrac{T_N}{n}$. The units corresponding to the numbers $R + jk, j = 0, 1, ..., (n - 1)$ constitute the sample. As an illustration, let us consider the data in Example 5.1. Here $T_N = 350$ and $k = 70$. Suppose the random number chosen between 1 and 70 is 51. Then, the units corresponding to systematic sample numbers 51, 121, 191, 261 and 331 are 2, 3, 6, 8 and 10. It is clear from the procedure that the ith unit is included in the sample if $T_{i-1} < R + jk \leqslant T_i$ for $j = 0, 1, ..., (n - 1)$, and the probability of inclusion of the ith unit is np_i. This method can be applied even when $\dfrac{T_N}{n}$ is not an integer. The sampling interval k can be taken as the integer nearest to $\dfrac{T_N}{n}$. In this case, however, sample size will vary.

An unbiased estimator of the population total Y, derived by Hartley and Rao (1962), is given by

$$\hat{Y}_{H-R} = \sum_i^n \frac{y_i}{\pi_i} = \sum_i^n \frac{y_i}{np_i} \tag{5.10.1}$$

The main drawback of this method, as in systematic random sampling, is that an unbiased variance estimator on the basis of a single sample is not possible. Hartley and Rao (1962) have further considered this selection procedure when the units are arranged at random and asymptotic approach has been used in developing formulae for variance and estimated variance. For fairly large values of N and for values of n relatively small compared to N, the approximate sampling variance of the estimator can be written as

$$V(\hat{Y}_{H-R}) \cong \frac{1}{n} \sum_i^N (y_i/p_i - Y)^2 \, p_i \, [1 - (n - 1)p_i] \tag{5.10.2}$$

which shows that even when the units are arranged at random, selecting them with pps systematic sampling is more efficient than pps sampling, wr, for the terms $[1 - (n - 1)p_i]$ are acting as reduction terms. Further, this variance can be estimated by

$$v(\hat{Y}_{H-R}) = \frac{1}{n^2(n-1)} \sum_i^n \sum_{j>i}^n \left[1 - n(p_i + p_j) + \sum_{j'}^N p_{j'}^2 \right] \left(\frac{y_i}{p_i} - \frac{y_j}{p_j} \right)^2$$

(5.10.3)

5.11 RANDOM GROUP METHOD

Rao, Hartley and Cochran (1962) suggested a procedure of varying probabilities, wor, which consists of dividing the units of the population into n groups at random with N_i units in the ith group, $i = 1, 2, \ldots, n$ and then selecting one unit with pps from each of the n random groups. According to this method of selection, let

$$\begin{pmatrix} y_1, y_2, \ldots, y_n \\ p_1, p_2, \ldots, p_n \end{pmatrix}$$ be the varying probability sample.

An estimator of the population total Y is defined by

$$\hat{Y}_{\text{RHC}} = \sum_i^n y_i/p_i/\pi_i = \sum_i^n \pi_i \, y_i/p_i \qquad (5.11.1)$$

where $\pi_i = \sum_j^{N_i} p_{ij}$, p_{ij} being the initial probability of selection for the jth unit in the ith group ($j = 1, \ldots, N_i, i = 1, \ldots, n$).

THEOREM 5.11.1 In the random group selection method, the estimator $\hat{Y}_{\text{RHC}}$ is unbiased and its sampling variance is given by

$$V(\hat{Y}_{\text{RHC}}) = \frac{(\sum_i^n N_i^2 - N)}{N(N-1)} \left(\sum_i^{N_i} \frac{y_i^2}{p_i} - Y^2 \right) \qquad (5.11.2)$$

Proof We know that $E(\hat{Y}_{\text{RHC}}) = E_1 E_2(\hat{Y}_{\text{RHC}})$

where E_2 denotes the expectation for a given split of the population and E_1 the expectation over all possible splits of the population into n groups of sizes $N_1, N_2, \ldots, N_n$. Then, for any given split

$$E_2(\hat{Y}_{\text{RHC}}) = \sum_i^n Y_i = Y$$

where $\quad Y_i = \sum_t^{N_i} y_t$

Therefore,

$$E(\hat{Y}_{\text{RHC}}) = E_1(Y) = Y$$

This shows that $\hat{Y}_{\text{RHC}}$ is an **unbiased** estimator. Again, we have

$$V(\hat{Y}_{\text{RHC}}) = E_1 V_2(\hat{Y}_{\text{RHC}}) + V_1 E_2(\hat{Y}_{\text{RHC}})$$

$$= E_1 V_2(\hat{Y}_{\text{RHC}}) \quad \text{Since the second term on the right-hand side is zero.}$$

Now, $\quad V_2(\hat{Y}_{\text{RHC}}) = \sum_t^{N_i} \frac{p_t}{\pi_i} \left(\frac{y_t}{p_t/\pi_i} - Y_i \right)^2$

$$= \sum_{t < t'}^{N_i} p_t p_{t'} \left(\frac{y_t}{p_t} - \frac{y_{t'}}{p_{t'}} \right)^2$$

Since $N_i(N_i - 1)/N(N - 1)$ is the probability, in a random split, that a pair of observations fall into the ith group, we have

$$E_1 V_2(\hat{Y}_{\text{RHC}}) = \sum_j^n \frac{N_i(N_i - 1)}{N(N - 1)} \sum_{t < t'}^N p_t p_{t'} \left(\frac{y_t}{p_t} - \frac{y_{t'}}{p_{t'}} \right)^2$$

$$= \frac{(\sum_i^n N_i^2 - N)}{N(N - 1)} \left(\sum_t^N \frac{y_t^2}{p_t} - Y^2 \right)$$

COROLLARY 1 If the choice is that $N_1 = N_2 = \ldots = N_n = N/n$, then $V(\hat{Y}_{\text{RHC}})$ minimizes and its value is obtained by

$$\left. \begin{array}{l} V_{\min}(\hat{Y}_{\text{RHC}}) = \frac{1}{n} [1 - (n-1)/(N - 1)] \sum_i^N p_i(y_i/p_i - Y)^2 \\[2mm] \cong \frac{1}{n} \sum_i^N p_i \{1 - (n - 1)/N\} \left(\frac{y_i}{p_i} - Y \right)^2 \end{array} \right\} \quad (5.11.3)$$

A comparison with the Hartley-Rao estimator shows that the method brings a slightly larger variance.

COROLLARY 2 In sampling wr, the estimator of Y is

$$\hat{Y}' = \sum_n y_s/np_s \qquad (5.11.4)$$

with its variance $V(\hat{Y}') = \sum_t^N \dfrac{y_t^2}{np_t} - \dfrac{Y^2}{n}$

then,
$$V(\hat{Y}_{\text{RHC}}) = n \frac{(\sum_i^n N_i^2 - N)}{N(N-1)} V(\hat{Y}') \qquad (5.11.5)$$

COROLLARY 3 An unbiased estimator of $V(\hat{Y}_{\text{RHC}})$ is given by

$$v(\hat{Y}_{\text{RHC}}) = \frac{(\sum_i^n N_i^2 - N)}{(N^2 - \sum_i^n N_i^2)} \{\sum_i^n \pi_i (y_i/p_i - \hat{Y}_{\text{RHC}})^2\} \qquad (5.11.6)$$

When $N_1 = N_2 = \cdots = N_n = \dfrac{N}{n}$

$$v(\hat{Y}_{\text{RHC}}) = \frac{(N-n)}{N(n-1)} \sum_i^n \pi_i \left(\frac{y_i}{p_i} - \hat{Y}_{\text{RHC}}\right)^2 \qquad (5.11.7)$$

In case $n = 2$, the results are simple and can be derived easily without any complication.

EXAMPLE 5.5 If the yields (in 10 kg) of the 8 orchards in Example 5.4 are 60, 35, 30, 44, 30, 50, 22 and 40 respectively, estimate, along with the standard error, the total production of the 8 orchards, from samples selected in the example, by using

  (i) Des Raj ordered estimator,
 (ii) Murthy unordered estimator, and
(iii) Horvitz-Thompson estimator.

Since the sample selected in Example 5.4 includes the orchards at serial numbers 5 and 7, the yield (in 10 kg) of two selected orchards are 30 and 22, respectively.

Sl. No.	No. of trees (X_i)	Yield (Y_i)	$p_i = (X_i/X)$
1	50	60	0.185
2	30	35	0.111
3	25	30	0.093
4	40	44	0.148
5	26	30	0.096
6	44	50	0.163
7	20	22	0.074
8	35	40	0.130
Total	290	311	1.000

We have the following results from Example 5.4.

For the sample selected, we have $x_1 = 26$, $x_2 = 20$, $y_1 = 30$, $y_2 = 22$, $p_1 = 0.096$ and $p_2 = 0.074$.

(i) Des Raj Ordered Estimator

The Des Raj ordered estimator for total production of orchards is given by

$$\hat{Y}_D = \frac{1}{2}\left[\frac{y_1}{p_1}(1 + p_1) + \frac{y_2}{p_2}(1 - p_1)\right]$$

$$= \frac{1}{2}\left[\frac{30}{0.096}(1 + 0.096) + \frac{22}{0.074}(1 - 0.096)\right]$$

$$= 305.625 \text{ (in 10 kg units)}$$

The estimate of $V(\hat{Y}_D)$ is given by

$$v(\hat{Y}_D) = \frac{1}{4}(1 - p_1)^2 (y_1/p_1 - y_2/p_2)^2$$

$$= \frac{1}{4}(1 - 0.096)^2 (30/0.096 - 22/0.074)^2 = 47.2645$$

$$\therefore \quad \text{Standard error } \hat{Y}_D = \sqrt{v(Y_D)} = \sqrt{47.2645} = 6.87$$

(ii) Murthy Unordered Estimator

The Murthy unordered estimator for total production is given by

$$\hat{Y}_M = \frac{1}{(2 - p_1 - p_2)}\left[\frac{y_1}{p_1}(1 - p_2) + \frac{y_2}{p_2}(1 - p_1)\right]$$

$$= \frac{1}{(2 - 0.096 - 0.074)}\left[\frac{30}{0.096}(1 - 0.074) + \frac{22}{0.074}(1 - 0.096)\right]$$

$$= 304.98 \text{ (in 10 kg units)}$$

The estimate of the variance of the estimator $\hat{Y}_M$ is

$$v(\hat{Y}_M) = \frac{(1 - p_1)(1 - p_2)(1 - p_1 - p_2)}{(2 - p_1 - p_2)^2}\left[\frac{y_1}{p_1} - \frac{y_2}{p_2}\right]^2$$

$$= \frac{(1 - 0.096)(1 - 0.074)(1 - 0.096 - 0.074)}{(2 - 0.096 - 0.074)^2}$$

$$\cdot \left[\frac{30}{0.096} - \frac{22}{0.074}\right]^2 = 46.8386$$

$$\therefore \text{ Standard error of } \hat{Y}_M = \sqrt{v(\hat{Y}_M)} = \sqrt{46.8386} = 6.84$$

(iii) *Horvitz–Thompson Estimator*

The Horvitz–Thompson estimator for population total is given by

$$\hat{Y}_{HT} = \sum_{i}^{n} \frac{y_i}{\pi_i}$$

with

$$v\left(\hat{Y}_{HT}\right) = \sum_{i}^{n} \sum_{j>i}^{n} \frac{(\pi_i \pi_j - \pi_{ij})}{\pi_{ij}} \left(\frac{y_i}{\pi_i} - \frac{y_i}{\pi_i} \right)^2$$

Now, we have

$$S = \sum_{i}^{8} \frac{P_i}{1 - p_i} = 1.157$$

For the selected units,

$$\pi_1 = p_1 \left[S + 1 - \frac{p_1}{1 - p_1} \right] = 0.1969$$

$$\pi_2 = p_2 \left[S + 1 - \frac{p_2}{1 - p_2} \right] = 0.1538$$

$$\pi_{12} = p_1 p_2 \left[\frac{1}{1 - p_1} - \frac{1}{1 - p_2} \right] = 0.0155$$

Thus

$$\hat{Y}_{HT} = \frac{30}{0.1969} + \frac{22}{0.1538} = 295.403 \text{ (in 10 kg units)}$$

and

$$v\left(\hat{Y}_{HT}\right) = \left[\frac{0.1969 \times 0.1538 - 0.0155}{0.0155} \right] \left[\frac{30}{0.1969} - \frac{22}{0.1538} \right]^2$$

$$= 79.91$$

$$\therefore \quad \text{Standard error of } \hat{Y}_{HT} = \sqrt{v\left(\hat{Y}_{HT}\right)} = \sqrt{79.91} = 8.93$$

EXAMPLE 5.6 From the population of 8 orchards in Example 5.4, select a sample of 2 orchards using
  (i) Randomised pps systematic sampling
  (ii) Random group method and
  (iii) Probability proportional to size
and estimate the average production along with their respective standard errors.

(i) *Randomised pps Systematic Sampling* (Hartley and Rao procedure)

For selecting a sample of size 2 using randomised pps systematic sampling, we arrange the population units at random and the arrangement thus obtained is:

S. no.	No of trees (X_i)	Cumulative total
1	30	30
2	50	80
3	26	106
4	25	131
5	40	171
6	20	191
7	44	235
8	35	270

The value of k, the sampling interval in the above example, is

$$\frac{\sum_i^N X_i}{n} = \frac{270}{2} = 135$$

Selecting a random number between 1 and 135, we get 120. Another random number is then obtained by adding 135 to the random number selected. The units corresponding to the cumulative totals 120 and 255 are at serial numbers 4 and 8, respectively. Thus, the orchards selected in the sample are at serial numbers 4 and 8 in the original table with the number of trees being 25 and 35, respectively.

The estimate and the estimated variance of the estimate of average production, given by Hartley and Rao, are

$$\hat{\bar{Y}} = \frac{1}{nN} \left[\frac{y_1}{p_1} + \frac{y_2}{p_2} \right]$$

and

$$v(\hat{\bar{Y}}) = \frac{1}{N^2 n^2 (n-1)} [1 - n(p_1 + p_2)$$

$$+ n \sum_i^N p_i^2] \cdot \left(\frac{y_1}{p_1} - \frac{y_2}{p_2} \right)^2$$

In this case,

$$N = 8, \; y_1 = 30, \; x_1 = 25, \; p_1 = \frac{25}{270} = 0.093$$

$$n = 2, \; y_2 = 40, \; x_2 = 35, \; p_2 = \frac{35}{270} = 0.130$$

Thus, $\hat{\bar{Y}} = \frac{1}{16}\left[\frac{30}{0.093}+\frac{40}{0.130}\right] = 39.39$

and $v(\hat{\bar{Y}}) = \frac{1}{4 \times 1}[1 - 2(0.093 + 0.130) + 2 \times 0.13526]$

$$\times \left(\frac{30}{8 \times 0.093} - \frac{40}{8 \times 0.130}\right)^2 = 0.1783$$

∴ Standard error of $\hat{\bar{Y}} = \sqrt{v(\hat{\bar{Y}})} = \overline{0.1783} = 0.4224$

(ii) Random Group Method

As the units are randomised in (i), we take the first 4 units as forming group I and the remaining 4 units as group II, respectively. Now, from each group, one unit is to be selected by probability proportional to the number of trees. For this, let us prepare the following table:

	Group I			Group II	
S. No.	No. of trees	Probability	S. No.	No. of trees	Probability
1	30	0.111	1	40	0.148
2	50	0.185	2	20	0.074
3	26	0.096	3	44	0.163
4	25	0.093	4	35	0.130
Total	131	0.485	Total	132	0.515

Using Lahiri's method of selection and referring to the table of random numbers, the admissible pairs selected for group I and group II are (2, 41) and (3, 38). Hence, the units selected are 1 and 6 in the original table.

The RHC estimate and its estimate of the variance are

$$\hat{\bar{Y}} = \frac{1}{N}\left[\frac{y_1}{p_1/\pi_1} + \frac{y_2}{p_2/\pi_2}\right]$$

and

$$v(\hat{\bar{Y}}) = \left(1 - \frac{2}{N}\right)\left[\pi_1\left(\frac{y_1}{Np_1} - \hat{\bar{Y}}\right)^2 + \pi_2\left(\frac{y_2}{Np_2} - \hat{\bar{Y}}\right)^2\right]$$

For the units selected, we have

$$N = 8, y_1 = 60, x_1 = 50, p_1 = \frac{50}{270} = 0.185$$

$$n = 2, \, y_2 = 50, \, x_2 = 44, \, p_2 = \frac{44}{270} = 0.163$$

$$\pi_1 = (p_1 + p_2 + p_3 + p_4)$$
$$= 0.185 + 0.111 + 0.093 + 0.96 = 0.485$$

$$\pi_2 = (p_5 + p_6 + p_7 + p_8)$$
$$= 0.148 + 0.163 + 0.074 + 0.130 = 0.515$$

Thus
$$\hat{\bar{Y}} = \frac{1}{8}\left[\frac{60 \times 0.485}{0.185} + \frac{54 \times 0.515}{0.163}\right] = 39.41$$

and
$$v(\hat{\bar{Y}}) = (1 - 2/8)\left[0.485\left(\frac{60}{8 \times 0.185} - 39.41\right)^2 \right.$$
$$\left. + 0.515\left(\frac{50}{8 \times 0.163} - 39.41\right)^2\right]$$

$$= 0.0802$$

$\therefore$ Standard error of $\hat{\bar{Y}} = 0.2832$

(iii) *Probability Proportional to Total Size* (Midzuno–Sen Sampling procedure)

In this method, the first unit is selected by probability proprtional to the number of trees and the others by simple random sampling, without replacement, from the remaining units. Using Lahiri's method of selection and referring to the random number table, the pair selected is (6, 18). As the size of the 6th unit is greater than the second number of the random pair, i.e., 18, the unit at serial number 6 is selected. The remaining units are now numbered from 1 to 7 and, thereafter, one random number from 1 to 7 is drawn to select the second unit in the sample. The random number from the table happens to be 6 and, therefore, the unit at serial number 6 in the new arrangement is selected. Thus, the units selected by this method are at serial numbers, 6 and 7 in the original table.

The Horvitz-Thompson estimator for average and the Yates-Grundy estimate of the variance for the estimate under this selection procedure are

$$\hat{\bar{Y}} = \frac{1}{N}\left[\frac{y_1}{\pi_1} + \frac{y_2}{\pi_2}\right]$$

and
$$v(\hat{\bar{Y}}) = \frac{1}{N^2}\left(\frac{\pi_1 \pi_2 - \pi_{12}}{\pi_{12}}\right)\left(\frac{y_1}{\pi_1} - \frac{y_2}{\pi_2}\right)^2$$

We have, $\qquad N = 8, y_1 = 50, x_1 = 44$

$\qquad\qquad n = 2, y_2 = 22, x_2 = 20,$

$$\pi_1 = \frac{8-2}{8-1} \times 0.163 + \frac{2-1}{8-1} = 0.282$$

$$\pi_2 = \frac{8-2}{8-1} \times 0.074 + \frac{2-1}{8-1} = 0.206$$

$$\pi_{12} = \frac{2-1}{8-1}\left[\frac{(8-1)}{(8-2)}(0.163 + 0.0174) + \frac{(2-2)}{(8-2)}\right]$$

$$= .033$$

Therefore, $\qquad \hat{\bar{Y}} = \frac{1}{8}\,(50/0.282 + 22/0.206] = 35.51$

and $\qquad v(\hat{\bar{Y}} = \frac{1}{64}\,\frac{(0.282 \times 0.206 - 0.033)}{0.033}\,(50/0.282 - 22/0.206)^2$

$$= 58.2596$$

$\therefore$ Estimated standard error of $\hat{\bar{Y}} = \sqrt{58.2596}$

$$= 7.6328$$

5.12 INCLUSION PROBABILITY PROPORTIONAL TO SIZE (πps) SAMPLING PROCEDURES

There are several π ps sampling procedures in literature. Most of them are for sample size 2 alone. For sample sizes greater than 2, the expressions involved are complicated. We discuss here two sampling procedures, one due to Durbin (1967) and another Hanurav (1966), as these are the simplest of all the procedures. For selecting a πps sample of required size n, the pps systematic sampling which has already been discussed can also be used.

5.12.1 Durbin's πps Sampling Scheme

Durbin's πps sampling procedure for sample size 2 consists in selecting the first unit with probability p_i for the ith unit and the second unit with the conditional probability for the jth unit when the ith unit is drawn at the first draw as

$$p_{j/i} = p_i\left(\frac{1}{1-2p_i} + \frac{1}{1-2p_j}\right)\left(1 + \sum_{k=1}^{N}\frac{p_k}{1-2p_k}\right)^{-1}$$

where $j \neq i$

For this selection procedure, the expressions for inclusion probabilities are given by

$$\pi_i = 2\,p_i, \qquad i = 1, 2, \ldots, N$$

and
$$\pi_{ij} = 2\,p_i\,p_j \left(\frac{1}{1 - 2\,p_i} + \frac{1}{1 - 2\,p_j} \right) \left(1 + \sum_{k=1}^{N} \frac{p_k}{1 - 2\,p_k} \right)^{-1}$$

where $j \neq i = 1, 2, \ldots, N$

For Durbin's selection procedure, the Horvitz-Thompson estimator and Yates-Grundy variance estimator can be used for estimation. It may be noted here that the Yates-Grundy variance estimator is always non-negative and the procedure provides a more efficient estimator than sampling with replacement.

5.12.2 Hanurav's πps Sampling Scheme

The selection of 2 units under Hanurav's πps sampling scheme consists of selecting 2 units by pps with replacement. If the units selected are distinct, the sample is retained. Otherwise, it is rejected and another sample of 2 units is selected with probability proportional to p_i^2, for the ith unit, with replacement. If the units selected now are distinct, the sample is retained. Otherwise, a further sample of 2 units is drawn with probability proportional to p_i^4, with replacement, and so on. For this selection procedure

$$\pi_i = 2\,p_i$$

and
$$\pi_{ij} = 2\,p_i\,p_j \left(1 + \sum_{k=1}^{\infty} w_k \right)$$

where
$$w_k = \frac{(p_i\,p_j)\,2^k - 1}{S\,(1)\,S\,(2)\,\ldots\,S\,(k)}$$

and
$$S\,(t) = \sum_{j=1}^{N} p_j \cdot 2^t$$

It may be noted here that the Horvitz-Thompson estimator of the population total admits a non-negative Yates-Grundy variance estimator and this procedure provides an estimate more efficient than pps sampling with replacement. Srivastava and Singh (1981) have given a simple sampling procedure, for inclusion probabilities proportional to size, which permits a considerable degree of flexibility.

SET OF PROBLEMS

5.1 From a population of N units, a sample of two distinct units is drawn. The first unit of the sample is selected with probabilities $p_1 = (i = 1, \ldots, N)$, $\Sigma\, p_i = 1$, based on measures of size X_i. After dropping the unit already selected, the second unit is selected with probability proportionate to q_i, $\Sigma\, q_i = 1$, where q_i are obtained by

$$p_i = \sum_{i \neq j} p_j\, q_i / (1 - q_i), \text{ for } i = 1, 2, \ldots, N$$

The probability of the unit U_i, included in the sample, equals $2\, p_i$. Prove that the variance of the estimator $\Sigma\, (y_i / 2p_i)$ of this sampling scheme is less than or equal to the variance of the same estimator in the case of pps sampling, wr.

5.2 In sampling with unequal probabilities, wor, a sample of size 2 is drawn. The first unit is selected with pps and the second with pps of the remaining units. Show that Yates and Grundy's variance estimator is always positive for this sampling system.

5.3 If the first unit in the sample is selected with pps, the second with pps of the remaining units, while the remaining $(n - 2)$ units are selected with equal probabilities without replacement. Prove that Yates and Grundy's variance estimator would be positive for this sampling system.

(Rao 1961)

5.4 If $\begin{pmatrix} y_1, y_2, \ldots, y_n \\ p_1, p_2, \ldots, p_n \end{pmatrix}$ be the values of the units in the order which they are drawn, along with their initial probabilities of selection, and

$$z_i = y_1 / Np_1$$

$$z_i = \frac{y_i\,(1 - p_i)\,(1 - p_1 - p_2) \ldots (1 - p_1 - p_2 \ldots - p_{i-1})}{N\,(N-1) \ldots (N - i + 1)\, p_1\, p_2 \ldots p_i}$$

for $i = 2, \ldots, n$

Show that $\bar{z} = \sum_{i=1}^{n} z_i / n$ is an unbiased estimator of the mean. Further, show that

$$\hat{\bar{Y}}^2 = \frac{\Sigma\, y_i\, z_i}{nN} + \frac{2\,(N-1)}{Nn\,(n-1)} \sum_{i<j} y_i\, z_i$$

Hence, or otherwise, obtain an unbiased estimate of the variance of $\bar{z}$.

(Das, 1951)

5.5 (i) Prove that with the Midzuno's method of sample selection, the probability that any specific sample will be drawn as

$$\frac{(n-1)!\,(N-n)!}{(N-1)!}\, \sum_{i}^{n} X_i / X$$

(ii) Show that with the Midzuno's scheme of sampling,

$$t = \sum_{i=1}^{n} y_i / \sum_{i=1}^{n} p_i$$

where the summation is over the units in the sample and p_i are the initial probabilities of selection of the units selected in the sample, is an unbiased estimator of the population total. Give also an unbiased estimate of $V(t)$.

5.6 The results of a sample survey on the number of bearing lime trees and the area reported under lime, in each of the twenty-two villages growing lime in one of the tehsils of Nellore district, are given below:

S. No. of village	Area under lime (in acres)	No. of bearing lime trees
1	32.77	2328
2	7.97	754
3	0.62	105
4	15.61	949
5	42.85	3091
6	40.03	1736
7	9.39	840
8	6.33	311
9	5.05	0
10	94.55	3044
11	53.71	2483
12	0.67	128
13	0.82	102
14	2.15	60
15	0.43	0
16	123.36	11799
17	0.29	26
18	3.00	317
19	4.00	190
20	2.00	180
21	6.21	752
22	45.85	3091

From this population,

(i) select five samples of 10 villages each with probabilities proportional to area under lime, with replacement by Lahiri's method.

(ii) from one of these samples, estimate the total number of bearing lime trees in the above tehsil together with its standard error.

(iii) compare the efficiency of pps wr with that of simple random sampling.

5.7 A pilot scheme for the study of cultivation practices and yield of guava was carried out in Allahabad district of Uttar Pradesh during 1980-81. The villages were selected with probability proportional to area under fresh fruit. The data concerning the selected villages from the selected tehsil are given below:

S.No. of the selected village	Area under fresh fruit	Area under guava crop
1	127.00	166.15
2	32.00	24.73
3	74.99	100.77
4	57.00	87.14
5	68.00	116.28
6	61.01	60.22
7	6.00	13.59
8	27.00	41.70
9	68.00	10.52
10	13.17	13.85
11	8.00	12.92
12	8.00	10.73
13	22.00	38.64
14	10.00	15.92
15	30.00	9.09
16	98.99	155.51
17	7.38	10.34
18	62.99	95.16
19	13.00	22.40
20	14.00	10.97
21	30.00	39.07
22	10.00	13.70
23	13.00	26.40
24	7.00	1.50
25	8.00	12.57
26	24.00	2.00
27	3.00	6.72
28	24.00	20.75
29	38.00	51.65
30	13.00	16.42
31	4.00	3.90
32	12.00	2.44
33	1.00	3.90
34	5.00	15.31
35	1.00	1.44
36	13.00	14.88
37	19.00	23.01
38	5.00	3.44
39	12.00	14.32
40	12.00	24.39
41	7.38	9.88
42	7.00	17.66
43	2.00	3.26
44	8.00	6.89
45	1.00	0.84
46	10.00	13.02
47	22.00	32.85

(i) Considering the above 47 villages as constituting the population, divide it into three classes, one having area under fresh fruit less than or equal to 25 acres, the second between 25 and 60 acres and the third over 60 acres. From each of these classes, select a sample of two villages with probability proportional to the area under fresh fruit and without replacement, using the random numbers in the order given below:

9978   20873   15319   5736   19978   28865

(ii) Estimate the total area under guava and its standard error using
   (a) Horvitz-Thompson estimator
   (b) Des Raj ordered estimator
   (c) Unordered Murthy estimator

(iii) Divide the villages in each class into two equal groups at random and select one village from each group with probability proportional to the area under fresh fruits, with the help of the random numbers given above. Estimate the total area under guava with its standard error.

5.8 The data given below relate to 15 villages of one stratum, collected in a survey conducted in Andhra Pradesh for estimating the number of cashewnut trees. The area under cashewnut in different villages, as reported by Patwaris, and the actual number of cashewnut trees in different villages are given in Columns (2) and (3) of Table given below:

(1)	(2)	(3)
1	71.45	2614
2	46.28	1927
3	116.27	5162
4	178.11	6248
5	276.98	10842
6	22.72	1045
7	67.34	2359
8	26.21	1228
9	82.85	3107
10	0.36	1082
11	97.16	3247
12	178.16	8039
13	17.83	1013
14	521.46	19489
15	17.34	557

(i) Draw a pps sample of size 2 without replacement from this stratum and obtain an unbiased estimate of the total number of trees and the variance of the estimate by adopting (a) Des Raj estimator (b) Horvitz-Thompson estimator.

(ii) Divide the stratum into two groups at random and select one village from each group with probability to area under cashewnut trees. Estimate the total number of trees with its standard error. You may use the random numbers 20153 and 44227 for selecting units.

5.9 Data on the number of boats landing (x) and the catch of fish (y) at a particular centre in the Malabar Coast of India on a particular day of 12 hours (6.00 am to 6.00 pm) are given below:

Hours	1	2	3	4	5	6	7	8	9	10	11
x(in number)	42	52	19	6	23	56	36	59	14	14	26
y(in maunds)	568	887	223	88	352	1295	934	1265	486	443	980

Select a sample of size 3 with probability proportional to the number of boats landing, by the following methods:
 (i) Hartley-Rao's pps systematic sampling
 (ii) Rao, Hartley and Cochran's random group method
From the selected samples, estimate the total catch of fish in a day along with respective standard errors.

5.10 To obtain an idea about the relative performance of various sampling procedures for samples of size 2, two populations of size $N = 4$ were studied by Yates and Grundy, and Des Raj, which are given below:

Serial No. of the unit	p_i	Population A (y_i)	Population B (y_i)
1	0.1	0.5	0.8
2	0.2	1.2	1.4
3	0.3	2.1	1.8
4	0.4	3.2	2.0

Obtain the sampling variances of the following for sample size 2:
 (i) pps sampling with replacement
 (ii) Ordered Des Raj estimator
 (iii) Unordered Murthy estimator
 (iv) Horvitz-Thompson estimator
 (v) Midzuno system of sampling
 (vi) pps systematic sampling
 (vii) Rao, Hartley and Cochran method
Compare the efficiency of each method with that of sampling with replacement.

5.11 A population consists of 7 units of sizes 10, 20, 30, 40, 50, 60, and 70. A sample of 2 units is to be drawn by pps wor. Find the probabilities of inclusion in the sample for (i) each unit and (ii) each pair of units.

5.12 A sample of 10 villages was drawn from a tehsil with pps wr, size being 1951 census population $(X = 415,149)$.

The relevant data are given below:

Village	1951 census population (x)	Cultivated area (y) in acres	Village	1951 census population (x)	Cultivated area (y) in acres
1	5511	4824	6	7357	5505
2	865	924	7	5131	4051
3	2535	1948	8	4654	4060
4	3523	3013	9	1146	809
5	8368	7678	10	1165	1014

(i) Estimate the total cultivated area and its standard error.

(ii) Obtain the sample size required to ensure the relative standard error of 2 per cent.

REFERENCES

Basu, D., "On sampling with and without replacement," *Sankhya*, **20**, 287-294, (1958).

Brewer, K.R.W. and G.C. Undy, "Samples of two units drawn with unequal probabilities without replacement," *Aust. J. Statist.*, **4**, 89-100, (1962).

Das, A.C., "On two phase sampling and sampling with varying probabilities" *Bull. Int. Statist. Inst.*, **61**, 384-390, (1951).

Des Raj, "Some estimators in sampling with varying probabilities without replacement," *J. Amer. Statist. Assoc.*, **51**, 269-284, (1956).

Durbin, J., "Design of multi-stage surveys for the estimation of sampling errors," *Applied Statistics.*, **16**, 152-164, (1967).

Hanurav, T.V , "Optimum utilization of auxiliary information: πps sampling of two units from a stratum," *J. R.Statist. Soc.*, **29** B, 374-391, (1967).

Hartley, H.O., "Systematic sampling with unequal probabilities and without replacement," *J. Amer. Statist. Assoc.*, **61**, 739-748, (1966).

Hartley, H.O. and J.N.K. Rao, "Sampling with unequal probabilities and without replacement," *Ann. Math. Statist.*, **33**, 350-374, (1962).

Horvitz, D.G. and D.J. Thompson, "A generalization of sampling without replacement from a finite universe," *J. Amer. Statist. Assoc.*, **47**, 663-685, (1952).

Lahiri, D.B., "A method of sample selection providing unbiased ratio estimates," *Bull. Int. Statist.*, **33**, 133-140, (1951).

Madow, N.G., "On theory of systematic sampling," *Ann. Math. Statist.*, **20**, 333-354, (1949).

Midzuno, H., "On the sampling systems with probabilities proportional to sums of sizes," *Ann. Inst. Statist. Math.*, **3**, 99-107, (1952).

Murthy, M.N., "Ordered and unordered estimators in sampling without replacement;" *Sankhya*, **18**, 379-390, (1957).

Narain, R.D., "On sampling without replacement with varying probabilities," *J. Ind. Soc. Agr. Statist.*, **4**, 57-60, (1951).

Rao, J.N.K., "On the estimate of the variance in unequal probability sampling," *Ann. Inst. Statist. Math.*, **13**, 57-60, (1961).

Rao, J.N.K., H.O. Hartley and W.G. Cochran, "On a simple procedure of unequal probability sampling without replacement," *J.R. Statist. Soc.*, **24B**, 484-491, (1962).

Sen, A.R., "Present status of probability sampling and its use in estimation of farm characteristics," *Econometrica*, **20**, 130, (1952).

Singh, D., "On efficiency of sampling with varying probabilities without replacement," *J. Ind. Soc. Agr. Statist.*, **6**, 48-57, (1954).

Srivastava, A.K. and D. Singh, "A sampling procedure with inclusion probabilities proportional to size," *Biometrika*, **68**, 732-734, (1981).

Yates, F. and P.M. Grundy, (1953). "Selection without replacement from within strata probability proportional to size," *J.R. Statist Soc.*, **15 B**, 253-261, (1953).

6

Ratio Estimators

There is repetition everywhere, and nothing is found only once in
the world.   Goethe

6.1 INTRODUCTION

In many surveys, information on an auxiliary variate which is highly
correlated with the variable under study is readily available and can
be used for improving sampling design. Stratified sampling and pps
schemes are two examples of improved sampling designs, in which the
use of data on auxiliary variate has been made in Chapters 3 and 5.
However, in both the schemes, it implies that the information on
auxiliary variate on individual sampling units is available prior to
preparation of the sampling design. In case data on auxiliary
variate for individual sampling units are not available but only the
aggregate value for all units of auxiliary variate is available, the two
schemes cannot be used. In such a situation, the aggregated data on
auxiliary variate can still be used at the time of estimation of the parameters under study, provided the data on auxiliary variate for the
sampled units can be easily obtained while recording the values of the
study variate. Two such methods of estimation are known as *ratio*
method of estimation and *regression method of estimation*. In this
chapter, the ratio method of estimation will be discussed.

6.2 DEFINITIONS AND NOTATIONS

Let us denote

y_i = the value of the characteristic under study for the ith unit of the population

x_i = the value of the auxiliary characteristic on the same unit

Y = the total of y characteristic of the population

X = the total of x characteristic of the population

$R = Y/X = \overline{Y}/\overline{X}$ = the ratio of the population totals or means of character y and x

ρ = the correlation coefficient between x and y in the population.

Suppose it is desired to estimate Y, or $\overline{Y}$ or R by drawing a simple random sample of n units from the population. Let us assume that based on n pairs of observations, $\bar{y}$ and $\bar{x}$ are the sample means of the characteristics y and x, respectively, and the population total X or mean $\overline{X}$ is known. The ratio estimators of the population ratio $Y/X = R$, the total Y, and the mean $\overline{Y}$, may be defined by

$$\left.\begin{aligned}
\hat{R} &= \frac{\mathbf{y}}{\mathbf{x}} = \frac{\bar{y}}{\bar{x}} \\
\hat{\overline{Y}}_R &= \frac{\mathbf{y}}{\mathbf{x}} X = \frac{\bar{y}}{\bar{x}} X = \hat{R} X \\
\hat{Y}_R &= \frac{\mathbf{y}}{\mathbf{x}} \overline{X} = \hat{R}\,\overline{X}, \text{ respectively}
\end{aligned}\right\} \quad (6.2.1)$$

and

where $\mathbf{y}$ and $\mathbf{x}$ are the sample totals for y and x, respectively.

6.3 BIAS OF RATIO ESTIMATORS

In the previous chapters, we discussed unbiased estimators but many a time, biased estimators are to be used when these are comparatively more efficient. There can be many illustrations where one would like to use such estimators in order to obtain higher precision without changing the cost of the survey or sampling scheme. It will be shown here that $\hat{R}$, $\hat{Y}$ and $\hat{\overline{Y}}_R$ are all biased estimators but preferred to the unbiased estimators under certain conditions which we shall discuss in subsequent sections. Since the theory is the same for all these estimators, most results will relate to the problem of estimating a ratio.

THEOREM 6.3.1 In simple random sampling, the bias of the ratio estimator $\hat{R}$ is given by

$$B(\hat{R}) = - \frac{\text{Cov}(\hat{R}, \bar{x})}{\bar{X}} \tag{6.3.1}$$

Proof We know that

$$\text{Cov}(u, v) = E(uv) - E(u) E(v)$$

Hence,
$$\text{Cov}(\hat{R}, \bar{x}) = E(\hat{R}\bar{x}) - E(\bar{x}) E(\hat{R})$$
$$= E(\bar{y}) - E(\bar{x}) E(\hat{R})$$
$$= \bar{Y} - \bar{X} E(\hat{R})$$

or
$$\frac{\text{Cov}(\hat{R}, \bar{x})}{\bar{X}} = \frac{\bar{Y}}{\bar{X}} - E(\hat{R})$$
$$= R - E(\hat{R}) = - B(\hat{R})$$

COROLLARY 1 Prove that

$$\frac{|B(\hat{R})|}{\sigma_{\hat{R}}} \leqslant \text{CV}(\bar{x})$$

where CV stands for the coefficient of variation.

COROLLARY 2 Show that

$$B(\hat{Y}_R) = X B(\hat{R}) \text{ and } B(\hat{\bar{Y}}_R) = \bar{X} B(\hat{R})$$

COROLLARY 3 Prove that $\hat{R}$ is unbiased if $\rho_{(\hat{R}, \bar{x})} = 0$, where ρ stands for the correlation coefficient.

THEOREM 6.3.2 Show that the first approximation to the relative bias of the ratio estimator in simple random sampling, wor, is given by

$$\frac{B(\hat{R})}{R} \cong \frac{(1-f)}{n\bar{X}\bar{Y}}(RS_x^2 - \rho S_x S_y) \tag{6.3.2}$$
$$\cong \frac{(1-f)}{n}(C_x^2 - \rho C_x C_y)$$

where $C_x = S_x/\bar{X}$ and $C_y = S_y/\bar{Y}$ are the coefficients of variation of x and y, respectively.

Proof We know that

$$\hat{R} - R = \frac{\bar{y}}{\bar{x}} - R = \frac{\bar{y} - \bar{x}R}{\bar{x}} = \frac{\bar{y} - \bar{x}R}{\bar{X} + (\bar{x} - \bar{X})}$$

$$= \frac{1}{\bar{X}} (\bar{y} - R\bar{x}) \left(1 + \frac{\bar{x} - \bar{X}}{\bar{X}} \right)^{-1}$$

Expanding by a Taylor's series, we get

$$\hat{R} - R \cong \frac{1}{\bar{X}} (\bar{y} - R\bar{x}) \left(1 - \frac{\bar{x} - \bar{X}}{\bar{X}} + \ldots \right)$$

Ignoring the terms of the second and higher orders, we have

$$E(\hat{R} - R) = \frac{1}{\bar{X}} \left[E(\bar{y} - R\bar{x}) - \frac{1}{\bar{X}} E(\bar{y} - R\bar{x})(\bar{x} - \bar{X}) \right]$$

Since $\quad E(\bar{y} - R\bar{x}) = \bar{Y} - R\bar{X} = 0$, and

$$E(\bar{y} - R\bar{x})(\bar{x} - \bar{X}) = E\{\bar{y} \cdot (\bar{x} - \bar{X})\} - RE\{\bar{x} \cdot (\bar{x} - \bar{X})\}$$

$$= E(\bar{y} - \bar{Y})(\bar{x} - \bar{X}) - RE(\bar{x} - \bar{X})^2$$

$$= \frac{(1 - f)}{n} [\rho S_x S_y - R S_x^2]$$

Hence $\quad B(\hat{R}) = \left\{ \frac{(1 - f)}{n\bar{X}^2} \right\} (R S_x^2 - \rho S_x S_y)$

or $\quad \dfrac{B(\hat{R})}{R} = \dfrac{(1 - f)}{n\bar{X}\bar{Y}} (R S_x^2 - \rho S_x S_y)$

By substituting,

$$C_x = \frac{S_x}{\bar{X}} \text{ and } C_y = \frac{S_y}{\bar{Y}},$$

we get the result.

For large and sufficiently high value of ρ the bias will be negligible. It should be noted that the bias in the ratio estimator becomes zero when $R = \rho S_y / S_x$, which is satisfied only if the line of regression of y on x passes through the origin. It can also be seen that contribution of higher order terms depends on the values of the moments and product moments of two variates and these are worked out by Sukhatme (1944). To a second order of approximation, the relative bias in a ratio estimator for a large population can be expressed as

$$B' = B \left(1 + \frac{3 C_x^2}{n} \right) \tag{6.3.3}$$

COROLLARY 1 Show that an upper bond to the ratio of the bias to its standard error is given by

$$\frac{|B(\hat{R})|}{\sigma_{\hat{R}}} \leqslant \frac{\sigma_{\bar{x}}}{\bar{X}} = C_x \qquad (6.3.4)$$

COROLLARY 2 An estimator of the bais $B(\hat{R})$ is given by

$$\hat{B}(\hat{R}) = \frac{[\hat{R}\, v(\bar{x}) - \text{Cov}(\bar{x}, \bar{y})]}{\bar{X}^2} \qquad (6.3.5)$$

6.4 APPROXIMATE VARIANCE OF RATIO ESTIMATOR

Since ratio estimators are generally biased, we should consider their mean square errors for the purpose of comparing their efficiency with that of any other estimator. Although these estimators are biased but they are consistent, and with simple random sampling for moderately large samples the bias is negligible. Consequently, for most practical purposes, approximate variance results are equally valid for comparison of its precision.

THEOREM 6.4.1 In simple random sampling, wor, for large n, an approximation to the variance of $\hat{R}$ is given by

$$V(\hat{R}) \cong \frac{(1-f)}{n\,\bar{X}^2} \sum_{i}^{N} \frac{(y_i - Rx_i)^2}{(N-1)} \qquad (6.4.1)$$

where $f(= n/N)$ is the sampling fraction.

Proof $V(\hat{R}) = E[\hat{R} - E(\hat{R})]^2$

$$\cong E(\hat{R} - R)^2 = E\left(\frac{\bar{y} - R\bar{x}}{\bar{x}}\right)^2 \cong \frac{E(\bar{y} - R\bar{x})^2}{\bar{X}^2}$$

Now consider the variate $u_i = y_i - Rx_i$, we have

$$\bar{u} = \bar{y} - R\bar{x}, \text{ and } \bar{U} = \bar{Y} - R\bar{X} = 0.$$

By applying Theorem 2.3.1, we get

$$V(\hat{R}) = \frac{1}{\overline{X}^2} \cdot E(\bar{u} - \overline{U})^2$$

$$= \frac{(1 - f)}{n\overline{X}^2} \cdot \sum_i^N \frac{(u_i - \overline{U})^2}{(N - 1)}$$

$$= \frac{(1 - f)}{n\overline{X}^2} \sum_i^N \frac{(y_i - Rx_i)^2}{(N - 1)}$$

COROLLARY 1   Show that

$$V(\hat{\overline{Y}}_R) \cong \frac{(1 - f)}{n} \sum_i^N \frac{(y_i - Rx_i)^2}{(N - 1)}$$

COROLLARY 2   Show that

$$V(\hat{Y}_R) \cong \frac{N^2(1 - f)}{n} \sum_i^N \frac{(y_i - Rx_i)^2}{(N - 1)}$$

THEOREM 6.4.2   Show that, to the first order of approximation, the variance of $\hat{R}$ can be expressed as

$$\left. \begin{array}{l} V(\hat{R}) = \dfrac{(1 - f)}{n\overline{X}^2} [S_y^2 + R^2 S_x^2 - 2\rho R\, S_x S_y] \\[2mm] = \dfrac{(1 - f)R^2}{n} [C_y^2 + C_x^2 - 2\rho\, C_x C_y] \end{array} \right\} \tag{6.4.2}$$

Proof   Since $\overline{Y} = R\overline{X}$, we can write in the relation (6.4.1)

$$u_i = y_i - Rx_i = (y_i - \overline{Y}) - R(x_i - \overline{X})$$

Then, $$V(\hat{R}) = \frac{(1 - f)}{n\,\overline{X}^2(N - 1)} \sum_i^N \{(y_i - \overline{Y}) - R(x_i - \overline{X})\}^2$$

$$= \frac{(1 - f)}{n\,\overline{X}^2(N - 1)} [\sum_i^N (y_i - \overline{Y})^2 + R^2 \sum_i^N (x_i - \overline{X})^2$$

$$- 2R \sum_i^N (x_i - \overline{X})(y_i - \overline{Y})]$$

The correlation coefficient ρ between y and x in the finite population is defined by

$$\rho = \frac{\sum_i^N (y_i - \overline{Y})(x_i - \overline{X})}{(N - 1)\, S_x S_y}$$

Thus we get,

$$V(\hat{R}) = \frac{(1-f)}{n\overline{X}^2} [S_y^2 + R^2 S^2 - 2R\,\rho S_x S_y]$$

$$= \frac{(1-f)}{n} R^2 \left[\frac{S_y^2}{\overline{Y}^2} + \frac{S_x^2}{\overline{X}^2} - 2\rho\,\frac{S_x}{\overline{X}}\,\frac{S_y}{\overline{Y}}\right]$$

$$= \frac{(1-f)\,R^2}{n} [C_y^2 + C_x^2 - 2\rho C_x C_y]$$

COROLLARY 1 Show that, to the first order of approximation,

$$V(\hat{\overline{Y}}_R) = \frac{(1-f)}{n} [S_y^2 + R^2 S_x^2 - 2R\rho S_x S_y]$$
$$= \frac{(1-f)}{n} \overline{Y}^2 [C_y^2 + C_x^2 - 2\rho C_x C_y] \quad\Bigg\}\quad (6.4.3)$$

COROLLARY 2 Show that, to the first order of approximation,

$$V(\hat{Y}_R) = \frac{N^2(1-f)}{n} [S_y^2 + R^2 S_x^2 - 2R\,\rho S_x S_y]$$
$$= \frac{(1-f)Y^2}{n} [C_y^2 + C_x^2 - 2\,\rho C_x C_y] \quad\Bigg\}\quad (6.4.4)$$

COROLLARY 3 Show that, to the first order of approximation,

$$(CV)^2 = \frac{(1-f)}{n} [C_y^2 + C_x^2 - 2\rho C_y C_x] \quad (6.4.5)$$

The quantity $(CV)^2$ has been called the relative variance by Hansen, Hurwitz and Madow (1953). Since the coefficient of variation, i.e. the standard error divided by the parameter to be estimated is the same quantity for all the three estimates $\hat{R}$, $\hat{\overline{Y}}_R$, and $\hat{Y}_R$ and the formula can be used to calculate their relative variances.

COROLLARY 4 If $C_x = C_y = C$, show that the relative variance

$$V\left(\frac{\hat{R}}{R}\right) = \frac{(1-f)}{n} \cdot 2C^2(1-\rho) \quad (6.4.6)$$

COROLLARY 5 A sample estimate of variance of the ratio estimator is given by

$$v(\hat{R}) = \frac{(1-f)}{n\,\overline{X}^2} \sum_i^n \frac{(y_i - \hat{R}x_i)^2}{(n-1)}$$
$$= \frac{(1-f)}{n\overline{X}^2} [\sum_i^n y_i^2 + \hat{R}^2 \sum_i^n x_i^2 - 2\hat{R} \sum_i^n x_i y_i] \quad\Bigg\}\quad (6.4.7)$$
$$= \frac{(1-f)}{n\overline{X}^2} [s_y^2 + \hat{R}^2 s_x^2 - \hat{R}\,r\,s_y\,s_x]$$

where r is the estimate of correlation coefficient from the sample values.

Relation (6.4.7) is a convenient form for computations and can be easily handled with the survey data.

Another form of a sample estimate of variance is

$$v_1(\hat{R}) = \frac{(1 - f)}{n\bar{X}^2} [s_y^2 + \hat{R}^2 s_x^2 - 2\hat{R} r s_x s_y]$$

There arises a question whether one should use $v(\hat{R})$ or $v_1(\hat{R})$ if $\bar{X}$ is known. Rao and Rao (1971) have observed that $v_1(\hat{R})$ is less biased for certain populations.

6.5 COMPARISON OF RATIO ESTIMATOR WITH MEAN PER UNIT

The conditions under which the ratio estimator is superior to the mean per unit will be worked out with a comparison of their variances. In simple random sampling, wor, the variance of the mean per unit is given by

$$V(\hat{\bar{Y}}_{\text{SR}}) = \frac{(1 - f) S_y^2}{n}$$

Also, we have seen that to the first order approximation, the variance of the mean based on the ratio method is given by

$$V(\hat{\bar{Y}}_R) = \frac{(1 - f)}{n} (S_y^2 + R^2 S_x^2 - 2R \rho S_x S_y)$$

Hence the ratio estimate will have smaller variance if

$$S_y^2 + R^2 S_x^2 - 2R \rho S_x S_y < S_y^2$$

or if

$$\rho > \frac{R S_x}{2 S_y} = \frac{1}{2}\left(\frac{S_x}{\bar{X}}\right)\Big/\left(\frac{S_y}{\bar{Y}}\right) = \frac{C_x}{2C_y} \qquad (6.5.1)$$

Thus, it depends on the value of correlation between y and x. If $C_x = C_y$, i.e. the value of coefficients of variation of x and y are the same, the ratio estimator is superior if ρ exceeds 0.5. The variability of the auxiliary variate x is an important factor. If cv of x is more than twice that of y, the ratio estimate is always less precise.

EXAMPLE 6.1 For studying milk yield, feeding and management practices of milch animals in the year 1977-78, the whole of Haryana

State was divided into 4 zones according to agro-climatic conditions. The total number of milch animals in 17 randomly selected villages (in 1977-78) of zone A, along with their livestock census data in 1976, are shown below:

S. No. of village	1	2	3	4	5	6	7
Number of milch animals in survey (y)	1129	1144	1125	1138	1137	1127	1163
Number of milch animals in census (x)	1141	1144	1127	1153	1117	1140	1153
S. No. of village	8	9	10	11	12	13	14
Number of milch animals in survey (y)	1153	1164	1130	1153	1125	1116	1115
Number of milch animals in census (x)	1146	1189	1137	1170	1115	1130	1118
S.No. of village	15	16	17				
Number of milch animals in survey (y)	1112	1112	1123				
Number of milch animals in ceusus (x)	1122	1113	1166				

Estimate the total number of milch animals in 117 villages of zone A
(i) by ratio method and
(ii) by simple mean per unit method
Also compare its precision, given the total number of milch animals in the census $= 143968$

Here,

$$N = 117, n = 17, \qquad X = 143968,$$
$$\sum x_i = 19381, \qquad \sum y_i = 19266$$
$$\bar{x} = 1143.06, \qquad \bar{y} = 1133.29$$

$$\hat{R} = 0.9939$$
$$s_y^2 = 287.85, \; s_x^2 = 458.56, \; s_{xy} = 262.86$$

(i) *Ratio Estimate* The total number of milch animals by the ratio method of estimation is given by

$$\hat{Y}_R = \hat{R} \, X = 0.9939 \times 143968 = 143090$$

and an estimate of the variance of $\hat{Y}_R$ is given as

$$v(\hat{Y}_R) = \frac{(1 - f)N^2}{n} [s_y^2 + R^2 s_x^2 - 2R\, s_{xy}]$$

$$= \frac{100 \times 117}{17} [287.85 + 403.99 - 522.51]$$

$$= 118{,}890$$

(ii) *Mean Per Unit Estimate*   The total number of milch animals by mean per unit estimate is given by

$$\hat{Y} = N\bar{y} = 117 \times 1133.29 = 132594.93$$

and an estimate of variance of $\hat{Y}$ is given by

$$v(\hat{Y}) = \frac{(1 - f)\, N^2}{n}\, s_y^2 = \frac{100 \times 117}{17} \times 287.85$$

$$= 129{,}285$$

Hence, the relative precision of ratio estimate is given as

$$\text{R.P.} = \frac{v(\hat{Y}) - v(\hat{Y}_R)}{v(\hat{Y}_R)} \times 100 = 18.8\%$$

6.6 CONFIDENCE LIMITS

For large samples, the estimate of the mean or total may be assumed to follow approximately normal distribution. The confidence limits for the total will be written as

$$\hat{Y}_R \pm z\sqrt{ v(\hat{Y}_R)} \tag{6.6.1}$$

where z is the value of the normal variate for a given level of confidence coefficient.

Similarly, the limits for R may be written as

$$\hat{R} \pm z\sqrt{ v(\hat{R})} \tag{6.6.2}$$

For small samples, let us assume that (x, y) follows a bivariate normal distribution and let $u_i = y_i - Rx_i$. We have

$$\bar{u} = \bar{y} - R\bar{x}, \quad \bar{U} = \bar{Y} - R\bar{X} = 0$$

and
$$V(\bar{u}) = \frac{(1 - f)}{n} (S_y^2 + R^2 S_x^2 - 2RS_{yx})$$

If u_i is normally distributed with mean 0 and variance V, then $\bar{u} = (\bar{y} - R\bar{x})$ will be normally distributed with mean 0 and variance V/n. Now we define,

$$t_{(\alpha, n-1)} = \frac{|\bar{u} - \bar{U}|}{\sqrt{v(\bar{u})}} \qquad (6.6.3)$$

where $v(\bar{u})$ is an unbiased estimate of $V(\bar{u})$, which is given by

$$v(\bar{u}) = \frac{(1-f)}{n}(s_y^2 + R^2 s_x^2 - 2R s_{xy}),$$

Substituting $s_{xy} = r s_y s_x$ and using $c_x = s_x/\bar{x}$ and $c_y = s_y/\bar{y}$ we have

$$v(\bar{u}) = \frac{(1-f)}{n}(c_y^2 + R^2 c_x^2 - 2R r\, c_x c_y)$$

The confidence limits of R with confidence coefficient $(1 - \alpha)$ are determined by the two roots of the quadratic in R, which is given after substituting these values in Eq. (6.6.3). The roots thus obtained are somewhat cumbersome. After some manipulations, the confidence limits of R may be expressed in an easier form as

$$\hat{R} \pm \hat{R} t_{(\alpha, n-1)} \left[\frac{(1-f)}{n} \cdot (c_y^2 + c_x^2 - 2r\, c_x c_y) \right]^{1/2} \qquad (6.6.4)$$

6.7 RATIO ESTIMATORS IN STRATIFIED SAMPLING

When the population is stratified and units are drawn by simple random sampling method from each stratum, there are two ways of obtaining a ratio estimate of the population total Y (1) *separate* ratio estimate, and (2) *combined* ratio estimate. We shall discuss them in the following sections.

6.7.1 Separate Ratio Estimator

If y_m, x_m are the sample totals in the mth stratum and X_m is the mth stratum total, we may define the estimator $\hat{Y}_{Rs}$ (s for *separate*) as

$$\hat{Y}_{Rs} = \sum_m \frac{y_m}{x_m} X_m = \sum_m \frac{\bar{y}_m}{\bar{x}_m} X_m \qquad (6.7.1)$$

THEOREM 6.7.1 If the sample sizes, $n'_m s$ are large in all strata and simple random sampling, wor, is done independently within each stratum, show that $\hat{Y}_{Rs}$ is a biased estimator with negligible bias and its sampling variance is given by

$$V(\hat{Y}_{Rs}) = \sum_m N_m^2 \frac{(1 - f_m)}{n_m}(S_{y_m}^2 + R_m^2 S_{x_m}^2 - 2 R_m \rho_m S_{y_m} S_{ym})$$

$$(6.7.2)$$

where R_m ($= Y_m / X_m$) and ρ_m, are the true ratio and the coefficient of correlation, respectively in the mth stratum.

Proof Let us define the ratio estimator for the mth stratum.

$$\hat{Y}_{Rm} = \frac{y_m}{x_m} X_m$$

Then $\hat{Y}_{Rs} = \sum_m \hat{Y}_{Rm}$

Here it can be seen that the estimator $\hat{Y}_{Rs}$ is built up from stratum level ratio estimators. It follows that $\hat{Y}_{Rs}$ is not an unbiased but a consistent estimator of the population total and will have the same properties as the usual ratio estimator. Using Eq. (6.3.2), the relative bias is obtained as

$$\frac{B(\hat{Y}_{Rs})}{Y} = \sum_m \frac{(1 - f_m)}{n_m} (C_{x_m}^2 - \rho C_{x_m} C_{ym})$$

If we assume teat that in each stratum approximates to unity, $n_m = n/k$ and that C_{x_m}, C_{y_m} and ρ_m are the same over all strata, say C_x, C_y and ρ, respectively, then relative bias reduces to $\frac{k}{n} (C_x^2 - \rho C_x C_y)$. This shows that if sample size within each stratum is sufficiently large $\hat{Y}_{Rs}$ provides a satisfactory estimate of the population total.

With similar assumptions, it can be shown that the variance of $\hat{Y}_{Rs}$, to a first approximation, is given by

$$V(\hat{Y}_{Rs}) = V(\sum_m \hat{Y}_{Rm})$$

$$= \sum_m V(\hat{Y}_{Rm}) \quad \text{since sampling is independent in the different strata, the product term vanishes.}$$

$$= \sum_m N_m^2 \frac{(1 - f_m)}{n_m} (S_{y_m}^2 + R_m^2 S_{x_m}^2 - 2R_m \, \rho_m \, S_{y_m} \, S_{x_m})$$

since $\hat{Y}_{Rm}$ is the ratio estimator within the mth stratum as the sample size is large in all the strata.

COROLLARY In stratified random sampling, wor, almost unbiased estimator of $V(\hat{Y}_{Rs})$ is given by

$$v(\hat{Y}_{Rs}) = \sum_m N_m^2 \frac{(1 - f_m)}{n_m} (s_{y_m}^2 + R_m \, s_{x_m}^2 - 2 \, \hat{R}_m \, s_{xy_m})$$

(6.7.3)

where s_{xy_m} stands for the estimated covariance in the mth stratum.

6.7.2 Combined Ratio Estimator

It was assumed, in case of separate estimator, that n_m's were large in each stratum. However, it may not hold good always in practice. To overcome this difficulty, Hansen, Hurwitz and Grurney (1946) suggested a *combined* ratio estimator $\hat{Y}_{Rc}$ (c for *combined*) for a sample from a stratified population as

$$\hat{Y}_{Rc} = \frac{\hat{Y}_{st}}{\hat{X}_{st}} \cdot X = \frac{\bar{y}_{st}}{\bar{x}_{st}} \cdot X,$$

(6.7.4)

where $\bar{y}_{st} = \dfrac{\hat{Y}_{st}}{N} = \sum_m \dfrac{N_m}{N} \, \bar{y}_m$

and $\qquad \bar{x}_{st} = \dfrac{\hat{X}_{st}}{N} = \sum_m \dfrac{N_m}{N} \, \bar{x}_m$

are estimated population means from a stratified sample, and X is the overall total of x.

THEOREM 6.7.2 If the total sample size n is large and simple random sampling, wor, is done in each stratum independently, then $\hat{Y}_{Rc}$ is a consistent estimator and its sampling variance is given by

$$V(\hat{Y}_{Rc}) = \sum_m N_m^2 \frac{(1 - f_m)}{n_m} (S_{y_m}^2 + R^2 S_{x_m}^2 - 2R \rho_m S_{y_m} S_{x_m})$$

(6.7.5)

Proof Since $\hat{R}$ is a consistent estimator, $\hat{Y}_{Rc}$ is a ratio estimator and therefore, also a consistent estimator of the population total. Further, to derive its bias, we can write

$$\hat{Y}_{Rc} - Y = \frac{N\bar{X}}{\bar{x}_{st}} (\bar{y}_{st} - R\bar{x}_{st}) \cong N(\bar{y}_{st} - R\bar{x}_{st})$$

Let us define a variate $u_{mj} = y_{mj} - R x_{mj}$

Hence, it can be shown that $\bar{u}_{st} = \bar{y}_{st} - R\bar{x}_{st}$ and,

$$\bar{U} = \bar{Y} - R\bar{X} = 0$$

Proceeding on lines similar to Eq. (6.3.2), the relative bias is obtained as

$$\frac{B(\hat{Y}_{Rc})}{Y} = \sum_m \frac{(1 - f_m)}{n_m} W_m^2 \left(\frac{S_{x_m}^2}{\bar{X}} - \frac{\rho S_{x_m} S_{y_m}}{\bar{X}\bar{Y}} \right)$$

If we assume that n_m is proportional to N_m and $S_{x_m}/\bar{X}$, $S_{y_m}/\bar{Y}$ and ρ_m are the same over all strata, say C_x, C_y and ρ, respectively, the relative bias reduces to $\frac{(1 - f)}{n} (C_x - \rho C_x C_y)$. It follows that the relative bias in $\hat{Y}_{Rc}$ diminishes rapidly with total sample size even when sample size is small in the individual stratum. Similarly, the sampling variance to the first approximation

$$V(\hat{Y}_{Rc}) = V(N\bar{u}_{st}) = N^2 V(\bar{u}_{st})$$

$$= \sum_m N_m^2 \frac{(1 - f_m) S_u^2}{n_m^2}$$

$$= \sum_m N_m^2 \frac{(1 - f_m)}{n_m} (S_{y_m}^2 + R^2 S_{x_m}^2 - 2R \rho_m S_{y_m} S_{x_m})$$

Corollary In stratified random sampling, wor, an almost unbiased estimator of $V(\hat{Y}_{Rc})$ is given by

$$v(\hat{Y}_{Rc}) = \sum_m N_m^2 \frac{(1 - f_m)}{n_m} (s_{y_m}^2 + \hat{R}^2 s_{x_m}^2 - 2 \hat{R} s_{xy_m})$$
(6.7.6)

where s_{xy_m} has its usual meaning.

6.7.3 Comparison of Separate and Combined Ratio Estimators

To make a comparative study, we can write

$$
V(\hat{Y}_{Rc}) - V(\hat{Y}_{Rs}) = \sum_m N_m^2 \frac{(1 - f_m)}{n_m} [(R^2 - R_m^2) S_{xm}^2
$$

$$
- 2 (R - R_m) \rho_m S_{ym} S_{xm})]
$$

$$
= \sum_m N_m^2 \frac{(1 - f_m)}{n_m} [(R - R_m)^2 S_{xm}^2
$$

$$
+ 2(R - R_m) (\rho_m S_{ym} S_{xm} - R_m S_{xm})^2] \qquad (6.7.7)
$$

Since the last term in Eq. (6.7.7) is small, the right side is likely to be positive. Thus, the separate ratio estimator is expected to be more precise provided the sample in each stratum is large enough for the approximate variance formula to be applied. The combined estimator will have a large variance as compared to the separate estimator, but the bias in the former is expected be smaller than in the latter. Unless the population ratio of y to x in different strata varies considerably, the combined estimator will usually have negligible bias and precision will be as high as the separate estimator. In cases when the line of regression of y on x passes through the origin within each stratum, the separate ratio estimator will be more precise than the combined one. Hence, the rules for choice between the two methods should be

1. If the sample taken from each stratum is small, the combined ratio estimator should be used unless there is a wide difference between the strata ratios R_m. When there are such differences and regrouping of the strata is possible so that each group will not differ much and has a large sample size, the separate ratio estimator should be used.

2. If the sample size in each stratum is large so that the approximate variance formula can be applied, then it is better to use the separate ratio estimator unless it involves extra calculation work. If it does, then one should make sure of the gain before actually using it.

3. If X_m are known independently from stratum to stratum, the separate ratio estimator may be used as a method of estimation. When $X_m = N_m$, the estimator becomes equivalent to the stratified estimator and there is no problem of bias from these estimates.

4. If the sampling units are the elementary units and the denominator of the ratio is simply the number of elementary units in the sample, both the estimators are unbiased. In such cases, the estimator should be chosen by consideration of the variance and gain to be obtained by using it.

EXAMPLE 6.2 The following data were collected in a pilot survey for estimating the extent of cultivation and production of fresh fruits in three districts of Uttar Pradesh in the year 1976–77.

Stratum number	Total no. villages of (N_m)	Total area (in hect.) under orchard (X_m)	No. of villages in sample (n_m)	Area under orchards in ha. (x_m)	Total no. of trees (y_m)
1	985	11253	6	10.63, 9.90, 1.45, 3.38, 5.17, 10.35,	747, 719, 78, 201, 311, 448,
2	2196	25115	8	14.66, 2.61, 4.35, 9.87, 2.42, 5.60, 4.70, 36.75,	580, 103, 316, 739, 196, 235, 212, 1646,
3	1020	18870	11	11.60, 5.29, 7.94, 7.29, 8.00, 1.20, 11.50, 1.70, 2.01, 7.96, 23, 15	488, 227, 374, 491, 499, 50, 455, 47, 879, 115, 115

Estimate the total number of trees in the three districts by various methods and compare their precision. The calculations have been shown in the table given below:

Stratum	W_m	$\left(\dfrac{1}{n_m} - \dfrac{1}{N_m}\right)$	$\bar{x}_m$	$\bar{y}_m$	$\hat{R}_m$	$W_m \times \bar{x}_m$	$W_m \times \bar{y}_m$	$s_{x_m}^2$	$s_{y_m}^2$	$s_{x_m y_m}$
1	0.2345	0.16598	6.81	417.33	61.28	1.60	97.66	16.03	74778.80	1008.75
2	0.5227	0.12454	10.07	503.38	49.99	5.26	263.12	129.64	259107.98	5643.81
3	0.2428	0.08902	7.97	340.00	42.66	1.94	82.55	38.39	65885.60	1403.69

Here $\hat{R} = \sum W_m \bar{y}_m / \sum W_m \bar{x}_m = 443.53/8.80 = 50.40$

(i) *Combined Ratio Estimate* The estimate of the total number of trees is given by

$$\hat{Y}_{Rc} = \frac{\sum W_m \bar{y}_m}{\sum W_m \bar{x}_m} X = \frac{443.53}{8.80} \times 55238 = 2783995$$

$$v(\hat{Y}_{Rc}) = \sum N_m^2 \left(\frac{1}{n_m} - \frac{1}{N_m}\right)(s_{y_m}^2 + \hat{R}s_{x_m}^2 - 2\hat{R}s_{xy_m})$$

$$= (985)^2 (0.16598) [74778.80 + (50.40)^2 \times 16.03$$

$$- 2 \times 50.40 \times 1008.75] + (2196)^2 \times (0.12454) [259107.90$$

$$+ (50.40)^2 \times 129.64 - 2 \times 50.40 \times 5643.81]$$

$$+ (1020)^2 \times 0.08902 [65885.60 + (50.40)^2 \times 38.39$$

$$- 2 \times 50.40 \times 1403.81]$$

$$= 161057.35 \times 13815.57 + 602802.00 \times 19518.23$$

$$+ 92595.60 \times 21910.39 = 6019519627.34$$

(ii) *Separate Ratio Estimate* Another estimate of the total number of trees is given by

$$\hat{Y}_{Rs} = \sum \hat{R}_m X_m$$

$$= 61.28 \times 11253 + 49.99 \times 25115 + 42.66 \times 188.70$$

$$= 2750076.89 = 2750077$$

Estimated variance of $\hat{Y}_{Rs}$ is given by

$$v(\hat{Y}_{Rs}) = \sum N_m^2 \left(\frac{1}{n_m} - \frac{1}{N_m}\right)(s_{y_m}^2 + \hat{R}_m^2 s_{x_m}^2 - 2\hat{R}_m s_{xy_m})$$

$$= (985)^2 \times (0.16598) [74778.80 + (61.28)^2 \times 16.03$$

$$- 2 \times (61.28) \times 1008.75]$$

$$+ (2196)^2 \times (0.12454) [259107.90 + (49.99)^2 \times 16.03$$

$$- 2 \times 49.99 \times 5643.81] + (1020)^2 \times (0.08902 \times [65885.60$$

$$+ (42.66)^2 \times 38.39 - 2 \times 42.66 \times 1403.69]$$

$$= 161057.35 \times 11342.88 + 602802.00 \times 18810.18$$

$$+ 92595.60 \times 15937.79 = 2441137855.48$$

The efficiency of separate ratio estimate $(\hat{Y}_{Rs})$ over the combined ratio estimate $(\hat{Y}_{Rc})$ is given by

$$R.P. = \frac{6019519627.34}{2441137855.48} \times 100$$
$$= 246.58\%.$$

6.8 UNBIASED RATIO ESTIMATOR AND ITS VARIANCE

Under simple random sampling, the ratio estimator $\hat{Y}_R$ is seen to be biased. There are, however, some methods of modifying the sampling scheme or estimating procedure to make the estimator unbiased. One approach is to modify the sampling scheme by drawing units with varying probabilities. Lahiri (1951) suggested a procedure of selecting samples with *probability proportionate to aggregate size* $_n$(ppas), where the first unit in the sample is included with pp to $\sum x_i$ and the remaining $(n - 1)$ units with equal probabilities. Midzuno [i] (1951) suggested another method in which the first unit is selected with probability proportional to x_i and the remaining $(n - 1)$ units are selected from the remaining $(N - 1)$ units with equal probability and without replacement. In this section, we shall discuss in brief, sampling with varying probability without replacement methods.

Let us suppose that, from a finite population of size N, the $_N$first unit in the sample is selected with probability p_i $(i = 1, \ldots, N)$, $\sum p_i = 1$, and the remaining $(n - 1)$ units with equal probability without[i]replacement. The probability that unit U_i is the first one to be selected and subsequent units form a simple random sample, wor, of size $(n - 1)$ is $p_i/\binom{N-1}{n-1}$. When p_i is proportional to x_i so that $p_i = x_i/X$, the probability of selecting a specified sample $s \equiv s(y_1, y_2, \ldots, y_n)$ is given by

$$p(s) = \frac{\sum_i^n p_i}{\binom{N-1}{n-1}} = \frac{\sum_i^n x_i}{\binom{N-1}{n-1}X} \tag{6.8.1}$$

Now let us consider $\hat{Y}_R = \frac{\bar{y}}{\bar{x}}X = \frac{y}{x}X$

Then, $$E(\hat{Y}_R) = XE\left(\frac{y}{x}\right) = X\sum{}'\frac{y}{x}\frac{x}{\binom{N-1}{n-1}X}$$

$$= \sum{}'\frac{y}{\binom{N-1}{n-1}} = Y$$

where $\sum'$ stands for summation over all possible samples.

Thus, the usual ratio estimator becomes unbiased in this sampling scheme. The variance of the estimator is given by

$$V(\hat{Y}_R) = E(\hat{Y}_R^2) - Y^2$$

$$= \frac{X}{\binom{N-1}{n-1}}\sum{}'(y^2/x) - Y^2 \qquad (6.8.2)$$

where $\sum'$ stands for summation over all possible combinations.

An unbiased estimator of the variance of $\hat{Y}_R$ is given by

$$v(\hat{Y}_R) = \hat{Y}_R^2 - \frac{NX}{x}\left[\bar{y}^2 - \left(\frac{1}{n} - \frac{1}{N}\right)s_y^2\right] \qquad (6.8.3)$$

Estimates of acreage of some principal crops in India are obtained by this method and reasonably good forecasts are made before crop harvest. Data on crop areas are collected field by field and maintained by the *Patwari* or *Lekhpal* (village accountant) for all villages within his jurisdiction. This information is used as a basis for selection of villages with probability proportional to the area under the crop.

EXAMPLE 6.3 The following data are related to the total cultivated area during 1978-79 and the area under wheat in the two consecutive years 1978-79 and 1979-80 for a sample of 16 villages selected in the Baghpat Tehsil of Meerut District (UP) for the survey on fertilizer practices. The villages were selected with replacement, with probability proportional to the cultivated area, as recorded in 1978-79. The total cultivated area and area under wheat in 1978-79 for all 304 villages of the Baghpat Tehsil are 191648 and 61100 hectares, respectively.

(i) Estimate the area under wheat for the Tehsil for the year 1979-80 using the ratio method of estimation and calculate its standard error.

(ii) What would be the estimate and its standard error if the information for the previous year was not used.

S. N. of villages	Total cultivated area during 1978–79 (in ha)	Area under wheat during 1978–79 (in ha)	1979–80 (in ha)
1	376	100	85
2	747	200	239
3	1149	423	406
4	1487	503	503
5	674	258	217
6	4163	1275	1191
7	595	89	69
8	1873	699	584
9	515	243	294
10	2534	672	745
11	2541	597	611
12	1307	455	421
13	1017	282	361
14	1343	421	277
15	291	80	96
16	1163	465	489

Here, we have $N = 364$, $X = 61100$ and $n = 16$

Calculations have been done in the tabular form given as below:

S. N.	Total cultivated area in 1978–79 (in ha) (A_i)	Area under wheat during 1978–79 (in ha) (x_i)	1979–80 (in ha) (y_i)	$l_i = \dfrac{y_i}{A_i}$	$l'_c = \dfrac{x_i}{A_i}$
1	376	100	85	0.226064	0.265957
2	747	200	239	0.319946	0.267738
3	1149	423	406	0.353351	0.368146
4	1487	503	503	0.338265	0.338265
5	674	258	217	0.321958	0.382789
6	4163	1275	1191	0.286092	0.306270
7	595	89	69	0.115966	0.149580
8	1873	699	584	0.311799	0.373198
9	515	243	294	0.570874	0.571875
10	2534	672	745	0.294002	0.265193
11	2541	597	611	0.240557	0.234947
12	1307	455	421	0.322112	0.348125
13	1017	282	361	0.354966	0.313477
14	1343	421	277	0.280715	0.277286
15	291	80	96	0.329897	0.274914
16	1163	465	489	0.420464	0.399828
Total	191,648			5.086928	5.037558

(i) The ratio estimate of the total area under wheat for the year 1979-80 is given by

$$\hat{Y}_R = \frac{\bar{z}}{\bar{v}} X$$

where $\quad X = 61,100 \quad \bar{z} = \frac{1}{n} \sum_{i}^{n} z_i = \frac{A}{n} \sum_{i} l_i$

and $\qquad\qquad \bar{v} = \frac{1}{n} \sum_{i}^{n} v_i = \frac{A}{n} \sum_{i}^{n} l_i'$

Thus $\qquad\qquad \hat{Y}_R = \frac{\sum l_i}{\sum l_i'} X = \hat{R} X$

$$= 1.009804 \times 61,100 = 61699.02$$

An estimate of the variance of $\hat{Y}_R$ is given by

$$v(\hat{Y}_R) = \frac{A^2}{n(n-1)} [\sum l_i^2 + \hat{R}^2 \sum l_i'^2 - 2\hat{R} \sum l_i l_i']$$

We have $\qquad\qquad \hat{R} = \frac{\sum l_i}{\sum l_i'} = 1.006778,$

$$\hat{R}^2 = 1.013605117$$

$$\sum l_i^2 = 1.7532604136, \qquad \sum l_i'^2 = 1.6743255075$$

$$\sum l_i l_i' = 1.6056407512$$

Putting these values in the above formula $v(\hat{Y}_R)$, we have

$$v(\hat{Y}_R) = \frac{(1916448)^2}{16 \times 15} [0.035153243] = 537975.6757$$

Standard error of $\hat{Y}_R = \sqrt{v(\hat{Y}_R)} = \sqrt{537,975.6757} = 733.47$

(ii) If the information for the previous year is not used, the estimate of the total is given by

$$\hat{Y} = \bar{z} = \frac{1}{n} A \sum_{i}^{n} l_i$$

and the estimate of its variance is

$$v(\hat{Y}) = \frac{A^2}{n(n-1)} \left[\sum_i^n l_i^2 - \frac{(\sum_i^n l_i)^2}{n} \right]$$

Substituting values, we get

$$\hat{Y}_R = \frac{191648}{16} \times 5.086928 = 60931.22$$

and

$$v(\hat{Y}) = \frac{(191648)^2}{16 \times 15}(0.13597601) = 20,799,919.6547$$

Estimated standard error of $\hat{Y} = \sqrt{20,799,919.6547} = 4560.69$

Thus, the efficiency of the ratio method with varying probabilities as compared to simple estimate with varying probabilities may obtained as

$$\frac{v(\hat{Y})}{v(\hat{Y}_R)} = \frac{20,799,919.6547}{537,975.6757} \times 100 = 386.63\%$$

6.9 UNBIASED RATIO-TYPE ESTIMATOR

An unbiased ratio type estimator has been proposed by Hartley and Ross (1954). They suggested average of the ratios y_i/x_i, i.e. an estimator

$$\bar{r} = \frac{1}{n} \sum_i^n \frac{y_i}{x_i} = \frac{1}{n} \sum_i^n r_i$$

was derived and was corrected for bias. We shall discuss some important results in brief here.

THEOREM 6.9.1 In simple random sampling, wor, an unbiased estimator of $R (= Y/X)$ is given by

$$\hat{R}_1 = \bar{r} + \frac{n(N-1)}{N(n-1)}(\bar{y} - \bar{r}\bar{x})\bar{X} \tag{6.9.1}$$

For large samples, an approximation to sampling variance is given by

$$V(\hat{R}_1) \cong \frac{(S_y^2 + \bar{R}^2 S_x^2 - 2\bar{R}\rho S_y S_x)}{n\bar{X}^2} \tag{6.9.2}$$

where

$$\bar{R} = E\left(\frac{y_i}{x_i}\right)$$

Proof In simple random sampling,

$$E(\bar{r}) = \frac{1}{n} \sum_{i}^{n} E(r_i) = E(r_i) = \frac{E(r_i) E(x_i)}{\bar{X}}$$

Hence, $B(\bar{r}) = E(\bar{r}) - R = -\dfrac{\text{Cov}(r_i, x_i)}{\bar{X}} = -\dfrac{N-1}{N\bar{X}} S_{rx}$ (6.9.3)

Since an unbiased estimate of S_{rx} is

$$s_{rx} = \sum_{i}^{n} \frac{r_i(x_i - \bar{x})}{(n-1)} = \frac{n}{(n-1)}(\bar{y} - \bar{r}\,\bar{x})$$ (6.9.4)

It follows that an unbiased ratio-type estimator of R is given by relation (6.9.1). An exact expression for the variance of this estimator has been given by Robson (1957), and Goodman and Hartley (1958). From these results, it can be shown that a large sample approximation to variance is given by relation (6.9.2).

COROLLARY An unbiased estimator of the population total is

$$\hat{Y}_{R_1} = \bar{r} X + \frac{n(N-1)}{(n-1)}(\bar{y} - \bar{r}\,\bar{x})$$ (6.9.5)

Mickey (1959) and Williams (1961) have given generalized ratio-type estimators. The estimator given by Hartley and Ross is a particular case.

In these methods, there is a search for an adjustment of bias in $\hat{R}$. Quenouille (1956) gave a broad class of estimators, which has been given the name of the *jackknife* method. A lot of work has been done on these lines, but most are of a theoretical nature and difficult to use in practice. Therefore, their discussion has been postponed.

6.10 PRODUCT ESTIMATOR

In Section 6.5, it was observed that the ratio estimator $\hat{Y}_R = \dfrac{\bar{y}}{\bar{x}} X$ of the population total, Y is more precise than the mean per unit estimator $N\bar{y}$, provided the correlation coefficient ρ is greater than $C_x/2C_y$ and R is positive. If the correlation coefficient between the main variate y and auxiliary variate x is negative, we cannot make use of the ratio estimator. In such situations, Goodman (1960) has proposed another type of estimators for the mean $\bar{Y}_P$ and the total Y, defined as

$$\hat{\bar{Y}}_P = \frac{\bar{y}\,\bar{x}}{\bar{X}}$$

and

$$\hat{Y}_P = \frac{\hat{Y}\,\hat{X}}{\bar{X}} = \frac{N\bar{y}\,\bar{x}}{\bar{X}}$$ (6.10.1)

which may be termed as the product estimators (subscript p for *product*).

These estimators are suggested as they are complementary to the ratio estimators and, hence, expected to be useful in situations, where ratio estimators are not efficient. The product estimator is analogous to the ratio estimator and its theory and treatment are similar.

With assumptions similar to those for the ratio estimator $\hat{Y}_R$, we derive expressions for the expected value, bias and variance of the estimator $\hat{Y}_P = \hat{Y}\,\hat{X}/X$. Similar results can be derived for $\hat{\overline{Y}}_P = \bar{y}\,\bar{x}/\overline{X}$ also.

THEOREM 6.13.1 Show that the product estimator $\hat{Y}_P = \hat{Y}\,\hat{X}/X$ is a biased estimator and its bias is given by

$$\frac{B(\hat{Y}_P)}{Y} \cong \frac{(N-n)}{N\,n}\,\rho\,C_x\,C_y \tag{6.10.2}$$

where C_x and C_y are the coefficients of variation of x and y, respectively.
The proof is obvious.

COROLLARY In large samples, the relative bias is less than the product of the coefficients of variation of x and y, i.e.

$$\frac{|B(\hat{Y}_P)|}{Y} \leqslant C_x\,C_y \tag{6.10.3}$$

Hence, it should be noted that if the terms of second and higher degrees are negligible, the bias of the product estimator becomes zero.

THEOREM 6.10.2 If the sample size is moderately large, the sampling variance of the product estimator $\hat{Y}_P$, to the first order of approximation, is given by

$$V(\hat{Y}_P) = V(\hat{Y}) + 2R\,\text{Cov}\,(\hat{X},\hat{Y}) + R^2\,V(\hat{Y}) \tag{6.10.4}$$

Proof By using relation (1.3.8), the sampling variance of the product estimator $\hat{Y}_P$ can be written as

$$V(\hat{Y}_P) \cong Y^2 \left[\frac{V(\hat{Y})}{Y^2} + 2\,\text{Cov}\,\frac{(\hat{X},\hat{Y})}{XY} + \frac{V(\hat{X})}{X^2} \right]$$

$$\cong V(\hat{Y}) + 2R\,\text{Cov}\,(\hat{X},\hat{Y}) + R^2\,V(\hat{X})$$

COROLLARY If cv is the coefficient of variation of either $\hat{\bar{Y}}_P$ or $\hat{Y}_P$ and the sample size is large, then

$$(cv)^2 = \frac{(1-f)}{n}(C_y^2 + C_x^2 + 2\,\rho\,C_y\,C_x) \qquad (6.10.5)$$

Estimators of the bias and variance of the product estimator can be obtained by substituting the sample estimates of R, C_y, C_x and ρ in the corresponding formulae.

For a comparative study of the product estimator $\hat{Y}_P$, with the mean per unit estimator $\hat{Y}$, let us consider that the sample is large and bias is negligible. The estimator $\hat{Y}_P$ will be more efficient than the mean per unit estimator $\hat{Y}$ if $V(\hat{Y}_p) < V(\hat{Y})$, which leads to the conditions

$$\left.\begin{array}{l} \rho < -\dfrac{C_x}{2C_y} \\[4mm] \text{and} \\[2mm] \rho > \dfrac{C_x}{2C_y} \end{array}\right\} \begin{array}{l}\text{in case both } Y \text{ and } X \text{ are}\\ \text{positive or negative.}\\[4mm] \text{in case either } Y \text{ or } X \text{ is}\\ \text{negative.}\end{array} \qquad (6.10.6)$$

It is assumed that both Y and X are non-zero.

Thus, the result is of great interest because it shows that, for any given supplementary variate, one can decide whether to use mean per unit or ratio or product estimator on the basis of the correlation coefficient between y and x. For the whole range of the correlation coefficient between y and x, the conditions and a choice of the estimator to be used are as follows:

*Use product estimator $(\hat{Y}_P)$ — (i) if $-1 < \rho < -C_x/2C_y$ and both Y and X are positive or negative or (ii) if $C_x/2C_y < \rho \leqslant +1$ and either of Y or X is negative.

*Use mean per unit (unbiased) estimator $(\hat{Y})$ — if $-C_x/2C_y \leqslant \rho \leqslant C/2C_y$ and both Y and X are positive/negative or either Y or X is negative.

*Use ratio estimator $(\hat{Y}_R)$ — (i) if $C_x/2C_y < \rho \leqslant +1$ and both Y and X are positive/negative (ii) if $-1 \leqslant \rho < -C_x/2C_y$ and either Y or X is negative.

Goodman (1960) has derived the exact variance of $\hat{Y}_P$ along with variance estimators. Murthy (1964) has discussed the use of product estimators and has proposed a technique of obtaining unbiased product estimators based on interpenetrating sub-samples.

6.11 MULTIVARIATE RATIO ESTIMATOR

Olkin (1958) has proposed the ratio estimator for k auxiliary variates. Let the k-auxiliary characters $(x_1, x_2, \ldots, x_k)$ be positively correlated with the characteristic under study, y. For the population total, a multivariate ratio estimator is defined by

$$\hat{Y}_{\text{MR}} = \sum W_i \frac{\bar{y}_i}{\bar{x}_i} X_i = \sum_i^k W_i \hat{Y}_{Ri} \qquad (6.11.1)$$

where $\hat{Y}_{R_i} = \frac{\bar{y}_i}{\bar{x}_i} X_i$, W_i are the weights taken from a weighting function such that $\sum W_i = 1$, and the population totals X_i are known.

This type of estimator is appropriate only when the regression of y on $x_1, x_2, \ldots, x_k$ is linear, the line passes through the origin and the weight function is optimal, i.e. those W_i which maximize the precision.

We shall consider here the case when two x-variates are used to maximize the precision of $\hat{Y}_{\text{MR}}$. Generalization to several auxiliary variates is straight-forward and can be done without any difficulty.

Let y denote the variate under study and x_1 and x_2 the auxiliary variates. Then, the estimator $\hat{Y}_{\text{MR}}$ is given by

$$\hat{Y}_{\text{MR}} = W_1 \hat{Y}_{R_1} + W_2 \hat{Y}_{R_2} \qquad (6.11.2)$$

It can be easily shown that, to the first order of approximation, the expected value is

$$E(\hat{Y}_{\text{MR}}) = Y + (1 - f)\frac{Y}{n} [W_1(C_{x_1}^2 - \rho_{yx_1} C_y C_{x_1})$$

$$+ W_2(C_{x_2}^2 - \rho_{yx_2} C_y C_{x_2})] \qquad (6.11.3)$$

where C_y, C_{x_1}, and C_{x_2} are the coefficients of variation of y, x_1 and x_2, respectively, and ρ_{yx_1} and ρ_{xy_2} are the correlation coefficients between y and x_1 and between y and x_2, respectively. It shows that the estimator $\hat{Y}_{\text{MR}}$ is biased and that the bias, to the first order of approximation, will be negligible if the sample size is sufficiently large.

To obtain the sampling variance, we have

$$V(\hat{Y}_{\text{MR}}) = W_1^2 V(\hat{Y}_{R_1}) + W_2^2 V(\hat{Y}_{R_2}) + 2W_1 W_2 \text{ Cov }(\hat{Y}_{R_1}, \hat{Y}_{R_2})$$

$$= W_1^2 A_{11} + W_2^2 A_{22} + 2W_1 W_2 A_{12} \qquad (6.11.4)$$

where
$$A_{11} = \frac{(1-f)}{n} Y^2 [C_y^2 + C_{x1}^2 - 2\rho_{yx1} C_y C_{x1}]$$

$$A_{22} = \frac{(1-f)}{n} Y^2 [C_y^2 + C_{x2}^2 - 2\rho_{yx_2} C_y C_{x2}]$$

$$A_{12} = \frac{(1-f)}{n} Y^2 [C_y^2 - \rho_{yx1} C_y C_{x1} - \rho_{yx_2} C_y C_{x2}$$
$$+ \rho_{x1x2} C_{x1} C_{x2}]$$

Putting $W_2 = 1 - W_1$ in relation (6.12.4), differentiating w.r.t. W_1 and equating to zero, we find that the optimum weights W_1 and W_2 are given by

$$W_1 = \frac{A_{22} - A_{12}}{A_{11} + A_{22} - 2A_{12}},$$
$$W_2 = \frac{A_{11} - A_{12}}{A_{11} + A_{22} - 2A_{12}} \quad\quad (6.11.5)$$

and the minimum variance is

$$V_{\min} (\hat{Y}_{MR}) = \frac{A_{11} A_{22} - A_{12}^2}{A_{11} + A_{22} - 2A_{12}} \quad\quad (6.11.6)$$

For the sake of simplicity, assume that

$$\rho_{yx1} = \rho_{yx2} = \rho, \quad \rho_{x1x2} = \rho'$$

and
$$C_{x1} = C_{x2} = C_x$$

Then, the minimum variance in relation (6.11.6) reduces to

$$V(\hat{Y}_{MR}) \cong \frac{(1-f) Y^2}{n} \left[\frac{C_y^2 - 2 \rho C_y C_x + C_x^2}{2(1+\rho')} \right] \quad\quad (6.11.7)$$

If no auxiliary variate is used, the variance of the total estimator $\hat{Y}_{SR}$ is given by

$$V(\hat{Y}_{SR}) = \frac{(1-f)}{n} Y^2 C_y^2$$

Comparing their variances, we can say that the estimator $\hat{Y}_{MR}$, based on two auxiliary variates, will be more precise than the estimator Y_{SR}, if

$$\frac{\rho}{(1+\rho')} \frac{C_y}{C_x} > \frac{1}{4} \quad\quad (6.11.8)$$

If x_1 coincides with x_2, i.e. there is a perfect correlation between x_1 and x_2, the above result reduces to $\rho C_y/C_x > \frac{1}{2}$. Here, it can also be seen that the method is capable of giving more precise results than the one based on single auxiliary variate. Only limitations of computations involved restrict its application to surveys. In practice, the weights are estimated from $\hat{A}_{11}, \hat{A}_{22}$ and $\hat{A}_{22.}$, we have

$$\hat{A}_{11} = \frac{(1 - f)\,\hat{Y}^2}{n}\,(c_y^2 + c_{x1}^2 - 2r_{01}\,c_y c_{x1})$$

$$\hat{A}_{22} = \frac{(1 - f)\,\hat{Y}^2}{n}\,(c_y^2 + c_{x_2}^2 - 2r_{01} c_y c_{x_2})$$

and $\qquad \hat{A}_{12} = \dfrac{(1 - f)\,\hat{Y}^2}{n}\,(c_y^2 + r_{12}\,c_{x1}\,c_{x2} - r_{01}\,c_y c_{x1} - r_{02}\,c_y c_{x2})$

where $\qquad\qquad c_y^2 = \dfrac{s_y^2}{\bar{y}^2}$ etc., $\qquad$ and $\qquad r_{01} = r_{yx1}$ etc.

For further investigation, the reader is referred to Singh (1965, 1967) who has suggested ratio-cum-product estimators for two auxiliary variates, and Rao and Mudholkar (1967) who have extended the multivariate ratio estimator to a weighted combination of ratio estimators when x_i and y are positively correlated, and product estimators when x_i and y are negatively correlated.

SET OF PROBLEMS

6.1 If y and x are unbiased estimators of the population totals Y and X of the main variate and the auxiliary variate, respectively, based on any sample design, show that the ratio of the exact bias of the ratio estimator $\dfrac{y}{x} X$ to its standard error is not greater than the relative standard error of the estimator x. Derive an approximate expression for the bias and the mean square error stating clearly the assumptions involved.

6.2 If y_i and x_i are unbiased estimators of the population totals Y and X of the main variate and the auxiliary variate, respectively, based on the ith sample selected with probability p_i, show that the ratio estimator $\dfrac{y_i}{x_i} X$ will be unbiased for the population total Y if the sample i is selected with probability proportional to $x_i\,p_i$. Derive an expression for its sampling variance.

6.3 If y and x are unbiased estimators of the population totals Y and X, respectively, based on the same set of sample units drawn according to any sample design, derive the expressions correct to the second degree of approximation for bias and mean square error of the ratio estimator $\dfrac{y}{x} X$ for estimating the population total y. State clearly the underlying assumptions and the degree of the approximations taken. Also state under what conditions the above ratio estimator is more efficient than the estimator $\hat{Y}$ for large samples?

6.4 (a) Define ratio estimator for estimating the population total of a character y and derive an expression for the standard error of the estimator. State the conditions under which the ratio estimator is 'blue' (best linear unbiased estimator).

(b) If the coefficient of variation of the auxiliary variate x is more than twice the coefficient of variation of the character y, show that, in large samples with simple random sampling, the ratio estimator is less precise than the mean per unit estimator. Is the converse true?

6.5 In simple random sampling, wor, the ratio estimator of $R = Y/X$ is given by $\bar{y}/\bar{x}$. Obtain an exact expression for the variance of $\bar{y}/\bar{x}$.

6.6 If y_i, x_i $(i = 1, \ldots, m)$ are unbiased estimators of Y and X respectively, based on m interpenetrating sub-samples of the same size, prove that

$$\bar{r}\bar{X} + m(\bar{y} - \bar{r}\bar{x})/(m - 1)$$ is an unbiased estimator of Y,

where $$\bar{r} = \frac{1}{m} \sum_{i}^{m} \left(\frac{y_i}{x_i}\right)$$

6.7 Values of y and x are measured for each unit in a simple random sample to estimate the population ratio, $R = \bar{Y}/\bar{X}$. Which of the following estimators would you recommend to estimate R?

(i) Always use $\bar{}/\bar{X}$.

(ii) Always use $\bar{y}/\bar{x}$.

(iii) Either use $\bar{y}/\bar{X}$ or $\bar{y}/\bar{x}$, depending upon the conditions (Given that $\bar{X}$ is known).

Give reasons for your choice.

6.8 Suppose a finite population of N individuals has NP_1 individuals as agriculturists, of which NP_2 individuals are literates. In order to estimate the population ratio P_2/P_1, a random sample, wor, is drawn where p_1 and p_2 are the sample proportions corresponding to P_1 and P_2, respectively.

(i) Show that the estimator p_2/p_1 is more efficient than p_2/P_1, when P_1 is known.

(ii) Derive the condition for p_2/p_1 to be more efficient than P_2/p_1 when P_2 is known. (Elkin, 1953).

6.9 If a population is divided into L strata, one unit is selected randomly from each stratum with definite probabilities and this gives rese to $\bar{y}, \bar{x}$ as estimates of $\bar{Y}, \bar{X}$, respectively. By taking k interpenetrating samples, $\bar{y}_i \ \bar{x}_i$ $(i = 1, \ldots, k)$, the following estimates are defined as

$$y_{sts} = \sum_{i}^{k} \frac{\bar{y}_i}{k}, \ \bar{x}_{st} = \sum_{i}^{k} \frac{\bar{x}_i}{k},$$

$$\bar{r}_{st} = \sum_{i}^{k} \frac{(y_i/x_i)}{k} = \sum_{i}^{k} \frac{r_i}{k}$$

Prove that an unbiased estimator of $\bar{Y}$ is given by

$$\bar{r}_{st}\,\bar{X} + \frac{(\bar{r}-r)\,(\bar{x}_i - \bar{x}_{st})}{k-1} = \bar{r}_{st}\,\bar{X} + \frac{1}{(k-1)}\,(\bar{y}_{st} - \bar{r}_{st}\,\bar{x}_{st})$$

Derive the sampling variance of this estimator.

6.10 A sample survey for the study of yield and cultivation practices of guava was conducted in District Allahabad during 1971-72. Out of a total of 146 guava-growing villages in Phulpur–Saran Tehsil, 13 were selected by the method of simple random sampling. The data for the total number of guava trees and area under guava orchards for the 13 selected villages are given below:

S. N. of villages	1	2	3	4	5	6	7
Total N. of guava trees	492	1008	714	1265	1889	784	294
Area under guava orchards (in acres)	4.80	5.99	4.27	8.43	14.39	6.53	1.88

S. N. of villages	8	9	10	11	12	13
Total N. of guava trees	798	780	619	403	467	197
Area under guava orchards (in acres)	6.35	6.58	9.18	2.00	2.20	1.00

Given that the total area under guava orchards of 146 villages is 354.78 acres, estimate the total number of guava trees in the tehsil along with its standard error, using the area under guava orchards as the auxiliary variate. Discuss the efficiency of your estimate with the one which does not make any use of the information on the auxiliary variate.

6.11 The number of cows in milk enumerated (y) from a random sample of 20 villages from a tehsil having 84 villages, as also the corresponding census figures (x) in the previous year, are given below:

Villages	y	x
1	237	155
2	1060	583
3	405	205
4	1085	738
5	666	526
6	542	284
7	1337	758
8	1166	681
9	399	143
10	228	111
11	813	616
12	666	576
13	681	540
14	2743	2242
15	1228	940
16	472	387
17	643	675
18	180	220
19	583	654
20	1195	1787

Given that the census estimate of the number of cows in milk in the tehsil was 74488, estimate the number of cows in milk in the current year with and without using the census information and compare the efficiencies of the estimates.

6.12 A sample of 34 villages was selected from a population of 170 villages, with pps wr for estimating the area under wheat in the region during 1974. The cultivated area was 78000 acres in 1971. Later it was found that the figures for area under wheat (21288 acres) in 1973 were also available for all villages in the region. The relevant data are given below:

S.N.	Area under wheat (in acres)		
	(1971) (x_1)	(1973) (x_2)	(1974) (y)
1	401	70	50
2	630	163	149
3	1194	320	284
4	1170	440	381
5	1065	250	278
6	827	125	111
7	1737	558	634
8	1060	254	278
9	360	101	112
10	946	359	355
11	4170	109	99
12	1625	481	498
13	827	125	111
14	96	5	6
15	1304	427	339
16	377	78	80
17	259	75	105
18	186	45	27
19	1767	564	515
20	604	238	249
21	700	92	85
22	524	247	221
23	571	134	133
24	962	131	144
25	407	129	103
26	715	190	175
27	845	363	335
28	1016	235	219
29	184	73	62
30	282	62	79
31	194	71	60
32	439	137	100
33	854	196	141
34	820	255	263

(i) Estimate the area under wheat in 1974, by the method of ratio estimation, using the information on wheat area for 1973 and estimate its standard error.

(ii) Determine the efficiency of the ratio estimate as compared to that of the usual unbiased estimate.

REFERENCES

Elkin, J. M., "Estimating the ratio between the proportions of two classes when one is a subclass of the other", *J. Amer. Statist. Assoc.*, **48**, 128-130, (1953).

Goodman, L.A., "On the exact variance of products," *J. Amer. Statist. Assoc.*, **55**, 708-713, (1960).

_____and H.O. Hartley, "The precision of unbiased ratio-type estimators, *J. Amer. Statist. Assoc.*, **53**, 491-508, (1958).

Hansen, M.H., W.N. Hurwitz, and M. Gurney, "Problems and methods of the sample survey of business," *J. Amer. Statist. Assoc.*, **47**, 173-189, (1946).

_____W.N. Hurwitz, and W.G. Madow, *Sample Survey Methods and Theory*, John Wiley & Sons, New York, (1953).

Hartley, H.O. and A. Ross, "Unbiased ratio estimators," *Nature*, **174**, 270-271, (1954).

Lahiri, D.B., "A method of sample selection providing unbiased ratio estimates," *Bull. Int. Statist. Inst.*, **32**, 133-140, (1951).

Mickey, M.R., "Some finite population unbiased ratio and regression estimators," *J. Amer. Statist. Assoc.*, **54**, 594-612, (1959).

Midzono, M., " On the sampling system with probability proportionate to sum of sizes," *Ann. Inst. Statist. Math*, **2**, 99-106, (1951).

Murthy, M.N., "Product method of estimation," *Sankhya*, **26**, 69-74, (1964).

Olkin, I., "Multi-variate ratio estimation for finite population," *Biometrika*, **45**, 154-165, (1958)

Quenoille, M.H., "Note on bias in estimation," *Biometrika*, **43**, 353-360, (1956).

Rao, P.S.R.S. and G.S. Mudholkar, "Generalized multivariate estimations for the mean of finite populations," *J. Amer. Statist. Assoc.*, **62**, 1008-1012, (1967).

_____and J.N.K. Rao, "Small sample results for ratio estimators," *Biometrika*, **58**, 625-630, (1971).

Robson, D.S , "Applications of multivariate polykays to the theory of unbiased ratio type estimation," *J. Amer. Statist. Assoc.*, **52**, 511-522, (1957).

Singh, M.P., "On the estimation of ratio and product of the population parameters," *Sankhya*, **27**, 321-328, (1965).

_____"Ratio cum product method of estimation," *Metrika*, **12**, (1967).

Sukhatme, P.V., (1944). "Moments and product moments of moment statistics for samples of finite and infinite populations," *Sankhya*, **6**, 363-382, (1944).

Williams, W.H., "Generalizing unbiased ratio and regression estimators", *Biometrics*, **17**, 267-274, (1961).

7

Regression Estimators

At thirty man suspects himself a fool;
Knows it at forty, and reforms his plan:
At fifty chides his infamous delay,
Pushes his prudent purpose to resolve,
In all the magnanmity of thought
Resolves, and re-resolves: then dies the same.

J. Young

7.1 INTRODUCTION

Like ratio estimators, linear regression estimators also make use of auxiliary information for increasing precision. It was seen that the ratio estimator provides a precise estimate of the population mean if regression is linear and the line passes through the origin. When regression is linear and the line does not go through the origin, it is better to use estimators based on linear regression. In other words, if the study variate (y) is approximately a constant and a multiple of the auxiliary variate, it is more precise to estimate the population mean or total by fitting a linear regression. Such an estimator is called a *regression estimator*. We shall now discuss the main characteristics of such estimators and compare them with the mean per unit estimator and ratio estimator.

7.2 DIFFERENCE ESTIMATOR

Let y and x be correlated characteristics. We wish to estimate $\overline{Y}$. If, from a simple random sample, we obtain the unbiased estimators $\bar{y}$ and $\bar{x}$ of $\bar{Y}$ and $\bar{X}$, respectively, then we can improve upon the estimator $\bar{y}$ by introducing a *difference function*. Thus, a simple difference estimator is defined by

$$\bar{y}_D = \bar{y} + (\bar{X} - \bar{x}) \tag{7.2.1}$$

It is assumed that there is a unit change in y when a unit change is made in x, x and y having equal variances. This assumption will not be valid if the relationship is of the type $y = cx + k$, where c and k are constants. In this situation, a more *generalized difference* estimator is defined as

$$\bar{y}_D = \bar{y} + c\,(\bar{X} - \bar{x}) \tag{7.2.2}$$

where c and $\bar{X}$ are known quantities.

In simple random sampling, the difference estimator and its variance are discussed in the following theorems.

THEOREM 7.2.1   In simple random sampling, wor, the difference estimator $\bar{y}_D$ is unbiased and its sampling variance is given by

$$V(\bar{y}_D) = V(\bar{y}) + c^2\,V(\bar{x}) - 2c\,\text{cov}(\bar{y}, \bar{x}) \tag{7.2.3}$$

$$= V(\bar{y})(1 + \alpha^2 - 2\,\alpha\,\rho)$$

where ρ is the correlation coefficient between y and x, and $\alpha = c\,S_x/S_y$.

The proof is simple, and its development is left as an exercise.

From this result, it may be noted that $\bar{y}_D$ has smaller variance than $\bar{y}$ if $\rho > \alpha/2$.

COROLLARY   Show that $V(\bar{y}_D)$ will have its minimum value when

$$c = \text{cov}(\bar{y}, \bar{x})/V(\bar{x}) \tag{7.2.4}$$

This proves that the optimum value to be given to c is β, the value of the regression coefficient of y on x. In this case, $\alpha = \rho$ and the minimum value of variance is $V(\bar{y})(1 - \rho^2)$.

THEOREM 7.2.2   In simple random sampling, wor, the sampling variance of $\bar{y}_D$ is obtained by

$$V(\bar{y}_D) = \frac{(1 - f)}{n}(S_y^2 + c^2\,S_x^2 - 2\,\rho\,c\,S_y\,S_x) \tag{7.2.5}$$

where c is some specified quantity.

Proof Since c is a given constant and, $E(\bar{X} - \bar{x}) = 0$, therefore, the unbiasedness of $\bar{y}_D$ is proved. Further, denoting

$$u_i = y_i - c(x_i - \bar{X}),$$

we have

$$\bar{u} = \bar{y} - c(\bar{x} - \bar{X}),$$

and

$$\bar{U} = \bar{Y} - c(\bar{X} - \bar{X}) = \bar{Y}$$

Hence, $\quad V(\bar{y}_D) = V(\bar{u}) = \dfrac{(1-f)}{n} S_u^2$

where

$$S_u^2 = \frac{\sum_i^N (u_i - \bar{U})^2}{(N-1)}$$

$$= \frac{\sum_i^N [(y_i - \bar{Y}) - c(x_i - \bar{X})]^2}{(N-1)}$$

$$= S_y^2 + c^2 S_x^2 - 2c\rho S_y S_x$$

This proves the theorem.

COROLLARY 1   For the case $c = R(= Y/X)$, the variance of the difference estimator $\bar{y}_D$ is exactly the same as the first order approximation of $V(\hat{\bar{Y}}_R)$ given in relation (6.4.4).

COROLLARY 2   The difference estimator $\bar{y}_D$ will be more precise than the mean per unit estimator $\bar{y}$ if $c < 2\rho S_y/S_x$.

This gives the condition that the difference estimator will be more efficient than the mean per unit estimator if $c < 2\beta$, otherwise, the mean per unit estimator will be efficient. Hence, it may be concluded that the difference estimator will be simple to apply as its results are exact. If one can guess c from a pilot survey, then a proper decision about its use can be taken.

An illustration of some characteristic situations in the application of difference estimates has been given by Zarkovic (1956).

COROLLARY 3   In simple random sampling, wor, an unbiased estimator of $V(\bar{y}_D)$ is obtained by

$$v(\bar{y}_D) = v(\bar{y}) + c^2 v(\bar{x}) - 2c \, \text{cov}(\bar{y}, \bar{x}) \qquad (7.2.6)$$

$$= \frac{(1-f)}{n} (s_y^2 + c^2 s_x^2 - 2c s_{yx})$$

where

$$s_y^2 = \frac{\sum_i^n (y_i - \bar{y})^2}{(n-1)}, \quad s_x^2 = \frac{\sum_i^n (x_i - \bar{x})^2}{(n-1)}$$

and

$$s_{yx} = \frac{\sum_i^n (y_i - \bar{x})(x_i - \bar{x})}{(n-1)}$$

7.3 REGRESSION ESTIMATOR

While discussing the difference estimator, it was seen that the optimum value to be given to c is β, where β is the regression coefficient of y on x. Generally, β is not known in advance and its value is estimated from the sample. Suppose y_i and x_i are obtained for each unit in the sample, then a least square estimate of β is

$$b = \frac{\sum_{i}^{n} (y_i - \bar{y}) (x_i - \bar{x})}{\sum_{i}^{n} (x_i - \bar{x})^2}$$

Thus, linear regression estimators of the population mean $\bar{Y}$ and population total Y are given by

$$\left.\begin{aligned} \bar{y}_l &= \bar{y} + b\,(\bar{X} - \bar{x}) \\ \hat{Y}_l &= N\,[\bar{y} + b\,(\bar{X} - \bar{x})] \end{aligned}\right\} \tag{7.3.1}$$

and

Here, $\qquad \hat{Y}_l = N\,\bar{y}_l$

Therefore, we shall present the theory of the linear regression estimator $\bar{y}_l$, only. Since b is a random variate, exact expressions for the bias and variance of the regression estimator are difficult to derive. In this section, large sample approximations to its bias and sampling variance will be given.

7.3.1 Bias of Regression Estimator

The regression estimator is consistent because, by increasing the sample size n to the population size N, we have $\bar{x} = \bar{X}$. Ultimately, the regression estimator reduces to $\bar{Y}$.

The regression estimator $\bar{y}_l$ is biased since

(i) β is generally estimated by taking the ratio of the estimate of cov $(\bar{y}, \bar{x})$ to that of $V(\bar{x})$, and

(ii) it involves the product of two estimates, viz. $b\,\bar{x}$. The bias of the regression estimator will usually be trivial and will decrease as sample size increases.

THEOREM 7.3.1 In simple random sampling, the bias of $\bar{y}_l$ is approximated by

$$B\,(\bar{y}_l) \cong -\,\text{cov}\,(\bar{x},\,b) \qquad (7.3.2)$$

which will be negligible if the sample size is large.

Proof Suppose
$$\bar{y} = \bar{Y}\,(1 + e),\, \bar{x} = \bar{X}\,(1 + e_1),\, \text{and}\, b = \beta\,(1 + e_2)$$

where $E(e) = E(e_1) = E(e_2) = 0$

Substituting these values in Eq. (7.3.1), we get

$$\bar{y}_l = \bar{Y} + (e\bar{Y} - e_1\,\beta\bar{X}) - e_1\,e_2\,\beta\bar{X} \qquad (7.3.3)$$

Therefore, the bias of the regression estimator is

$$B\,(\bar{y}_l) = \bar{Y}\,E\,(e) - \beta\bar{X}\,E\,(e_1) - \beta\bar{X}\,E\,(e_1\,e_2) \qquad (7.3.4)$$
$$= -\,\text{cov}\,(\bar{x},\,b)$$

since $E(e) = E\,(e_1) = 0$

For large samples, usually cov $(\bar{x},\,b)$ decreases. It becomes zero if the joint distribution of y and x is a bivariate normal.

COROLLARY 1 If $b = \hat{R}\,(= \bar{y}/\bar{x})$, the regression estimator $\bar{y}_l$ reduces to the ratio estimator $\hat{\bar{Y}}_R$ and its bias is given as in Eq. (7.3.2).

COROLLARY 2 If the sample is selected in the form of k independent sub-samples, then the bias can be unbiasedly estimated by

$$\hat{B}\,(\bar{y}_l) \cong \frac{\sum\limits_{i}^{k} (\bar{x}_i - \bar{y}_i)\,(b_i - b)}{(k - 1)} \qquad (7.3.5)$$

This estimator of bias can be used to get the unbiased regression estimator. A similar method has been discussed by Mickey (1959) and Williams (1961). The technique developed by Murthy (1962) can also be applied to make unbiased regression estimators.

7.3.2 Sampling Variance of Regression Estimator

When the sample size is large enough, the contribution of the term involving $e_1\,e_2$ in relation (7.3.3) is expected to be very small and the bias in the regression estimator $\bar{y}_l$ reduces to zero. Therefore, if the terms of the second and higher orders are neglected, we can derive the sampling variance of the regression estimator.

THEOREM 7.3.2   In simple random sampling, the large sample variance of the regression estimator is given by

$$V(\bar{y}_l) = V(\bar{y}) + \beta^2 V(\bar{x}) - 2\beta \operatorname{cov}(\bar{y}, \bar{x}) \qquad (7.3.6)$$

Proof   If the term involving $e_1 e_2$ in relation (7.3.3) is ignored, we have

$$\bar{y}_l - \overline{Y} = e\overline{Y} - e_1 \beta \overline{X}$$

Therefore, $V(\bar{y}_l) = V(e\overline{Y} - e_1 \beta \overline{X})$

$$= V(e\overline{Y}) + \beta^2 V(e_1 \overline{X}) - 2\beta \operatorname{cov}(e\overline{Y}, e_1 \overline{X})$$

$$= V(\bar{y}) + \beta^2 V(\bar{x}) - 2\beta \operatorname{cov}(\bar{y}, \bar{x})$$

COROLLARY 1   In simple random sampling, wor, the sampling variance of the regression estimator $\bar{y}_l$ is given by

$$\left.\begin{aligned}
V(\bar{y}_l) &= \frac{(1-f)}{n}(S_y^2 + \beta^2 S_x^2 - 2\beta S_{xy}) \\
&= \frac{(1-f)}{n} S_y^2 (1 - \rho^2)
\end{aligned}\right\} \qquad (7.3.7)$$

This shows that if regression is linear and b is the least square estimate of β then the regression estimator $\bar{y}_l$ is more precise than the difference estimator $\bar{y}_D$. If the regression of y on x is perfectly linear, i.e. $|\rho| = 1$, the variance of the regression estimator becomes zero. Also if y and x are uncorrelated, the variance of $\bar{y}_l$ reduces, to that of the mean per unit estimator $\bar{y}$.

COROLLARY 2   In large samples, an almost unbiased estimator of $V(\bar{y}_l)$ is given by

$$\left.\begin{aligned}
v(\bar{y}_l) &= \frac{(1-f)}{n}(1 - r^2) s_y^2 \\
&= \frac{(1-f)}{n}(s_y^2 - b^2 s_x^2) \\
&= \frac{(1-f)}{n(n-1)}\left[\sum_i^n (y_i - \bar{y})^2 - \frac{\{\sum_i^n (y_i - \bar{y})(x_i - \bar{x})\}^2}{\sum_i^n (x_i - \bar{x})^2}\right]
\end{aligned}\right\} \qquad (7.3.8)$$

The last form in the above relation is good for computational purposes. According to standard regression theory, it is suggested that the divisor $(n-2)$ be used instead of $(n-1)$.

Williams (1963) has suggested a method for getting some precise unbiased regression estimators. A detailed discussion of these estimators is given by Rao (1969).

7.4 COMPARISON WITH THE MEAN PER UNIT AND RATIO ESTIMATORS

For a large sample size n, the variance for estimators of the population mean $\bar{Y}$ can be written as

$$V(\bar{y}) = \frac{(1-f)}{n} S_y^2 \quad \text{(mean per unit)}$$

$$V(\bar{y}_R) = \frac{(1-f)}{n} (S_y^2 - 2R \rho S_y S_x + R^2 S_x^2) \quad \text{(ratio)}$$

$$V(\bar{y}_l) = \frac{(1-f)}{n} (1 - \rho^2) S_y^2 \quad \text{(regression)}$$

Firstly, comparing with the mean per unit estimator, we observe that the variance of the regression estimator is always smaller unless $\rho = 0$. In case $\rho = 0$, the variances for both are equal. The reduction in variance is large when ρ is high, and small when ρ is low.

Comparing next with the ratio estimator, it can be seen that the variance of the regression estimator is less than that of the ratio estimator if

$$(\rho S_y - R S_x)^2 > 0$$

or

$$(\beta - R)^2 > 0 \tag{7.4.1}$$

which is always true unless $\beta = R$. In this situation, both estimates have the same variance and this occurs only when the regression of y on x is a straight line through the origin.

This comparison suggests that all these estimators belong to the class of estimators

$$\bar{y}_g = \bar{y} + z(\bar{X} - \bar{x}) \tag{7.4.2}$$

where z is a random variate having some values in a finite range.

By a suitable choice of z, one can generate different types of estimators. If the variate z is zero, $\bar{y}_g$ conforms to the mean per unit estimator, $\bar{y}$. If z is a constant, $\bar{y}_g$ becomes a difference estimator. When

z is $\bar{y}/\bar{x}$, the estimator $\bar{y}_g$ becomes the ratio estimator $\hat{\bar{Y}}_R$. Similarly, when z is equal to the sample regression coefficient b, it becomes the linear regression estimator $\bar{y}_l$. In large samples, the regression estimators is the most efficient and can be used if computational labour is not heavy.

The above comparisons suggest that the regression estimator is not always a proper choice although it provides a variance equal to or less than that of others. Some rules for a choice among alternatives are as follows:

1. When advance information on an appropriate value of β ($= c$) is available, then, with simple computations good results can be obtained with such values of c with the difference estimator.

2. When the correlation coefficient between variates is nearly equal to the ratio of their standard deviations, i.e. $\rho \cong \sigma_x/\sigma_y$, the difference estimator with $c=1$ will attain equally precise results as the regression estimator.

3. When the correlation coefficient between variates is nearly equal to the ratio of their coefficients of variation, i.e. $\rho \cong C_x/C_y$, the ratio estimator will provide equally precise results as the regression estimator.

4. When ρ is different from σ_x/σ_y, the regression estimator should be preferred. In this situation, the difference estimator with $c = 1$ will not provide precise results.

5. When ρ is different from C_x/C_y, the regression estimator should be preferred. In this situation, the variance of the ratio estimator will be larger than that of the regression estimator.

6. When computations for the regression estimator are heavy, time consuming and expensive, its use is recommended only if the gains from such computations are much more significant than the additional costs.

EXAMPLE 7.1   Using the data given in Example 6.1, estimate the total number of milch animals in 117 villages of zone A by the method of regression estimation. Also, compare its precision with the ratio estimate and mean per unit estimate.

Here,
$$N = 117, \quad n = 17 \quad X = 143968$$
$$\sum x_i = 19381, \qquad \sum y_i = 19266$$
$$\bar{x} = 1140.06, \qquad \bar{y} = 1133.29$$
$$s_x^2 = 458.56, \quad s_y^2 = 287.85, \quad s_{xy} = 262.86$$
Hence,
$$b = s_{xy}/s_x^2 = 0.5732, \qquad r = s_{xy}/s_x s_y = 0.7235$$

Regression estimate of the total number of milch animals is obtained as

$$\hat{Y}_l = N\bar{y}_l = N\,[\bar{y} + b\,(\bar{X} - \bar{x})]$$
$$= 117\,[1130.29 - 0.5732\,(1140.06 - 1230.49)]$$
$$= 139263$$

and an estimate of the variance of $\hat{Y}_l$ is given as

$$v\,(\hat{Y}_l) = \frac{(1 - f)\,N^2}{n}\,(1 - r^2)\,s_y^2$$
$$= \frac{100 \times 117}{17}\,150.6757 = 103{,}700$$

From Example 6.1,

$$v(\hat{Y}_R) = 118{,}890 \quad \text{and} \quad v(\hat{Y}) = 129{,}285$$

Hence, the relative precision of the regression estimate over the mean per unit estimate is

$$\text{R.P.} = \left\{\frac{v\,(\hat{Y})}{v\,(\hat{Y}_l)}\right\} \times 100 = 124.67\%$$

Similarly, the relative precision of the regression estimate over the ratio estimate is given by

$$\text{R. P.} = \left\{\frac{v(\hat{Y}_R)}{v\,(\hat{Y}_l)}\right\} \times 100 = 114.64\%$$

7.5 REGRESSION ESTIMATORS IN STRATIFIED SAMPLING

When the population is stratified into k strata and, in the mth stratum, n_m units are selected from N_m units with simple random sampling, wor, like the ratio estimators, two regression estimators are possible. One is the *separate regression estimator* which is obtained by estimating the regression coefficients for each stratum separately. Another is the *combined regression estimator* which is obtained by getting a common (pooled) regression coefficient for all the strata.

7.5.1 Separate Regression Estimator

A separate regression estimator $\bar{y}_{ls}$ (s for separate) in stratified sampling may be defined as

$$\bar{y}_{ls} = \sum_i^k W_m \left[\bar{y}_m + b_m \left(\bar{X}_m - \bar{x}_m \right) \right] \qquad (7.5.1)$$

where W_m, $\bar{y}_m$, $\bar{x}_m$, $\bar{X}_m$ and b_m are the corresponding terms for the mth stratum. Like the usual regression estimator, $\bar{y}_{ls}$ is biased. If b_m is chosen in advance, and not based on the sample values, $\bar{y}_{ls}$ is an unbiased estimator of $\bar{Y}$.

THEOREM 7.5.1 If sampling is independent in different strata and sample size is large enough in each stratum, then $\bar{y}_{ls}$ is an almost unbiased estimator. To the first order of approximation, its variance is given by

$$V(\bar{y}_{ls}) = \sum_m^k W_m^2 \frac{(1 - f_m)}{n_m} \left(S_{y_m}^2 - 2b_m \, \rho_m \, S_{y_m} S_{x_m} + b_m^2 S_{x_m}^2 \right) \qquad (7.5.2)$$

The proof is obvious.

COROLLARY 1 If $b_m = \beta_m$, the true regression coefficient in the mth stratum is satisfied and the variance may be written as

$$V(\bar{y}_{ls}) = \sum_m^k \frac{W_m^2 (1 - f_m)}{n_m} (1 - \rho_m^2) \, S_{y_m}^2 \qquad (7.5.3)$$

COROLLARY 2 Provided the sample size n_m is large in all strata, an estimator of variance is given by

$$v(\bar{y}_{ls}) = \sum_m^k \frac{W_m^2 (1 - f_m)}{n_m (n_m - 1)} \left[\sum_j^{n_m} (y_{mj} - \bar{y}_m)^2 - b_m^2 \sum_j^{n_m} (x_{mj} - \bar{x}_m)^2 \right] \qquad (7.5.4)$$

The estimator $\bar{y}_{ls}$ also suffers from the same drawback as the corresponding ratio estimator, the ratio of the bias to its standard error becoming quite large. Since the leading terms in bias come from the quadratic regression of y_m on x_m, this may be of the same sign in all strata. Hence, the overall bias in the estimator $\bar{y}_{ls}$ may be very large.

7.5.2 Combined Regression Estimator

A combined regression estimator $\bar{y}_{lc}$ (c for combined) in stratified sampling may be defined as

$$\bar{y}_{lc} = \bar{y}_{st} + b_c \, (\bar{X} - \bar{x}_{st}) \qquad (7.5.5)$$

where $\bar{y}_{st}$ and $\bar{x}_{st}$ are the stratified sample means of y and x variates, and b_c is the pooled estimate obtained by

$$b_c = \frac{\sum\limits_{m}^{k} \sum\limits_{j}^{n_m} (y_{mj} - \bar{y}_m)(x_{mj} - \bar{x}_m)}{\sum\limits_{m}^{k} \sum\limits_{j}^{n_m} (x_{mj} - \bar{x}_m)^2}$$

The important point is to obtain the value of b_c. An easier method is to select a few extreme values of x, some high and some low. A straight line through the average values is drawn. The slope of this line is taken as the estimate of b_c, which is given by $(y_1 - y_2)/(x_1 - x_2)$, where (x_1, y_1) and (x_2, y_2) are two points on the line.

It can be shown that the combined regression estimator $\bar{y}_{lc}$ is also a biased estimator of $\bar{Y}$. It is analogous to $\hat{\bar{Y}}_{R_c}$. All the results can be drawn on similar lines.

THEOREM 7.7.2 If sampling is independent in different strata and sample size is large enough in each stratum, the variance of $\bar{y}_{lc}$ is given by

$$V(\bar{y}_{lc}) = \sum_{m}^{k} W_m^2 \frac{(1 - f_m)}{n_m} (S_{ym}^2 - 2b \, \rho_m \, S_{ym} \, S_{xm} + b^2 \, S_{xm}^2) \quad (7.5.6)$$

The proof is obvious.

COROLLARY 1 If $\beta_m = \dfrac{\rho_m \, S_{ym}}{S_{xm}}$ and weight $a_m = \dfrac{W_m^2 \, (1 - f_m)}{n_m} \, S_{xm}^2$

a combined regression coefficient, $\beta_c = \dfrac{\sum\limits_{m}^{k} a_m \, \beta_m}{\sum\limits_{m}^{k} a_m}$ minimizes the variance of $\bar{y}_{lc}$.

COROLLARY 2 Provided sample size n_m is large in all strata, the estimator of variance is given by

$$v(\bar{y}_{lc}) = \sum_m^k \frac{W_m^2 (1 - f_m)}{n_m (n_m - 1)} \sum_j^{n_m} [(y_{mj} - \bar{y}_m) - b_c (x_{mj} - \bar{x}_m)]^2 \quad (7.5.7)$$

The estimator $\bar{y}_{lc}$ also suffers from the same defect as the corresponding ratio estimator. This estimator will not be appropriate when there is reason to believe that the regression coefficient b_m varies considerably from stratum to stratum.

7.5.3 Comparison of Combined and Separate Estimators

Some caution is required when deciding whether the separate or combined regression estimator is to be used at any specific situation. The bias in the separate regression estimator is likely to be larger when sample size in each stratum is small. On the other hand, the defect of the combined estimator is that its variance is inflated if the population regression coefficients differ from stratum to stratum. If the regressions are approximately linear and the regression coefficients appear to be the same in all strata, use of the combined regression estimator is preferred. If the regressions are linear but the coefficients differ from stratum to stratum, it is advisable to adopt the separate estimator. Though the separate regression estimator is likely to be more efficient than the combined regression estimator, the latter is likely to have smaller bias.

EXAMPLE 7.2 With the data given in Example 6.2, estimate the total number of trees in the districts by the regression method of estimation and compare its precision.

We can use the calculations and results obtained in the above mentioned example. Also,

$$b_1 = 62.93, \quad b_2 = 43.53, \quad b_3 = 36.56$$

(i) *Separate Regression Estimate* The estimate of the total number of trees is given by

$$\hat{Y}_{ls} = \sum_m^k N_m[\bar{y}_m + b_m (\bar{X}_m - \bar{x}_m)] = 2,672,911$$

and $\quad v(\hat{Y}_{ls}) = \sum\limits_{m}^{k} N_m^2 \dfrac{(1 - f_m)}{n_m} (s_{ym}^2 - b_m^2 s_{xm}^2)$

$$= \sum\limits_{m}^{k} N_m^2 \left(\dfrac{1}{n_m} - \dfrac{1}{N_m}\right) (s_{ym}^2 - b_m^2 s_{xm}^2)$$

$$= 1{,}870{,}633{,}332$$

Therefore, efficiency over ratio estimate is given by

$$\left(\dfrac{2{,}441{,}137{,}835}{1{,}870{,}633{,}332}\right) \times 100 = 130.50$$

(ii) *Combined Regression Estimate* An estimate of the total number of trees is given by

$$\hat{Y}_{lc} = N\,\bar{y}_{lc} = N\,[\bar{y}_{st} + b\,(\bar{X} - \bar{x}_{st})] = 2{,}643{,}949$$

and $\quad v(\hat{Y}_{lc}) = \sum\limits_{m}^{k} \dfrac{W_m^2\,(1 - f_m)}{n_m(n_m - 1)} \sum\limits^{n_m} [(y_{mj} - \bar{y}_m) - b_c\,(x_{mj} - \bar{x}_m)]^2$

$$= 2{,}020{,}917{,}640$$

Therefore, efficiency over ratio estimate

$$= \dfrac{6{,}019{,}519{,}627}{2{,}020{,}917{,}640} \times 100 = 297.86$$

Similarly, the efficiency of the separate regression estimate $(\hat{Y}_{ls})$ over the combined regression estimate $(\hat{Y}_{lc})$ is

$$\text{RE} = \dfrac{2{,}020{,}917{,}640}{1{,}870{,}633{,}332} \times 100 = 108.03$$

7.6 MULTI-VARIATE REGRESSION ESTIMATOR

Frequently, data on more than one auxiliary variate are available and it is considered important to make use of all available information. If $\bar{y}$ is the usual estimator of the mean of the main variate, $\bar{x}_i$ is an unbiased estimator of the mean $\bar{X}_i$ of the auxiliary variate x_i and β_i is the regression coefficient of y on x_i, $i = 1, 2, \ldots, k$.

An estimator of the form defined by

$$\bar{y}_{ml_1} = \bar{y} + \sum_i^k \beta_i (\bar{X}_i - \bar{x}_i) \tag{7.6.1}$$

is used as a multi-variate estimator.

Des Raj (1965) proposed another difference estimator of the form

$$\bar{y}_{ml_2} = \sum_i^k w_i [\bar{y} + \beta_i (\bar{X}_i - \bar{x}_i)] \tag{7.6.2}$$

where β_i's are given regression coefficients and w_i are the weights adding up to unity.

For simplicity, we shall consider the case of two x-variates. In this case, the estimator may be written as

$$\bar{y}_{l_2} = \bar{y} + w_1 \beta_1 (\bar{X}_1 - \bar{x}_1) + w_2 \beta_2 (\bar{X}_2 - \bar{x}_2) \tag{7.6.3}$$

with its variance given by

$$\begin{aligned}
V(\bar{y}_{l_2}) &= \frac{(1-f)}{n} [V_y + w_1^2 \beta_1^2 V_{x_1} + w_2^2 \beta_2^2 V_{x_2} - 2\beta_1 w_1 \operatorname{cov}(y, x_1) \\
&\quad - 2\beta_2 w_2 \operatorname{cov}(y, x_2) + 2\beta_1 \beta_2 w_1 w_2 \operatorname{cov}(x_1, x_2)] \\
&= \frac{(1-f)}{n} [A_{00} + w_1^2 A_{11} + w_2^2 A_{22} - 2 w_1 A_{01} \\
&\quad - 2 w_2 A_{02} + 2 w_1 w_2 A_{12}]
\end{aligned} \tag{7.6.4}$$

where $A_{00}, A_{11}, \ldots$ are given in relation (6.11.4).

Putting $w_2 = 1 - w_1$ in the variance expression above, differentiating with respect to w_1 and equating the result to zero, we have the best weights given by

$$w_1 = \frac{(A_{22} - A_{12})}{(A_{11} + A_{22} - A_{12})}$$

and
$$w_2 = 1 - w_1 \tag{7.6.5}$$

By substituting the values of these weights in the variance expression, we have the minimum variance

$$V_{\min}(\bar{y}_{l_2}) = \frac{(1-f)}{n} \frac{A_{11} A_{12} - (A_{12})^2}{A_{11} + A_{22} - 2A_{12}} \tag{7.6.6}$$

Let us assume that $\rho_{yx_1} = \rho_{yx_2} = \rho_0$ and the correlation coefficient between x_1 and x_2 is ρ. If $C_{x_1} = C_{x_2} = C_0$ and the coefficient of variation of y is c, the optimum values of weights are given by

$$w_1 = w_2 = 1/2$$

By substituting these values in the variance, we have

$$V(\bar{y}_{l_2}) = \frac{(1-f)\,\overline{Y}^2}{n}\left[(1+\rho)\,\frac{c^2}{2} + C_0(C_0 - 2c\,\rho_0)\right] \qquad (7.6.7)$$

Further, if only one x variate is used, the variance is given by

$$V(\bar{y}_l) = \frac{(1-f)\,\overline{Y}^2}{n}\left[C^2 + C_0^2 - 2\rho_0\,C_0\right] \qquad (7.6.8)$$

Comparing relations (7.6.7) and (7.6.8), we find it always better to use the second variate if $\rho \neq 1$. Similarly, when no x variate is used and a comparison is made with the case of one x variate, it is better to use one x variate if $\rho_0 > c/2\,C_0$.

SET OF PROBLEMS

7.1 Show that, if the proportional increase in the variance of the regression estimator y_l arising due to the use of b instead of β is to be less than α, then the relative deviation is

$$\left|\frac{b}{\beta} - 1\right| < \left[\frac{\alpha(1-\rho^2)}{\rho^2}\right]^{1/2} \qquad \text{(Cochran, 1963)}$$

7.2 A random sample, wr, is selected from a finite population of N units, which itself is assumed to be a sample from a super population by following the model

$$E(y_i|x_i) = \alpha + \beta x_i,\ V(y_i|x_i) = a\,x_i^b,\ \mathrm{cov}\!\left(\frac{y_i,\,y_j}{x_i,\,x_j}\right) = 0,\ \ \text{for}\ j \neq i$$

where a and b are positive constants.

Show that the regression estimator of the population mean $\overline{Y}$ is always more precise than the ratio estimator when x is used as an auxiliary variate.

(Des Raj, 1958)

7.3 Using the difference estimator $t_1 = \bar{y} + \beta(\overline{X} - \bar{x})$ and the regression estimator $t_2 = \bar{y} + b\,(\overline{X} - \bar{x})$, an estimator is defined as $t_2 = t_1 + (b-\beta)\,(\overline{X} - \bar{x})$. Show that

$$\mathrm{MSE}(t_2) = V(t_1) + E(b-\beta)^2\,(\overline{X} - \bar{x})^2 + 2E(b-\beta)\,(\overline{X} - \bar{x})\,(t_1 - \overline{Y}),$$

$$[E(b-\beta)\,(\overline{X} - \bar{x})]^2 = \mathcal{O}\,(n^{-2})$$

and $$[E(b-\beta)\,(\overline{X} - \bar{x})\,(t_1 - \overline{Y})] = \mathcal{O}\,(n^{-1})$$

Hence, or otherwise, prove that

$$\mathrm{MSE}(t_2) \simeq V(t_1) \qquad \text{(Des Raj, 1968)}$$

7.4 An eye estimate of the fruit weights (x_i) on each tree in an orchard having 100 trees was made. The total weight (X) was found to be 12,500 kg. A random sample of 10 trees was taken and the actual weights of fruits (y_i) along with the eye estimates were as below:

Actual weight (y_i)	51	42	46	39	71	61	58	57	58	67
Eye est weight (x_i)	56	47	48	40	78	59	52	58	55	67

(i) Estimate the total actual fruit weight Y by taking the estimator

$$\hat{Y}_d = N[\bar{y} + (\bar{X} - \bar{x})]$$

and find its sampling variance.

(ii) Estimate the total actual fruit weight Y by taking the linear regression estimator

$$\hat{Y}_l = N[\bar{y} + b(\bar{X} - \bar{x})]$$

Also find the sampling variance and compare the results.

7.5 In Exercise 6.12, estimate the total number of guava trees by regression estimate and calculate the standard error. Compare the efficiency of the regression estimate with the ratio and mean per unit estimates.

7.6 Will the linear regression estimator with least square estimate b give a more precise estimate in Exercise 6.13?

7.7 Using two auxiliary variates x_1 and x_2, the multi-variate estimator given in Eq. (7.6.3) is used for estimating the population mean $\bar{Y}$. If the coefficients of variation of x_1 and x_2 are equal to C, the correlation coefficients between y and x_i, $i = 1, 2$, equal to ρ_0, and β_i are taken as $\bar{Y}/\bar{X}_i$, $i = 1, 2$.

Show that the optimum weights are given by $w_1 = w_2 = \frac{1}{2}$. Derive the variance of the estimator in this case and show that the estimator would be more efficient than the mean per unit estimator $\bar{y}$, if $\rho_0 > \frac{1}{4}(1 + \rho)/(C/C_0)$, where ρ is the correlation coefficient between x_1 and x_2 and C_0 is the coefficient of variation of y.

(Des Raj, 1965)

7.8 For estimating the total cattle population, a random sample, wr, of 24 villages was selected from the total 1238 villages. The number of cattle obtained in the survey is given below for each sample village, together with the corresponding census figure relating to a previous period:

S. N. of villages	Number of cattle Census	survey	S. N. of villages	Number of cattle Census	survey	S. N. of villages	Number of cattle Census	survey
1	623	654	9	161	210	17	330	375
2	690	696	10	298	555	18	218	212
3	534	530	11	2045	2110	19	160	147
4	293	315	12	1069	592	20	210	297
5	69	78	13	706	707	21	262	401
6	842	640	14	1795	1890	22	204	252
7	475	692	15	1406	1123	23	185	199
8	371	292	16	118	115	24	574	564

Compare the efficiency of the regression estimator with the ratio estimator. It is given that the number of cattle for the previous period of 1238 villages is 680,900.

7.9 In Exercise 6.14, estimate the area by the method of regression estimation and compare its efficiency with the ratio estimate obtained therein.

REFERENCES

Cochran, W.G., *Sampling Techniques*, John Wiley and Sons, New York, (1963).

Des Raj, "On the relative accuracy of some sampling techniques," *J. Amer. Statist. Assoc.*, **53**, 98-101, (1958).

———— "On a method of using multi-auxiliary in sample surveys," *J. Amer. Statist. Assoc.*, **60**, 270-277, (1965).

———— *Sampling Theory*, Tata McGraw-Hill Publishing Co., New Delhi, (1968).

Mickey, M.R., "Some finite population unbiased ratio and regression estimators," *J. Amer. Statist. Assoc.*, **54**, 594-612, (1959).

Murthy, M.N., "Almost unbiased estimators based on interpenetrating sub-samples," *Sankhya*, 24, 303-314, (1962).

Rao, J.N.K., "Ratio and regression estimators," *New Development in Survey Sampling*, John Wiley and Sons, New York, (1969).

Williams, W.H. "Generating unbiased ratio and regression estimators," *Biometrics*, **17**, 267-274, (1961).

———— "The precision of some unbiased regression estimators," *Biometrics*, **19**, 352-361 (1963).

Zarkovic, S.S., "An illustration of some characteristic situations in the application of the difference estimates," *Rev. Inter. Statist. Inst.*, **24**, 52-63, (1956).

8

Cluster Sampling

A Mathematician knows how to solve a problem but he can't do it.
W.E. Milne

8.1 INTRODUCTION

In random sampling, it is presumed that the population has been divided into a finite number of distinct and identifiable units defined as sampling units. The smallest unit into which the population can be divided is called an *element* of the population. A group of such elements is known as a *cluster*. When the sampling unit is a cluster, the procedure is called *cluster sampling*. If the entire area containing the population under study is divided into smaller segments and each element in the population belongs to one and only one segment, the procedure is sometimes called *area sampling*.

Generally, identification and location of an element requires considerable time. However, once an element has been located, the time taken for surveying a few neighbouring elements is small. Thus, the main function in cluster sampling is to specify clusters or to divide the population into appropriate clusters. Clusters are generally made up of neighbouring elements and, therefore, the elements within a cluster tend to have similar characteristics. As a simple rule, the number of elements in a cluster should be small and the number of clusters should be large. After dividing the population into specified clusters, the required number of clusters can be selected either by equal or unequal probabilities of selection. All the elements in selected clusters are enumerated.

For a given number of sampling units, cluster sampling is more convenient and less costly. The advantages of cluster sampling are that
(i) collection of data for neighbouring elements is easier, cheaper, faster and operationally more convenient than observing units spread over a region,
(ii) it is less costly than simple random sampling due to the saving of time in journeys, identification, contacts, etc.
(iii) when the sampling frame of elements may not be readily available.

Even if such a frame is made available it would be expensive to base an enquiry on a simple random sample of elements. From the point of view of statistical efficiency, cluster sampling is generally less efficient than simple random sampling due to the usual tendency of units in a cluster to be similar. In fact, the efficiency of cluster sampling is likely to decrease with increase in cluster size. For a given sample size, a smaller sampling unit will bring more precise results than a larger sampling unit. In most practical situations, the loss in efficiency may be balanced by the reduction in cost. Therefore, the efficiency per unit cost may be more in cluster sampling than in simple random sampling.

The selection of clusters can be random or by first selecting a unit called a *key unit* at random and then randomly taking the required number of neighbouring units to form the cluster. For example, to estimate the milk production, a cluster of three villages can be formed by first selecting a key village at random and then taking two more villages from a block of some specified area. Clusters may be overlapping or non-overlapping. Various sampling procedures, viz. simple random, stratified or systematic sampling procedures, discussed in earlier chapters, can be applied to cluster sampling by treating the clusters themselves as sampling units.

8.2 NOTATIONS

We shall first consider the case of equal clusters. Suppose the population consists of N clusters, each of M elements, and that a sample of n clusters is drawn by the method of simple random sampling.

N = number of clusters in the population

n = number of clusters in the sample

M = number of elements in the cluster

y_{ij} = the value of the characteristic under study for the jth element, $(j = 1, 2, \ldots, M)$ in the ith cluster, $(i = 1, 2, \ldots, N)$

$\bar{y}_{i.} = \sum\limits_{j}^{M} y_{ij}/M$ = the mean of per element of the ith cluster

$\bar{y}_n = \sum\limits_{i}^{n} \bar{y}_{i.}/n$ = the mean of cluster means in a sample of n clusters

$\bar{Y}_N = \sum\limits_{i}^{N} \bar{y}_{i.}/N$ = the mean of cluster means in the population

$\bar{Y} = \sum\limits_{i}^{N} \sum\limits_{j}^{M} y_{ij}/NM$ = the mean per element in the population

$S_i^2 = \sum\limits_{j}^{M} (y_{ij} - \bar{y}_{i.})^2/(M-1)$

= the mean square between elements within the ith cluster $(i = 1, 2, \ldots, N)$

$S_w^2 = \sum\limits_{i}^{N} S_i^2/N$ = the mean square within clusters,

(w for within)

$S_b^2 = \sum\limits_{i}^{N} (\bar{y}_{i.} - \bar{Y}_N)^2/(N-1)$

= the mean square between clusters means in the population
(b for between)

$S^2 = \sum\limits_{i}^{N} \sum\limits_{j}^{M} (y_{ij} - \bar{Y})^2/(N-1)$

= the mean square between elements in the population

$\rho = \dfrac{E(y_{ij} - \bar{Y})(y_{ik} - \bar{Y})}{E(y_{ij} - \bar{Y})^2} = \dfrac{\sum\limits_{i}^{N} \sum\limits_{j \neq k}^{M} (y_{ij} - \bar{Y})(y_{ik} - \bar{Y})}{(M-1)(NM-1)S^2}$

= the intracluster correlation co-efficient between elements within clusters

8.3 EQUAL CLUSTER SAMPLING: ESTIMATOR OF MEAN AND ITS VARIANCE

No new principles are involved in making estimates when a probability sample of n equal sized clusters has been taken and each cluster is enumerated completely. Since the clusters are of equal size, it is clear that $\overline{Y}_N = \overline{Y}$. For the sampling variance of the estimator $\bar{y}_n$ in the form developed by Hansen and Hurwitz (1942), we shall begin by proving the following theorem.

THEOREM 8.3.1 In simple random sampling, wor, of n clusters each containing M elements from a population of N clusters, the sample mean $\bar{y}_n$ is an unbiased estimator of $\overline{Y}$ and its variance is given by

$$V(\bar{y}_n) = \frac{(1-f)}{n} S_b^2 \tag{8.3.1}$$

$$\cong \frac{(1-f)}{n} S_M^2 [1 + (M-1)\rho] \tag{8.3.2}$$

where ρ is the intracluster correlation coefficient.

Proof By applying Theorem 2.3.1, it can be shown easily that $\bar{y}_n$ is an unbiased estimator of $\overline{Y}$ and its variance is given by

$$V(\bar{y}_n) = \frac{(1-f)}{n} \sum_{1}^{N} \frac{(\bar{y}_{i.} - \overline{Y}_N)^2}{(N-1)} = \frac{(1-f)}{n} S_b^2$$

Using

$$\sum_{i}^{N} (\bar{y}_{i.} - \overline{Y}_N)^2 = \frac{1}{M^2} \sum_{i}^{N} \sum_{j}^{M} (\bar{y}_{ij} - \overline{Y})^2 + \frac{1}{M^2} \sum_{i}^{N} \sum_{j \neq k}^{M} (y_{ij} - \overline{Y})(y_{jk} - \overline{Y})$$

after substituting the value of ρ in the above relation, we have

$$V(\bar{y}_n) = \frac{(1-f)}{n} \frac{NM-1}{M^2(N-1)} S^2 [1 + (M-1)\rho]$$

$$\cong \frac{(1-f)}{nM} S^2 [1 + (M-1)\rho] \quad \text{for large } N.$$

It has been shown that the variance in cluster sampling depends on the number of clusters in the sample, the size of the cluster, the intracluster correlation coefficient ρ and the variance S^2. If $M = 1$ it gives the sampling variance of a simple random sample of nM elements taken individually. Both procedures are equally good in this situation. If $M > 1$ and ρ is positive, cluster sampling will give a higher variance than the mean per element. If ρ is negative, there is reason to use cluster sampling.

COROLLARY 1 In simple random sampling, wor, of n clusters each containing M elements from a population of N clusters, the population total is estimated unbiasedly by

$$\hat{Y}_C = NM\, \bar{y}_n$$

with its sampling variance,

$$
\left.
\begin{aligned}
V(\hat{Y}_C) &= (1 - f)\, \frac{N^2}{n}\, M^2 S_b^2 \\
&\cong MN^2\, \frac{(1 - f)}{n}\, S^2\, [1 + (M - 1)\, \rho]
\end{aligned}
\right\}
\quad (8.3.3)
$$

COROLLARY 2 Instead of sampling in clusters, if a simple random sample of nM elements be directly taken from the population of NM elements, the variance of the mean per element would be

$$V(\bar{y}_n) = \frac{(1 - f)}{nM}\, S^2 \qquad (8.3.4)$$

COROLLARY 3 In simple random sampling, wor, an estimator of variance of $\bar{y}_n$ is given by

$$v(\bar{y}_n) = \frac{(1 - f)}{n}\, s_b^2 \qquad (8.3.5)$$

where

$$s_b^2 = \frac{\sum_i^n (\bar{y}_{i\cdot} - \bar{y}_n)^2}{(n - 1)}$$

COROLLARY 4 In simple random sampling, wr, the cluster mean $\bar{y}_n$ is an unbiased estimator of $\overline{Y}$ and its variance is given by

$$V(\bar{y}_n) = \frac{S_b^2}{n} \qquad (8.3.6)$$

An estimator of variance $V(\bar{y}_n)$ can be written as

$$v(\bar{y}_n) = \frac{s_b^2}{n} \qquad (8.3.7)$$

8.4 RELATIVE EFFICIENCY OF CLUSTER SAMPLING

In sampling nM elements from the population by simple random sampling, the variance of the sample mean $\bar{y}$ is given by

$$V(\bar{y}) = \frac{(1 - f)\, S^2}{nM}$$

Thus, the relative efficiency of cluster sampling compared with simple random sampling is given by

$$\text{Rel Efficiency} = \frac{V_{SR}(\bar{y})}{V_C(\bar{y}_n)} = \frac{S^2}{MS_b^2} \tag{8.4.1}$$

This shows that the efficiency of cluster sampling increases as the mean square between clusters decreases.

Also $\quad (N - 1)\, MS_b^2 = (NM - 1)\, S^2 - N\,(M - 1)\, S_w^2 \tag{8.4.2}$

Therefore, relative efficiency will increase with increase in the mean square within clusters. These results suggest that the clusters should be so formed that variation within clusters is maximum while variation between clusters is minimum. Another way to express relative efficiency is to make use of the concept of the intracluster correlation coefficient. For large N, the relative efficiency of cluster sampling in terms of intra-cluster coefficient ρ is given by

$$\text{Rel Efficiency }(E) = [1 + (M - 1)\,\rho]^{-1} \tag{8.4.3}$$

(i) In case of complete homogeneity of clusters, $S_w^2 = 0$ and so $\rho = 1$ and $E = 1/M$, i.e cluster sampling is not efficient.

(ii) In case of complete heterogeneity, $S_w^2 = S^2$, so $S_b^2 = 0$ and $\rho = -1/(M - 1)$, i.e. cluster sampling is very efficient.

Hence, it should be noted that ρ lies in the range $-1/(M - 1)$ to 1. It also shows that cluster sampling will be more efficient if ρ is negative. In practice, ρ is usually positive as neighbouring elements are grouped to form clusters. Generally, ρ decreases with increase in M. The efficiency of cluster sampling increases as the factor $(M-1)\,\rho$ increases with cluster size.

The efficiency can easily be calculated by estimating the value of ρ from the sample. An estimator of ρ can be written as

$$\hat{\rho} = \frac{(n - 1)\, Ms_b^2 - ns_w^2}{(n - 1)\, Ms_b^2 + n(M - 1)\, s_w^2} \tag{8.4.4}$$

where $\qquad s_w^2 = \sum_i^n \sum_j^M \frac{(y_{ij} - \bar{y}_{i.})^2}{n\,(M - 1)} \tag{8.4.5}$

Thus, for large N, an estimate of the relative efficiency of cluster sampling can be written as

$$\text{Est. Rel Efficiency }(e) = \frac{1}{M} + \frac{(M - 1)\, s_w^2}{M^2\, s_b^2} \tag{8.4.6}$$

And accordingly ρ can be estimated by

$$\hat{\rho} = \frac{(1 - e)}{(M - 1)e} \qquad (8.4.7)$$

For a sample of n clusters, the results in the form of analysis of variance can be written in the form of Table 8.1.

Table 8.1 Analysis of Variance for Sample of n Clusters

Source of variation	Degrees of freedom	Mean square
Between clusters	$n - 1$	$Ms_b^2 = \sum_i^n \dfrac{(\bar{y}_{i\cdot} - \bar{y}_n)^2}{(n - 1)}$
Within clusters	$n(M - 1)$	$s_w^2 = \sum_i^n \sum_j^M \dfrac{(\bar{y}_{ij} - \bar{y}_{i\cdot})^2}{n(M - 1)}$
Total sample	$nM - 1$	$s^2 = \sum_i^n \sum_j^M \dfrac{(\bar{y}_{ij} - \bar{y}_n)^2}{(nM - 1)}$
		$= \dfrac{(n - 1)\,Ms_b^2 + n(M - 1)\,s_w^2}{(nM - 1)}$

It should be noted clearly that in a random sample of n clusters, s_b^2 and s_w^2 will provide unbiased estimates of S_b^2 and S_w^2, respectively while s^2 will not be an unbiased estimate of S^2. The reason is that a sample of nM elements is not taken randomly from the population of NM elements. However, an unbiased estimate may be obtained easily by substituting the values in the relation, when

$$\hat{S}^2 = \frac{(N - 1)\,M\,s_b^2 + N(M - 1)\,s_w^2}{(NM - 1)} \qquad (8.4.8)$$

EXAMPLE 8.1 A pilot sample survey for study of cultivation practices and yield of guava was conducted by IASRI in Allahabad district of Uttar Pradesh (India). From Umerpur-Neerna village, out of a total of 412 bearing trees, 15 clusters of size 4 trees each were selected and yields (in kg) were recorded as given on next page.

Cluster	1st tree	2nd tree	3rd tree	4th tree
1	5.53	4.84	0.69	15.79
2	26.11	10.93	10.08	11.18
3	11.08	0.65	4.21	7.56
4	12.66	32.52	16.92	37.02
5	0.87	3.56	4.81	57.54
6	6.40	11.68	40.05	5.15
7	54.21	34.63	52.55	37.96
8	1.94	35.97	29.54	25.98
9	37.04	47.07	16.94	28.11
10	56.92	17.69	26.24	6.77
11	27.59	38.10	24.76	6.53
12	45.98	5.17	1.17	6.53
13	7.23	34.35	12.18	9.86
14	14.23	16.89	28.93	21.70
15	3.53	40.76	5.15	1.25

(i) Estimate the average yield (in kg) per tree of guava in the Umerpur–Neerna village of Allahabad along with its standard error.

(ii) Estimate the intracluster correlation coefficient between trees within clusters and efficiency of cluster sampling as compared to simple random sampling.

Here we have, $M = 4$, $N = 103$ and $n = 15$.

We prepare the following results:

Table 8.2 Calculations of Cluster Means and Their Variances

Cluster	Mean $(\bar{y}_i)$	$\sum_j \bar{y}_{ij}^2$	$M\bar{y}_i^2$	s_i^2
1	6.71	303.8067	180.0954	41.237
2	14.58	1027.7958	854.3056	59.163
3	5.88	198.0666	138.2976	19.923
4	24.78	2874.5927	2456.1936	139.466
5	9.20	795.0185	338.5600	152.157
6	15.81	1807.5993	999.8244	269.258
7	44.84	834.4251	8042.5024	99.264
8	23.36	2845.1765	2182.7584	220.806
9	33.19	4830.9302	4406.3044	141.542
10	26.91	4287.1930	2896.5924	463.533
11	23.08	2828.4951	2130.7456	232.838
12	14.71	2184.8911	865.5364	439.787
13	15.88	1476.3314	1008.6975	155.878
14	20.44	1995.5999	1671.1744	41.475
15	12.67	1701.9235	642.1156	353.269
Total	292.04			2829.341

Let y_{ij} be the yield of the jth tree in the ith cluster,

$$j = 1, 2, 3, 4 \text{ and } i = 1, 2, \ldots, 15.$$

(i) An estimate of the average yield per tree of guava is given as

$$\bar{y}_n = \frac{1}{nM} \sum_i^n \sum_j^M y_{ij} = \frac{1}{n} \sum_i^n \bar{y}_i.$$

$$= \frac{292.04}{15} = 19.47$$

Estimated variance of $\bar{y}_n$ is given by

$$v(\bar{y}_n) = \left(\frac{1}{n} - \frac{1}{N} \right) \frac{1}{n-1} \sum_i^n (\bar{y}_i. - \bar{y}_n)^2$$

$$= \left(\frac{1}{n} - \frac{1}{N} \right) \frac{1}{n-1} [\sum_i^n \bar{y}_i^2. - n \bar{y}_n^2]$$

$$= \left(\frac{1}{15} - \frac{1}{103} \right) \frac{1}{14} [7202.4262 - 15 \times 379.0809]$$

$$= \left(\frac{1}{15} - \frac{1}{103} \right) \times 108.3009 = 6.1686$$

Therefore, standard error of $\bar{y}_n = \sqrt{6.1686} = 2.48$

An estimate of the efficiency of cluster sampling as compared with simple random sampling can be written in the form of analysis of variance as below:

Source of variation	df	Mean square
Between clusters	14	108.3 ($= Ms_b^2$)
Within clusters	397	188.6 ($= s_b^2$)
Total sample	411	249.32 ($= \hat{S}^2$)

Thus
$$RE = \frac{249.32}{4 \times 108.3} = 0.556$$

and
$$\hat{\rho} = \frac{(1 - 0.576)}{(4 - 1) \times 0.576} = 0.245$$

8.5 OPTIMUM CLUSTER SIZE

For a given sample size, the sampling variance increases with cluster size and decreases with increasing number of clusters. On the other hand, the cost decreases with the cluster size and increases with the number of clusters. Hence, it is necessary to determine a balancing point by finding out the optimum cluster size and the number of clusters in the samples which can minimize the sampling variance for a given cost or, alternatively, minimize the cost for a fixed variance.

The efficiency of cluster sampling has been studied by Smith (1938) and Hansen and Hurwitz (1942). Mahalanobis (1940, 1942) have studied the problem of determination of the optimum cluster size from the points of view of both variance and cost. A comparative study of cluster sizes in cluster sampling and sub-sampling procedures has been made by Singh (1956). In this section we shall discuss the problem of optimum size of cluster for which maximum precision is attained with a given cost, or vice versa.

A good account of discussion on the construction of cost functions for cluster sampling has been given by Hansen, Hurwitz and Madow (1953). In brief, we have summed up the guidelines here.

The cost of a survey, apart from overhead cost, will be made up of two components:

(i) Cost due to expenses in enumerating the elements in the sample and in travelling within the cluster, which is proportional to the number of elements in the sample.

(ii) Cost due to expenses on travelling between clusters, which is proportional to the distance to be travelled between clusters. It has been shown empirically that the expected value of minimum distance between n points located at random is proportional to $n^{1/2}$.

The cost of a survey can be, therefore, expressed as

$$C = c_1 nM + c_2 n^{1/2} \tag{8.5.1}$$

where c_1 is the cost of enumerating an element, including the cost of travel between units within the cluster, and c_2 is the cost per unit distance travelled between clusters.

It has already been shown that the variance of the estimator $\bar{y}_n$ based on a sample of n clusters of size M each, is given by

$$V(\bar{y}_n) = \frac{(1-f) S_b^2}{n} \tag{8.5.2}$$

As shown by relation (8.4.2), S_b^2 can be obtained if we know (i) the variance S^2 between all elements in the population and (ii) the variance S_w^2 within clusters. Hence, an approach has always been made to predict S_w^2 as it is affected by the cluster size while S^2 remains unchanged by it.

Jessen (1942), Mahalanobis (1944) and Hendriks (1944) have attempted to develop a general law to predict how S_w^2 changes with size of cluster. On the basis of several agricultural surveys, they observed that S_w^2 appears to bear a relation with M which can be written by the empirical relation

$$S_w^2 = aM^b \quad (b > 0) \tag{8.5.3}$$

where a and b are positive constants and are to be determined from the survey data.

From the analysis of variance, we have

$$S_b^2 = \frac{(MN - 1)S^2 - N(M - 1) S_w^2}{M(N - 1)}$$

By substituting the value of S_w^2 from relation (8.5.3), for large N we have

$$S_b^2 = S^2 - (M - 1) a M^{b-1} \tag{8.5.4}$$

In relation (8.5.2), after ignoring the fpc and substituting the value of S_b^2 from relation (8.5.4), we have

$$V(\bar{y}_n) = \frac{1}{n} [S^2 - (M - 1) aM^{b-1}] \tag{8.5.5}$$

The problem under consideration is to calculate the values of n and M by minimizing V for given C. To minimize $\phi = C + \lambda V$ (λ being Lagrange's multiplier), differentiate ϕ w.r.t. n and M, respectively, and equate to zero. Solving the equations thus obtained, the value of n is given by

$$n = \left[\frac{- c_2 + (c_2^2 + 4c_1 CM)^{1/2}}{2 c_1 M} \right]^2 \tag{8.5.6}$$

It is difficult to get an explicit expression for M. However, M can be obtained from the equation

$$\frac{aM^{b-1} [bM - (b-1)]}{S^2 - (b - 1) aM^{b-1}} = 1 - \left(1 + \frac{4 c_1 CM}{c_2^2} \right)^{-1/2} \tag{8.5.7}$$

by the iterative method. On substituting the value of M thus obtained in relation (8.5.7), we can obtain the optimum value of n.

From relation (8.5.8), it follows that M will change according to changes in c_1, c_2 and C such that $c_1 CM/c_2^2$ is nearly constant. This leads to the conclusion that the optimum size of the unit will be smaller when

(i) the cost of enumeration of an element increases;
(ii) the cost of travel between units decreases; and
(iii) the cost of the survey is sufficiently large.

EXAMPLE 8.2   A survey for the estimation of the number of pepper standards and yield of pepper was conducted in the Irrikur Firka of Kerala State (India). The following calculations give the mean square between cluster means (S_b^2) and within mean square (S_w^2) for the number of bearing pepper standards in clusters of various sizes formed by grouping adjacent survey numbers of the seven villages selected from Irrikur Firka:

Cluster size	3	5	7	10	15	20
S_b^2	20.99	16.30	13.46	10.15	8.39	6.97
S_w^2	32.99	33.35	34.45	36.52	37.23	38.04

(i) Calculate the intra-class correlation coefficient and efficiencies of various clusters taken as sampling units, compared with that of an element.

(ii) Study the relationship between S_b^2 and size of cluster by fitting the following relations:

(a) $S_b^2 = (S^2/M^g)$,   where $S^2 = 42.259$
(b) $S_w^2 = aM^b$

In order to find the efficiency and intra-cluster correlation, we calculate S^2 using the relation

$$S^2 = \frac{M-1}{M} S_w^2 + S_b^2,$$

for N is large. The following calculations are prepared to get the estimates of E and ρ by using $E = (S^2/MS_b^2)$ and $\rho = (1 - E)/(M - 1) E$

Cluster size (M)	3	5	7	10	15	20
S_b^2	20.99	16.30	13.46	10.15	8.39	6.97
S^2	42.96	38.98	42.99	43.32	44.14	43.11
$E = \dfrac{S^2}{MS_b^2}$	0.682	0.478	0.456	0.424	0.343	0.309
$\rho = \dfrac{(1-E)}{(M-1)E}$	0.230	0.273	0.199	0.151	0.137	0.118

We prepare the following results for studying the relationships:

M	$\log M$	S_b^2	$\log S_b^2$	S_w^2	$\log S_w^2$	$\dfrac{\log M}{\log S_b^2}$	$\dfrac{\log M}{\log S_w^2}$
(1)	(2)	(3)	(4)	(5)	(6)	(7) = (2)/(4)	(8) =(2)/(6)
3	0.4771	20.99	1.3221	32.96	1.5180	0.6356	0.7242
5	0.6990	16.30	1.2122	33.35	1.5331	0.8394	1.0646
7	0.8451	13.46	1.1219	34.45	1.5372	0.9417	1.2990
10	1.0000	10.15	1.0021	36.52	1.5625	1.0244	1.8377
15	1.1761	8.39	0.9238	37.23	1.5709	1.0808	1.5709
20	1.3010	6.97	0.8432	38.04	1.5802	1.0918	2.0558

For the relationship $S_b^2 = S^2/M^g$, the estimate of g by the least square technique is given by

$$-g = \frac{\sum\limits^{6} \log M \log S_b^2 - 1/6 \sum\limits^{6} \log M \sum\limits^{6} \log S_b^2}{\sum\limits^{6} (\log M)^2 - \dfrac{(\sum\limits^{6} \log M)^2}{6}}$$

On substituting values, we have

$$-g = \frac{5.6179 - 5.8946}{5.5062 - 5.0386} = -0.592$$

Thus, the relationship between S_b^2 and M is

$$S_b^2 = \frac{S^2}{M^{0.592}}$$

Similarly, for the relationship $S_w^2 = aM^b$, the values of b and a are given by

$$b = \frac{\sum\limits^{6} \log S_w^2 \log M - \dfrac{\sum\limits^{6} \log S_w^2 \sum\limits^{6} \log M}{6}}{\sum\limits^{6} (\log M)^2 - \dfrac{(\sum\limits^{6} \log M)^2}{6}}$$

$$a = \text{antilog } \frac{(\sum\limits^{6} \log S_w^2 - b \sum\limits^{6} \log M)}{6}$$

Substituting the values, we have

$$b = \frac{8.5524 - 8.5150}{5.5062 - 5.0386} = \frac{0.0374}{0.4676} = 0.080$$

$$\log a = 1.5487 - 0.080 \times 0.9164 = 1.4754$$

Therefore, $a = 29.88$

Thus, the relationship between S_w^2 and M is

$$S_w^2 = 29.88 \, M^{0.080}$$

The values of s_b^2 and s_w^2 from the relationships

$$s_b^2 = S^2 M^{0.592} \quad \text{and} \quad s_w^2 = 29.88 \, M^{0.080},$$

are given below for various values of M along with their true values

M	s_b^2	S_b^2	s_w^2	S_w^2
3	22.03	20.99	32.63	32.96
5	16.28	16.30	33.99	33.35
7	13.34	13.46	34.91	34.45
10	10.80	10.15	35.92	36.52
15	8.49	8.39	37.11	37.23
20	7.16	6.97	38.04	38.04

8.6 CLUSTER SAMPLING FOR PROPORTIONS

Suppose it is required to estimate the proportion of elements belonging to a specified class when the population consists of N clusters, each of size M, and a random sample, wor, of n clusters is selected. Suppose the M elements in any cluster can be classified into two classes. Assuming that $y_{ij} = 1$, if the jth element of the ith cluster belongs to the class, and 0 otherwise. It can be easily seen that $p_i = a_i/M$ is the proportion in the ith cluster, a_i being the number of elements in the ith cluster belonging to the specified class. An unbiased estimator of the population proportion $P = \sum_i^N a_i/NM = \sum_i^N p_i/N$ is given by

$$\hat{P}_c = \sum_i^n \frac{p_i}{n} = \bar{p} \qquad (8.6.1)$$

where p_i is the proportion of elements belonging to the specified class in the ith cluster of the sample.

The sampling variance of $\hat{P}_c$ is given by

$$V(\hat{P}_c) = \frac{(1-f)}{n} S_b^2 \qquad (8.6.2)$$

where S_b^2 is the variance between cluster proportions and is given by

$$M S_b^2 = \sum_i^N \frac{(p_i - P)^2}{N} = PQ - \sum_i^N \frac{P_i Q_i}{N} \qquad (8.6.3)$$

For large N, we have

$$S^2 \cong S_b^2 + S_w^2 = PQ$$

and the within-variance S_w^2 is given by $\sum_i^N P_i Q_i/N$. Hence, the intracluster correlation coefficient ρ can be written as

$$\rho = 1 - \frac{\sum_i^N M P_i Q_i}{(M - 1) PQ} \qquad (8.6.4)$$

Therefore, the sampling variance, in terms of the intracluster correlation coefficient, can be expressed as

$$V(\hat{P}_c) = \frac{(1 - f) NPQ}{(N - 1) nM} [1 + (M - 1) \rho] \qquad (8.6.5)$$

An estimator of the total number of units belonging to the specified category can be obtained by multiplying $\hat{P}_c$ by NM and the expression for its sampling variance is $N^2 M^2$ times that given by Eq. (8.6.5).

If a simple random sample of nM elements could be taken, the variance of the sample proportion $\hat{P}$ would be given by

$$V(\hat{P}) = (1 - f) \frac{NPQ}{nM (N - 1)} \qquad (8.6.6)$$

The efficiency of cluster sampling as compared to simple random sampling, wor, can be obtained as

$$RE = \frac{(N - 1)}{(NM - 1)} \frac{NPQ}{NPQ - \sum_i^N P_i Q_i} \qquad (8.6.7)$$

An estimator of $V(\hat{P}_c)$ is given by

$$v(\hat{P}_c) = (1 - f) \frac{s_b^2}{n} = (1 - f) \frac{\sum_i^n (p_i - \bar{p})^2}{n (n - 1)} \qquad (8.6.8)$$

An estimator for the sampling variance of the total number of units belonging to a specified class can be obtained by multiplying $N^2 M^2$ to the value in relation (8.6.8).

8.7 UNEQUAL CLUSTER SAMPLING: ESTIMATORS OF MEAN AND THEIR VARIANCES

We have considered the case when the size of all the clusters is the same. But in many practical situations, cluster sizes vary. Now we shall discuss the case of unequal clusters.

Suppose there are N clusters. Let the ith cluster consist of M_i elements; $(i = 1, 2, \ldots, N)$ and $\sum_{i}^{N} M_i = M_0$. The population mean per element $\overline{Y}$ is defined by

$$\overline{Y} = \frac{\sum_{i}^{N} \sum_{j}^{M_i} y_{ij}}{\sum_{i}^{N} M_i} = \frac{\sum_{i}^{N} M_i \, \bar{y}_{i.}}{\sum_{i}^{N} M_i} = \frac{\sum_{i}^{N} M_i \, \bar{y}_{i.}}{M_0},$$

where $\bar{y}_{i.}$ is the mean per element of the ith cluster.

We may also define the pooled mean of the cluster means as

$$\overline{Y}_N = \sum_{i}^{N} \frac{\bar{y}_{i.}}{N}$$

Let a random sample, wor, of n clusters be drawn and all elements of the clusters surveyed. Three estimators of $\overline{Y}$, as given below, may be considered:

$$\bar{y}_n = \sum_{i}^{n} \frac{\bar{y}_{i.}}{n} \tag{8.7.1}$$

$$\bar{y}'_n = \frac{\sum_{i}^{n} M_i \, \bar{y}_{i.}}{\sum_{i}^{n} M_i} \tag{8.7.2}$$

$$\bar{y}_n^* = \frac{N}{nM_0} \sum_{i}^{n} M_i \, \bar{y}_{i.} = \frac{\sum_{i}^{n} M_i \bar{y}_{i.}}{n\overline{M}} \tag{8.7.3}$$

where

$$\overline{M} = \frac{\sum_{i}^{N} M_i}{N} = \frac{M_0}{N}$$

THEOREM 8.7.1 Show that the simple arithmetic mean given by relation (8.7.1) is not an unbiased estimator. The bias and sampling variance of the estimator are given by

$$B(\bar{y}_n) = -\frac{\text{cov}(\bar{y}_{i.}, M_i)}{\overline{M}} \tag{8.7.4}$$

and
$$V(\bar{y}_n) = \frac{(1-f)}{n} S_b^2 \qquad (8.7.5)$$

where
$$S_b^2 = \frac{\sum\limits_{i}^{N} (\bar{y}_i. - \overline{Y}_N)^2}{(N-1)}$$

Proof To prove that $\bar{y}_n$ is not unbiased, we can write

$$E(\bar{y}_n) = E\left[\frac{1}{n} \sum_i^n \bar{y}_i.\right] = \frac{1}{n} \sum_i^n E(\bar{y}_i.)$$

$$= \sum_i^N \frac{\bar{y}_i.}{N} = \overline{Y}_N \neq \overline{Y}$$

Thus, $\bar{y}_n$ is a biased estimator of the population mean $\overline{Y}$.

The bias of the estimator is given by

$$B(\bar{y}_n) = E(\bar{y}_n) - \overline{Y} = \frac{\sum\limits_{i}^{N} \bar{y}_i.}{N} - \frac{\sum\limits_{i}^{N} M_i \bar{y}_i.}{N \overline{M}}$$

$$= \frac{\sum\limits_{i}^{N} (\overline{M} - M_i) \bar{y}_i.}{N\overline{M}} = -\frac{\operatorname{cov}(\bar{y}_i., M_i)}{\overline{M}}$$

For a population in which M_i's do not appreciably vary from one cluster to another, the bias may not be materially significant. If M_i any $\bar{y}_i.$ are uncorrelated, the bias is zero and $\bar{y}_n$ is an unbiased estimator in this case.

This shows that bias is expected to be small when M_i and $\bar{y}_i.$ are not highly correlated. In such a case, it is advisable to use this estimator. Its sampling variance is given by

$$V(\bar{y}_n) = E[\bar{y}_n - E(\bar{y}_n)]^2 = E(\bar{y}_n - \overline{Y}_N)^2$$

$$= \frac{(1-f)}{n} \frac{\sum\limits_{i}^{N}(\bar{y}_i. - \overline{Y}_N)^2}{(N-1)} = \frac{(1-f)}{n} S_b^2$$

COROLLARY An unbiased estimator of $V(\bar{y}_n)$ is given by

$$v(\bar{y}_n) = \frac{(1-f)}{n} s_b^2 \qquad (8.7.6)$$

where
$$s_b^2 = \sum_i^n \frac{(\bar{y}_i. - \bar{y}_n)^2}{(n-1)}$$

THEOREM 8.7.2 Show that the estimator of the mean $\overline{Y}$ given by relation (8.7.2), which is a weighted mean of the cluster means and a ratio of two random variables, is consistent but not unbiased. Its sampling variance is given by

$$V(\bar{y}'_n) = \frac{(1-f)}{n} S_b'^2 \qquad (8.7.7)$$

where
$$S_b'^2 = \frac{\sum\limits_{i}^{N} M_i^2 (\bar{y}_{i.} - \overline{Y})^2}{\overline{M}^2 (N-1)}$$

Proof The estimator is given by replacing x_i by M_i and y_i by $M_i \bar{y}_{i.}$ in the ratio estimator. We have already seen that the ratio estimator is not unbiased, but consistent. A first order approximation to the variance of this estimator is given by

$$V(\bar{y}'_n) = \frac{(1-f)}{n} S_b'^2$$

COROLLARY An unbiased estimator of $V(\bar{y}'_n)$ is given by

$$v(\bar{y}'_n) = \frac{(1-f)\, s_b'^2}{n} \qquad (8.7.8)$$

where
$$s_b'^2 = \frac{\sum\limits_{i}^{n} M_i^2 (\bar{y}_{i.} - \bar{y}'_n)^2}{\overline{M}'^2 (n-1)} \quad \text{and} \quad \overline{M}' = \frac{\sum\limits_{i}^{n} M_i}{n}$$

THEOREM 8.7.3 Show that the estimator of the mean given by relation (8.7.3) is unbiased and its variance is given by

$$V(\bar{y}_n^*) = \frac{(1-f)}{n} S_b^2 \qquad (8.7.9)$$

where
$$S_b^{*2} = \frac{1}{(N-1)} \sum\limits_{i}^{N} \left(\frac{M_i}{\overline{M}} \bar{y}_{i.} - \overline{Y} \right)^2$$

Proof Let us consider the estimator $\bar{y}_n^* = \dfrac{\sum\limits_{i}^{n} M_i \bar{y}_{i.}}{n\overline{M}}$

$$E(\bar{y}_n^*) = E(\sum\limits_{i}^{n} M_i \bar{y}_{i.}/n\overline{M}) = \frac{\sum\limits_{i}^{n} E(M_i \bar{y}_{i.})}{n\overline{M}} = \overline{Y}$$

Hence $\bar{y}_n^*$ is an unbiased estimator. The sampling variance of the estimator is given by

$$V(\bar{y}_n^*) = \frac{(1 - f)}{n} \frac{\sum_i^N \left(\frac{M_i}{\bar{M}} \bar{y}_{i.} - \bar{Y} \right)^2}{(N - 1)}$$

$$= \frac{(1 - f)}{n} S_b^{*2}$$

It may be noticed that the estimator $\bar{y}_n^*$ will often be less precise. This occurs because the variance depends upon the variation of the product $M_i \bar{y}_{i.}$ and is likely to be larger than $\bar{y}_n$ unless $\bar{y}_{i.}$ and M_i vary in such a way that their product is almost constant.

COROLLARY   An unbiased estimator of $V(\bar{y}_n^*)$ is given by

$$v(\bar{y}_n^*) = \frac{(1 - f)}{n} s_n^{*2} \tag{8.7.10}$$

where
$$s_b^{*2} = \frac{\sum_i^n \left(\frac{M_i \bar{y}_{i.}}{\bar{M}} - \bar{y}_n^* \right)^2}{(n - 1)}$$

8.8   RELATIVE EFFICIENCY OF UNEQUAL CLUSTER SAMPLING

In a number of situations, it is easier to take some naturally formed groups of elements. Usually in such cases, cluster size would be unequal. For example, villages which are groups of households or households which are groups of persons, are usually taken as clusters for the purpose of sampling, on account of operational convenience. Thus unequal cluster sampling is the most practical situation and its relative efficiency with respect to simple random sampling should be worked out.

In unequal cluster sampling, the total number of elements $\sum_i^n M_i$ in the sample is a random variable with expected value $n\bar{M}$. If an equivalent simple random sample of size $n\bar{M}$ had been selected directly from the population of $N\bar{M}$ elements, the variance of the mean per element would be given by

$$V_{SR}(\bar{y}) = \frac{(N\bar{M} - n\bar{M})}{N\bar{M} n\bar{M}} S^2 = \frac{(1 - f)}{n\bar{M}} S^2 \tag{8.8.1}$$

Comparing this with the value given by Eq. (8.7.9), we have,

$$\text{R.E.} = \frac{S^2}{M \, S_b^{*2}} \qquad (8.8.2)$$

Hence, it is observed that the efficiency increases as the variation between clusters decreases. In general, cluster sampling will be efficient only when the variation between clusters is as small as possible.

8.9 VARYING PROBABILITY CLUSTER SAMPLING

In many practical situations, cluster size is positively correlated with the variable under study. In these cases it is advisable to select the clusters with probability proportional to the number of elements in the cluster. Hansen and Hurwitz (1943) gave a technique to select the clusters with probability proportional to their sizes M_i. In some cases the sizes M_i are known only approximately. There are several applications in which the size is not the number of elements in the cluster and some other measure is considered for probability selection. In this section, we shall confine ourselves to discussion of the case where the probability of selection is proportional to cluster size M_i.

Let p_i $(0 < p_i < 1)$ be the probability of selecting the ith cluster of size M_i $(i = 1, 2, \ldots, N)$ at each draw, such that $\sum_i^N p_i = 1$

Suppose that $\qquad z_{ij} = \dfrac{M_i \, y_{ij}}{M_0 \, p_i} \qquad (8.9.1)$

for $\qquad j = 1, 2, \ldots, M_i; \quad i = 1, 2, \ldots, N$

Further, suppose that n clusters are selected by pps with replacement, so that

$$\bar{z}_{i.} = \frac{M_i \, \bar{y}_{i.}}{M_0 \, p_i} \quad \text{for} \quad i = 1, 2, \ldots, N \qquad (8.9.2)$$

THEOREM 8.9.1 If a sample of n clusters is drawn with probabilities p_i and with replacement, then an unbiased estimator of $\bar{Y}$ is given by

$$\bar{z}_n = \frac{\sum_i^n \bar{z}_{i.}}{n} \qquad (8.9.3)$$

with its sampling variance

$$V(\bar{z}_n) = \frac{\sum\limits_{i}^{n} p_i (\bar{z}_{i.} - \bar{Y})^2}{n} \tag{8.9.4}$$

Proof To prove that $\bar{z}_n$ is unbiased, we have

$$E(\bar{z}_n) = E(\sum_{j}^{n} \bar{z}_{i.}/n) = \frac{1}{n} \sum_{i}^{n} E(\bar{z}_{i.})$$

$$= \frac{1}{n} \sum_{i}^{n} \sum_{i}^{N} p_i \frac{M_i \bar{y}_{i.}}{M_0 p_i} = \frac{1}{n} \sum_{i}^{n} \sum_{i}^{N} \frac{M_i \bar{y}_{i.}}{M_0} = \frac{1}{n} \sum_{i}^{n} \bar{Y} = \bar{Y}$$

To obtain the sampling variance of $\bar{z}_n$, we may proceed on similar lines as in Section 2 of Chapter 5. Thus, we get

$$V(\bar{z}_n) = \frac{1}{n} \sum_{i}^{n} p_i (\bar{z}_{i.} - \bar{Y})^2$$

COROLLARY 1 If units are drawn with probabilities proportional to size $p_i = M_i/M_0$ and with replacement, then an unbiased estimator of $\bar{Y}$ is given by

$$\bar{z}_n = \sum_{i}^{n} \bar{y}_{i.}/n \tag{8 9.5}$$

with variance $\quad V(\bar{z}_n) = \sum_{i}^{N} \frac{M_i}{M_0} (\bar{y}_{i.} - \bar{Y})^2/n \tag{8.9.6}$

COROLLARY 2 If units are drawn with probabilities p_i and with replacement, an unbiased estimator of variance $V(\bar{z}_n)$ is given by

$$v(\bar{z}_n) = \frac{\sum\limits_{i}^{n} (\bar{z}_{i.} - \bar{z}_n)^2}{n(n-1)} \tag{8.9.7}$$

COROLLARY 3 If units are drawn with probabilities proportional to size $p_i = M_i/M_0$ and with replacement, then an unbiased estimator of $V(\bar{z}_n)$ is given by

$$v(\bar{z}_n) = \frac{\sum\limits_{i}^{n} (\bar{y}_{i.} - \bar{y}_n)^2}{n(n-1)} \tag{8.9.8}$$

The efficiency of sampling n unequal clusters with pps, wr, as compared to simple random sampling, wr, by taking $n\overline{M}$ elements, can be obtained by comparing Eq. (8.9.5) with the variance

$$V_{SR}(\bar{y}) = \frac{S^2}{n\overline{M}}$$

which can be expanded by introducing $\bar{y}_i$. in the above relation. Thus,

$$V_{SR}(\bar{y}) = \frac{1}{n\overline{M}} \frac{1}{N\overline{M}} \left[\sum_i^N \sum_j^{Mi} (y_{ij} - \bar{y}_i.)^2 + \sum_i^N M_i (\bar{y}_i. - \overline{Y})^2 \right]$$

$$= \frac{\sigma_w^2}{n\overline{M}} + V(\bar{z}_n)$$

where
$$\sigma_w^2 = \sum_i^N \sum_j^{M_i} \frac{(y_{ij} - \bar{y}_i.)^2}{N\overline{M}} = \sum_i^N \frac{M_i \sigma_i^2}{N\overline{M}}$$

Hence, the efficiency of cluster sampling is given by

$$E = \frac{V_{SR}}{V_{pps}} = [\overline{M}(1 - \sigma_w^2/\sigma^2)]^{-1} \qquad (8.9.9)$$

It can be seen that E is always greater than 1. Other sampling schemes, such as pps, wor, sampling, and pps systematic sampling with suitable arrangement of clusters may also be used to increase the efficiency. Further comparisons have been made under super population model by Des Raj (1954, 1958), Zarcovic (1960), and Foreman and Brewer (1971).

SET OF PROBLEMS

8.1 If the NM elements in a population are grouped at random to form N clusters of M elements each, show that a random sample, wor, of n clusters would have the same efficiency as sampling nM elements in random sample, wor.

8.2 A finite population of M_0 elements is divided into N clusters with the ith cluster having M_i elements. A random sample, wor, of m elements is selected from the total population elements and the sample elements grouped according to the clusters to which they belong. If y is the sample total based on the values of the sample elements in the n selected clusters, show that $\hat{\overline{Y}} = Ny/mn$ is an unbiased estimator for the population mean $\overline{Y}$ and derive its sampling variance.

(Ghosh, 1963)

8.3 Let there be N clusters of M elements each. A sample of n clusters is taken systematically for estimating the population mean per element. Derive the sampling variance of the estimator in terms of the intracluster correlation coefficient (ρ_c) between pairs of elements in the clusters and (ρ'_c) between pairs of clusters at the samples, assuming N to be a multiple of n. (Madow, 1949)

8.4 A population consists of N clusters each containing M elements. A simple random sample of n clusters is selected to estimate the population mean per element. Assuming S_w^2, the variance within clusters, to be given by aM^b, $b < 0$ and the cost function to be $C = c_1 nM + c_2 \sqrt{n}$, find the optimum size of the cluster for which the variance of the estimator is minimum, when cost of survey is fixed.

8.5 For examining the efficiency of sampling households instead of persons for estimating the population of males in a given area, the following assumptions are made
   (i) each household consists of 4 persons (husband, wife and two children)
   (ii) the sexes of children are binomially distributed.
   Show that the intracluster correlation coefficient is 1/6 and efficiency of sampling households compared to that of sampling persons is 200%. (Sukhatme, 1954)

8.6 A population consists of N clusters, M_i being the size of the ith cluster ($i=1, 2, \ldots N$). Clusters are selected one-by-one by pps of clusters, wr. The cluster selected at the $(m + 1)$th draw is rejected if the number of distinct clusters selected in the first m draws is a pre-assigned number k. Suppose the ith cluster occurs m_i times in a sample of m draws, $m_i = 0, 1, 2, \ldots, i = 1, 2, \ldots, N$. If $\bar{y}_i$. is the mean of the ith cluster, show that $\bar{y} = \sum_i (m_i/m)\, \bar{y}_i.$ is an unbiased estimator of the population mean. Also show that an unbiased estimate of the variance of this estimator is obtained by

$$v(\bar{y}) = \sum_i^N \frac{m_i\,(\bar{y}_i. - \bar{y})^2}{m\,(m-1)}$$ (Sampford, 1962)

8.7 If the population is divided into N unequal clusters, the ith cluster consisting of M_i elements ($i = 1, 2, \ldots, N$), and r denotes the intracluster correlation coefficient defined by

$$r = \frac{\sum\limits_{i}^{N} \sum\limits_{j \neq k}^{M_i} (y_{ij} - \bar{Y})(y_{ik} - \bar{Y})}{M_0\,(\bar{M} - 1)\,S^2}$$

where M_0 and $\bar{M}$ have their usual meanings.

(i) Show that the limits of r are given as

$$-\frac{1}{M-1} < -\frac{M_0-1}{M_0\,(M-1)} \leqslant r \leqslant \frac{(N-1)\,S_b'^2}{N\,S^2}$$

where $$S_b'^2 = \sum_i^N \frac{M_i\,(M_i - 1)(\bar{y}_i. - \bar{Y})^2}{\bar{M}\,(\bar{M}-1)\,(N - 1)}$$

(ii) $M_i = m$ for $i = 1, 2, \ldots, N$ and $S_b'^2 = S_b^2$

Show that $\quad r = \dfrac{Nm \, \rho}{(Nm-1)}$

(iii) If $\bar{y}_n'$ is defined as in Eq. (8.7.2), show that

$$V(\bar{y}_n') \cong \frac{(1 - f) \, S^2}{n\bar{M}} [1 + r \, (\bar{M} - 1)]$$

where S^2 has its usual meaning. (Zasepa, 1962).

8.8 For studying the cultivation practices and yield of apple, a pilot sample survey was conducted in a district of Himachal Pradesh (India). The yield (in kilogrammes) of 15 clusters of 4 trees each, selected at random out of 308 bearing trees in a village, are given below:

Cluster\Tree	1	2	3	4
1	5.53	4.84	0.69	15.79
2	26.11	10.93	10.08	11.18
3	11.08	0.65	4.21	7.56
4	12.66	32.52	16.92	37.02
5	0.87	3.56	4.81	27.54
6	6.40	11.68	40.05	5.12
7	54.21	34.63	52.55	37.20
8	1 24	35.97	29.54	25.28
9	37.94	47.07	19.64	28.11
10	54.92	17.69	26.24	6.77
11	25.52	38.10	24.74	1.90
12	45.98	5.17	1.17	6.53
13	7.13	34.35	12.18	9.86
14	14.23	16.89	28.93	21.70
15	3.53	40.76	5.15	1.25

(i) Estimate the average yield per tree as well as the production of apple in the village and their standard errors.

(ii) Estimate the intracluster correlation coefficient between trees within clusters.

(iii) Estimate the efficiency of cluster sampling as compared to simple random sampling.

8.9 A pilot sample survey was conducted to study the management practices and yield of guava in a village in Uttar Pradesh (India). Of the total of 412 bearing trees, 16 clusters of size 4 each were selected and their yield records (kg/tree) are given on next page.

Cluster number	1st tree	2nd tree	3rd tree	4th tree
1	5.58	4.84	0.69	15.79
2	26.11	10.93	10.08	11.18
3	11.08	0.65	4.21	7.56
4	12.66	32.52	16.92	37.02
5	0.87	3.56	4.81	23.54
6	6.40	11.68	40.05	5.12
7	54.21	34.63	52.55	37.96
8	1.94	35.97	29.54	25.28
9	37.94	47.07	19.64	29.11
10	56.92	17.69	26 24	1.90
11	27.59	38.10	24.74	6.77
12	45.98	5.17	1.47	6.53
13	7.13	34.35	12.18	9.86
14	14.23	16.89	28.93	21.70
15	3.53	40.76	5.15	1.25
16	5.17	26.11	29.54	19.64

(i) Estimate the average yield per tree as well as the total yield of guava in the village along with their standard errors.

(ii) Estimate the efficiency of cluster sampling as compared to simple random sampling.

8.10 A survey on pepper was conducted to estimate the number of pepper standards and production of pepper in Kerala State (India). For this, 3 clusters from 95 were selected by srs, wor. The information on the number of pepper standards recorded is given below:

Cluster number	Cluster size	No. of pepper standards
1	12	41, 16, 19, 15, 144, 454, 212, 57, 28, 76, 199
2	12	39, 70, 38, 37, 161, 38, 27, 219, 46, 128, 30, 20
3	7	115, 59, 120

Estimate the total number of pepper standards along with standrad error for the region, given $\bar{M}$ the average cluster size for the population to be 10.

REFERENCES

Des Raj, "On sampling probabilities proportional to size," *Ganita*, **52**, 175-182, (1954).

——————— "On relative accuracy of some sampling technique," *J. Amer. Statist. Assoc.*, **53**, 98-101, (1958).

Foreman, E. K. and K. W. R. Brewer, "The efficient use of supplementary information in standard sampling procedures," *J. R. Statist. Soc.*, **33B**, 391-400, (1971).

Ghosh, S.P., "Post cluster sampling," *Ann. Math. Statist.*, **34**, 587-597, (1963).

Hansen, M.H. and W.N. Hurwitz, (1942). "Relative efficiencies of various sampling units in population enquiries," *J Amer. Statist. Assoc.*, **37**, 89-94, (1942).

Hansen, M.H. and W.N. Hurwitz, "On the theory of sampling from finite populations," *Ann. Math. Statist.*, **14**, 333-362, (1943).

Hansen, M.H., W. N. Hurwitz and W.G. Madow, *Sample Survey Methods and Theory*, John Wiley & Sons, New York, (1953).

Hendricks, W.A., "The relative efficiencies of groups of farms as sampling units," *J. Amer. Statist. Assoc.*, **39**, 366-376, (1944).

Jessen, R.J., "Statistical investigation of a sample survey for obtaining farm facts," *Iowa Agricultural Experiment Station, Research Bulletin*, **304**, (1942).

Madow, W.G., "On the theory of systematic sampling II," *Ann. Math. Stasist.*, **20**, 333-354. (1949).

Mahalanobis, P. C., "A sample survey of acreage under jute in Bengal," *Sankhya*, **4**, 511-530, (1940).

——————— *General Report on the Sample Census of Area under Jute in Bengal*, Indian Central Jute Committee, (1942).

——————— "On large-scale sample surveys," *Phil. Trans. Roy. Soc.*, **231 (B)**, 329-451, (1944).

Sampford, M. R. "Methods of cluster sampling with and without replacement for clusters of unequal sizes," *Biometrika*, **49**, 27-40, (1962).

Smith, H.F., "An empirical law describing heterogeneity in the yields of agricultural crops," *J. Agr. Sci.*, **28**, 1-23, (1938).

Singh, D., "On efficiency of cluster sampling," *J. Ind. Soc. Agr. Statist.*, **8**, 45-55, (1956).

Sukhatme, P.V., *Sampling Theory of Surveys with Applications*, Indian Soc. Agr. Statist., New Delhi, (1954).

Zarcovic, S.S., "On the efficiency of sampling with various probabilities and the selection of units with replacement," *Metrika*, **3**, 53-60, (1960).

Zasepa, R., "Badania Statystyezne Metoda Reprezentacying," Warszawa, (1962).

Multi-Stage Sampling

"Mine is a long and sad tale" said the Mouse, turning
to Alice and sighing.
"It is a long tail, certainly" said Alice, looking down with
wonder at the Mouse's tail, *"but why do you call it sad?"*

Lewis Carroll

9.1 SAMPLING PROCEDURE

In Chapter 8, we have discussed cluster sampling in which clusters were
considered sampling units and all the elements in the selected clusters
were enumerated completely. It has been stated there that cluster sampling is economical under certain circumstances but the method restricts
the spread of the sample over the population which results generally
in increasing the variance of the estimator. It is, therefore, logical to
expect that the efficiency of the estimator will be increased by distri-
buting elements over a large number of clusters and surveying only a
sample of units in each selected cluster instead of completely enumerating all the elements in the sample of clusters. *This type of sampling
which consists in first selecting the clusters and then selecting a
specified number of elements from each selected cluster is known as
sub-sampling or two-stage sampling.* In such sampling designs, clusters
which form the units of sampling at the first stage are called the *first
stage units (fsu)* or *primary sampling units (psu)* and the elements
within clusters are called *second-stage units (ssu)*. This procedure can

be generalized to three or more stages and is termed *multi-stage sampling*. For example, in crop surveys for estimating yield of a crop in a district, a block may be considered a primary sampling unit, the villages the second stage units, the crop fields the third stage units, and a plot of fixed size the ultimate unit of sampling.

Multi-stage sampling has been found to be very useful in practice and this procedure is being commonly used in large-scale surveys. Mahalanobis (1940) used this procedure in crop surveys and Ganguli (1941) has termed it *nested sampling*. Cochran (1939), Hansen and Hurwitz (1943), Sukhatme (1953) and Lahiri (1954) have discussed its use in agricultural and population surveys. Roy (1957) and Singh (1958) have considered the estimation of variance components for this sampling design. The multi-stage sampling procedure may be taken to be a better combination of random sampling and cluster sampling procedures. It can be expected to be (i) less efficient than single stage random sampling and more efficient than cluster sampling from the sampling variability point of view, and (ii) more efficient than single stage random sampling and less efficient than cluster sampling from the cost and operational point of view. The main advantage of this sampling procedure is that, at the first stage, the frame of fsu's is required which can be prepared easily. At the second stage, the frame of ssu's is required only for the selected fsu's and so on. This design is more flexible as it permits the use of different selection procedures in different stages. It should also be mentioned that multi-stage sampling may be the only choice in a number of practical situations where a satisfactory sampling frame of ultimate-stage units is not readily available and the cost of obtaining such a frame is large.

9.2 TWO-STAGE SAMPLING WITH EQUAL FIRST-STAGE UNITS: ESTIMATION OF MEAN AND ITS VARIANCE

Since, in two-stage sampling, the units are selected in stages by considering a probability structure at each stage, the selection procedures at both stages are to be considered in deriving the expected value and the variance of an estimator based on the number of observations taken on a sample of ssu's. For getting the expected vaule and sampling variance of estimators based on units selected through randomization at two stages, we may follow results given in Theorems 1.3.7 and 1.3.8 as summarized on next page.

$$E(t) = E_1 E_2(t) \tag{9.2.1}$$

$$V(t) = V_1 E_2(t) + E_1 V_2(t) \tag{9.2.2}$$

where E_1 and V_1 are expectation and variance over the first stage and E_2 and V_2 are the conditional expectation and variance over the second stage for a given sample of fsu's.

Let us assume that the population consists of NM elements grouped into N fsu's of M ssu's each. Let n be the number of fsu's in the sample and m the number of ssu's to be selected from each sampled first stage unit. Also we assume that the units at each stage are selected with equal probability. The following notations are used:

$y_{ij} =$ the value obtained for the jth ssu in the ith fsu

$$\bar{Y}_{i\cdot} = \frac{\sum_{i}^{M} y_{ij}}{M} = \text{Mean per element in the } i\text{th fsu}$$

$$\bar{Y} = \frac{\sum_{i}^{N} \bar{Y}_{i\cdot}}{N} = \text{mean per element in the population}$$

$$S_b^2 = \frac{\sum_{i}^{N} (\bar{Y}_{i\cdot} - \bar{Y})^2}{(N-1)} = \text{true variance between first stage unit means}$$

$$S_w^2 = \frac{\sum_{i}^{N} \sum_{i}^{M} (y_{ij} - \bar{Y}_{i\cdot})^2}{N(M-1)} = \text{true variance within first stage units}$$

$$\bar{y}_{i\cdot} = \frac{\sum_{j}^{m} y_{ij}}{m} = \text{sample mean per ssu in the } i\text{th fsu}$$

$$\bar{y} = \frac{\sum_{i}^{n} \bar{y}_{i\cdot}}{n} = \text{overall sample mean per element}$$

THEOREM 9.2.1 If the n fsu's and the m ssu's from each choosen fsu are selected by simple random sampling, wor, $\bar{y}$ is an unbiased estimator of $\bar{Y}$ with sampling variance

$$V(\bar{y}) = \frac{(N-n)}{N} \frac{S_b^2}{n} + \frac{(M-m)}{M} \frac{S_w^2}{mn} \tag{9.2.3}$$

Proof Applying relation (9.2.1) for getting expectation, we have

$$E(\bar{y}) = E_1 E_2(\bar{y}_{i\cdot}/i) = E_1(\bar{Y}_{i\cdot}) = \bar{Y}$$

It shows that the simple mean of all elements in the sample gives an unbiased estimator of the population mean.

To obtain the variance of the estimator, by relation (9.2.2), we have

$$V(\bar{y}) = V_1 [E_2 (\bar{y}/i)] + E_1 [V_2 (\bar{y}/i)]$$

$$= V_1 (\bar{Y}_{i.}) + E_1 \left[\frac{1}{n^2} \sum_{i}^{n} \left(\frac{1}{m} - \frac{1}{M} \right) S_i^2 \right]$$

$$= \frac{(N - n)}{nN} S_b^2 + \frac{(M - m)}{mM} \frac{S_w^2}{n}$$

where $\qquad S_w^2 = \frac{1}{N} \sum_{i}^{N} S_i^2$

If $f_1 = n/N$ and $f_2 = m/M$ are the sampling fractions in the first and second stages, the result can be written as

$$V(\bar{y}) = \frac{(1 - f_1)}{n} S_b^2 + \frac{(1 - f_2)}{nm} S_w^2 \qquad (9.2.4)$$

The variance given by relation (9.2.3) in two-stage sampling is made up of two components. One component comes from the variability of ssu's within fsu's and the second one arises from the variance of fsu's. If the selected fsu's are completely enumerated or, in other words, $m = M$, the variance of the sample mean will be given by the first component only and this situation has already been discussed in Chapter 8. If $n = N$ or, in other words, every fsu in the population is included in the sample, then this case corresponds to stratified sampling with fsu's as strata, and a simple random sampling of m ssu's is drawn from each of the strata.

COROLLARY 1   Under the conditions of Theorem 9.2.1, an unbiased estimator of $V(\bar{y})$ is given by

$$v(\bar{y}) = \frac{(1 - f_1)}{n} s_b^2 + \frac{f_1 (1 - f_2)}{nm} s_w^2 \qquad (9.2.5)$$

where $\qquad s_b^2 = \dfrac{\sum_{i}^{n} (\bar{y}_{i.} - \bar{y})^2}{(n - 1)}$

$$s_w^2 = \frac{\sum_{i}^{n} \sum_{j}^{m} (y_{ij} - \bar{y}_{i.})^2}{n (m - 1)}$$

COROLLARY 2   Show that an unbiased estimator of S_b^2 is given by

$$\hat{S}_b^2 = s_b^2 - \frac{(1 - f_2)}{m} s_w^2 \qquad (9.2.6)$$

COROLLARY 3   If the n fsu's are selected randomly with replacement and the m ssu's from each chosen unit are selected by simple random sampling, wor, $\bar{y}$ is an unbiased estimator of $\overline{Y}$ with sampling variance

$$V(\bar{y}) = \frac{S_b^2}{n} + (1 - f_2) \frac{S_w^2}{mn} \qquad (9.2.7)$$

COROLLARY 4 If the n fsu's are selected randomly, wor, and the m ssu's from each chosen unit are selected randomly, wr, $\bar{y}$, is an unbiased estimator of $\overline{Y}$ with its variance

$$V(\bar{y}) = (1 - f_1) \frac{S_b^2}{n} + \frac{S_w^2}{mn} \qquad (9.2.8)$$

COROLLARY 5   If the n fsu's and m ssu's from each chosen unit are selected by simple random sampling, wr, $\bar{y}$ is an unbiased estimator of $\overline{Y}$ with its variance

$$V(\bar{y}) = \frac{S_b^2}{n} + \frac{S_w^2}{mn} \qquad (9.2.9)$$

COROLLARY 6   If the n fsu's and m ssu's from each chosen unit are selected by simple random sampling, wor, the estimator

$$\hat{Y} = NM \sum_{i}^{n} \frac{\bar{y}_{i\cdot}}{n} \qquad (9.2.10)$$

is an unbiased estimator of the population total Y and its sampling variance is given by

$$V(\hat{Y}) = N^2 M^2 (1 - f_1) \frac{S_b^2}{n} + N^2 M^2 (1 - f_2) \frac{S_w^2}{mn} \qquad (9.2.11)$$

9.3 OPTIMUM ALLOCATION: EQUAL FIRST-STAGE UNITS

As the efficiency of two-stage sampling depends very much upon values of m and n, it is natural to think of their optimal values in a given practical situation. In a two-stage sampling design, a simple cost function of the survey can be written as

$$C = a + nc_1 + nmc_2 \qquad (9.3.1)$$

where a is the overhead cost, c_1 is the cost of including an fsu in the sample and c_2 is the cost of including an ssu in the sample.

In practice, c_1 is likely to be larger than c_2. Hence, a unit increase in n would increase the cost more than a unit increase in m. Let us consider the sampling variance of the estimator also. We find that the total variance in two-stage sampling consists of two parts (1) variance between fsu's and (2) variance within fsu's. In fact, the sampling variance of an estimator, in a two-stage design, can be written as

$$V = A_0 + \frac{1}{n}\left(A_1 + \frac{A_2}{m}\right) \tag{9.3.2}$$

where $$A_0 = \frac{S_b^2}{N}, \quad A_1 = S_b^2 - \frac{S_w^2}{M}, \quad A_2 = S_w^2$$

It can be seen that the variance component due to fsu's decreases with an increase in n only while the variance component due to ssu's decreases with an increase in both n and m.

Thus, the variance and cost functions behave in opposite direction for an increase in n and m and, therefore, it is necessary to determine an optimum allocation for both the values of n and m so that the efficiency per unit of cost is the maximum or the cost per unit is the minimum for a specified value of efficiency.

(a) *When cost is fixed* Suppose that cost C is fixed, say C_0. Using Lagrange's method of undetermined multipliers, we construct the function

$$L(n, m, \lambda) = V(\bar{y}) + \lambda(a + nc_1 + nmc_2 - C_0) \tag{9.3.3}$$

Differentiating L wrt n and m, equating the partial derivatives to zero and eliminating λ, we have

$$m_{opt} = \left(\frac{A_2 c_1}{A_1 c_2}\right)^{1/2} = \frac{S_w (c_1/c_2)^{1/2}}{[S_b^2 - S_w^2/M]^{1/2}} \tag{9.3.4}$$

Substituting the value of m in relation (9.3.1), we get the optimum value of n as

$$n_{opt} = (C_0 - a)\frac{(A_1/c_1)^{1/2}}{(A_1c_1)^{1/2} + (A_2c_2)^{1/2}} \tag{9.3.5}$$

Substituting the values of m_{opt} and n_{opt} in the expression for variance, we get the minimum variance as

$$V_{min}(\bar{y}) = A_0 + \frac{[\sqrt{A_1 c_1} + \sqrt{A_2 c_2}]^2}{(C_0 - a)} \tag{9.3.6}$$

(b) *When variance is fixed* Suppose the variance V of the estimator in two stage sampling is fixed, say V_0. Then, the values of n and m, which minimize the cost are given by Lagrange's method of undetermined multiplier. Applying a similar method as in (a) above, we get

$$m_{opt} = \left(\frac{A_2 c_1}{A_1 c_2}\right)^{1/2}$$

Substituting the values of m in relation (9.3.2), we get the optimum value of n as

$$n_{opt} = \frac{\sqrt{A_1 c_1} + \sqrt{A_2 c_2}}{V_0 - A_0} \sqrt{\frac{A_1}{c_1}} \tag{9.3.7}$$

Substituting the values of m_{opt} and n_{opt} in the expression for cost, we get the minimum cost as

$$C = a + \frac{[\sqrt{A_1 c_1} + \sqrt{A_2 c_2}]^2}{V_0 - A_0} \tag{9.3.8}$$

EXAMPLE 9.1 At an experimental station, there were 100 fields sown with wheat. Each field was divided into 16 plots of equal size (1/16th hectare). Out of 100 fields, 10 were selected by simple random sampling, wor. From each selected field, 4 plots were chosen by random sampling, wor. The yields in kg/plot are given below:

Selected field	Plots			
	1	2	3	4
1	4.32	4.84	3.96	4.04
2	4.16	4.36	3.50	5.00
3	3.06	4.24	4.76	3.12
4	4.00	4.84	4.32	3.72
5	4.12	4.68	3.46	4.02
6	4.08	3.96	3.42	3.08
7	5.16	4.24	4.96	3.84
8	4.40	4.72	4.04	3.98
9	4.20	4.66	3.64	5.00
10	4.28	4.36	3.00	3.52

(i) Estimate the wheat yield per hectare for the experimental station along with its standard error.

(ii) How can an estimate obtained from a simple random sample of 40 plots be compared with the estimate obtained above, in (i)?

(iii) Obtain optimum n and m under cost function $100 = 4n + nm$.

We are given, $N = 100$, $M = 16$, $n = 10$ and $m = 4$.

Calculations have been made as shown below:

S. N.	$\overset{4}{\underset{j}{\Sigma}} \bar{y}_{ij}$	$\bar{y}_{i.}$	$(\bar{y}_{i.} - \bar{y})^2$	$\overset{4}{\underset{j=1}{\Sigma}} y_{ij}^2$	$\bar{y}_{i.}^2$	$(\overset{4}{\underset{j}{\Sigma}} \bar{y}_{ij}^2 - m\bar{y}_{i.}^2)$
(1)	(2)	(3)	(4)	(5)	(6)	(7)
1	17.16	4.290	0.0267	74.091	18.404	0.475
2	17.02	4.255	0.0165	73.565	18.105	1.145
3	15.18	3.795	0.4469	59.733	14.402	2.125
4	16.88	4.220	0.0087	71.925	71.808	0.694
5	16.28	4.070	0.0143	67.009	16.545	0.749
6	14.54	3.635	0.2586	53.511	13.213	0.659
7	18.20	4.550	0.1794	83 950	20.703	1.138
8	17.14	4.285	0.0251	73.800	17.361	0.356
9	17.50	4.375	0.0618	77.605	19.141	1.041
10	15.16	3.790	0.4402	58.718	14.364	1.262
Total		41.265	1.4782	693.908		9.644

(i) An estimate of the average wheat yield, with usual notations, is given by

$$\bar{y} = \frac{1}{n} \overset{n}{\underset{i}{\Sigma}} \bar{y}_{i.} = \frac{41.265}{10} = 4.1265$$

The estimated variance of $\bar{y}$ is

$$v(\bar{y}) = \left(\frac{1}{n} - \frac{1}{N} \right) s_b^2 + \frac{1}{n} \left(\frac{1}{m} - \frac{1}{M} \right) s_w^2$$

Calculating, these values, we get

$$s_b^2 = \frac{1}{n-1} \overset{n}{\underset{i}{\Sigma}} (\bar{y}_{i.} - \bar{y})^2 = \frac{1.4782}{9} = 0.1642$$

and

$$s_w^2 = \frac{1}{n(m-1)} \sum_i^n \sum_j^m (y_{ij} - \bar{y}_{i.})^2 = \frac{9.644}{30} = 0.3215$$

Therefore,

$$v(\bar{y}) = \left(\frac{1}{10} - \frac{1}{100}\right) 0.1642 + \frac{1}{100}\left(\frac{1}{4} - \frac{1}{16}\right) 0.3215$$

$$= 0.0145$$

and standard error of $\bar{y} = \sqrt{0.0145} = 0.120$

(ii) In simple random sampling, the estimate of variance is given by

$$v_{\text{ran}}(\bar{y}) = \left(\frac{1}{nm} - \frac{1}{NM}\right) s^2$$

The estimate of S^2, using a two stage sampling design, can be written as

$$s^2 = \frac{1}{NM-1}\left[M(N-1) s_b^2 + \left\{ N(M-1) - (M-m)\frac{(N-1)}{m}\right\} s_w^2 \right]$$

$$= \frac{1}{1600-1}\left[16 \times 99 \times 0.1642 + \left\{ 100 \times 15 - \frac{99 \times 12}{4}\right\} 0.3215 \right]$$

$$= 0.4045$$

Thus,

$$v_{\text{ran}}(\bar{y}) = \left(\frac{1}{40} - \frac{1}{1600}\right) 0.4045 = 0.0099$$

(iii) The given cost function is of the form $C = c_1 n + c_2 nm$ with $c_1 = 4$, $c_2 = 1$, and $C = 100$. The optimum value of m is given by

$$m_{\text{opt}} = \left[\frac{c_1}{c_2} \frac{s_w^2}{s_b^2 - \frac{s_w^2}{m}} \right]^{1/2}$$

$$= \left[\frac{4}{1} \times \frac{0.3215}{0.1642 - (0.3215/4)} \right]^{1/2} = 4$$

Substituting the value of m in the given cost function, the optimum value of n is given by

$$n_{\text{opt}} = \frac{100}{8} \cong 13$$

9.4 THREE-STAGE SAMPLING WITH EQUAL PROBABILITY

The procedure of two-stage sampling can be carried to a third stage by sampling the ssu's instead of enumerating them completely. For example, in crop surveys for estimating the yield average, a village is considered the first stage sampling unit. Within a selected village, only some of the fields growing the crop are selected and taken as the second-stage units. When a field is selected, only certain parts (called plots) of it are sampled, which may be termed the third-stage units (tsu). Thus the results of three stage sampling can be obtained by extending those of two stage sampling, with further assumptions that each ssu has L third-stage units. It is also assumed that the units are selected with equal probability.

Let y_{ijk} be the value obtained for the kth third-stage unit in the jth second stage unit of the ith first stage unit. The relevant population means per element are as follows:

$$\overline{Y}_{ij} = \sum_k^L \frac{y_{ijk}}{L},$$

$$\overline{Y}_i = \sum_j^M \sum_k^L \frac{y_{ijk}}{LM},$$

$$\overline{Y} = \sum_i^N \sum_j^M \sum_k^L \frac{y_{ijk}}{LMN}$$

$\bar{y}_{ij}$, $\bar{y}_i$ and $\bar{y}$ will denote the corresponding values for the sample. Corresponding population variances will be

$$S_b^2 = \sum_i^N \frac{(\overline{Y}_i - \overline{Y})^2}{(N-1)}$$

$$S_w^2 = \sum_i^N \sum_i^M \frac{(\overline{Y}_{ij} - \overline{Y}_i)^2}{N(M-1)}$$

$$S_u^2 = \sum_i^N \sum_j^M \sum_k^L \frac{(y_{ijk} - \overline{Y}_{ij})^2}{NM(L-1)}$$

THEOREM 9.4.1 If the n fsu's, m ssu's and l ultimate units are chosen by simple random sampling, wor, $\bar{y}$ is an unbiased estimate of $\overline{Y}$ with variance

$$V(\bar{y}) = \frac{(1-f_1)}{n} S_b^2 + \frac{(1-f_2)}{nm} S_w^2 + \frac{(1-f_3)}{nml} S_u^2 \qquad (9.4.1)$$

where $f_1 = n/N, f_2 = m/M, f_3 = l/L$ are the sampling fractions at three stages, respectively. The proof is obvious.

The variance given by relation (9.4.1) is made up of three components corresponding to the three-stages of sampling. The first component is due to the variability of the fsu's, the second to variation of the ssu's and the third to tsu's. If $m = M$ and $l = L$, i.e. each of the n fsu's were completely enumerated, the variance of the sample mean will be given by the first component only, representing the variance of single stage sampling. Similarly, if each of the nm selected second stage units were completely enumerated, i.e. $l = L$, the variance of the sample mean will be given by the first two terms only, representing the variance of two-stage sampling design. In $n = N$ or, in other words, every fsu in the population is included in the sample, the variance of the sample mean will have the last two terms, i.e. it corresponds to a stratified two stage sampling design with the fsu's as strata.

COROLLARY 1   If sampling is done with replacement at every stage, $\bar{y}$ is an unbiased estimator of $\bar{Y}$ with sampling variance

$$V(\bar{y}) = \frac{S_b^2}{n} + \frac{S_w^2}{nm} + \frac{S_u^2}{nml} \qquad (9.4.2)$$

COROLLARY 2   An unbiased estimator of $V(\bar{y})$ is given by

$$v(\bar{y}) = \frac{(1 - f_1)}{n} s_b^2 + \frac{f_1(1 - f_2)}{nm} s_w^2 + \frac{f_1 f_2(1 - f_3)}{nml} s_u^2 \qquad (9.4.3)$$

where s_b^2, s_w^2, s_u^2 are the sample values corresponding to S_b^2, S_w^2, S_u^2, respectively.

COROLLARY 3   With cost function of the form

$$C = a + nc_1 + nmc_2 + nmlc_3, \qquad (9.4.4)$$

and variance function of the form

$$V(\bar{y}) = A_0 + \frac{A_1}{n} + \frac{A_2}{mn} + \frac{A_3}{mnl}$$

The optimum values of l, m and n which minimize the variance are given by

$$\left.\begin{array}{l} l_{\text{opt}} = \left(\dfrac{A_3 c_2}{A_2 c_3}\right)^{1/2} \\[3mm] m_{\text{opt}} = \left(\dfrac{A_2 c_1}{A_1 c_2}\right)^{1/2} \\[3mm] n_{\text{opt}} = \dfrac{(C_0 - a)\,(A_1/c_1)^{1/2}}{(A_1 c_1)^{1/2} + (A_2 c_2)^{1/2} + (A_3 c_3)^{1/2}} \end{array}\right\} \qquad (9.4.5)$$

with the minimum value of variance for fixed C, say C_0

$$V_{\min} = A_0 \frac{[(A_1 c_1)^{1/2} + (A_2 c_2)^{1/2} + (A_3 c_3)^{1/2}]^2}{(C_0 - a)} \qquad (9.4.6)$$

COROLLARY 4 With similar notations as in corrollary 3, the optimum values of l, m and n which minimize the cost, are given by

$$\left. \begin{array}{l} l_{\mathrm{opt}} = \left(\dfrac{A_3 c_2}{A_2 c_3}\right)^{1/2} \\[2ex] m_{\mathrm{opt}} = \left(\dfrac{A_2 c_1}{A_1 c_2}\right)^{1/2} \\[2ex] n_{\mathrm{opt}} = \dfrac{(A_1 c_1)^{1/2} + (A_2 c_2)^{1/2} + (A_3 c_3)^{1/2}}{(V_0 - A_o)(c_1/A_1)^{1/2}} \end{array} \right\} \qquad (9.4.7)$$

with the minimum cost for a specified variance, say V_0,

$$C = a + \frac{[(A_1 c_1)^{1/2} + (A_2 c_2)^{1/2} + (A_3 c_3)^{1/2}]^2}{(V_0 - A_0)} \qquad (9.4.8)$$

9.5 STRATIFIED MULTI-STAGE SAMPLING

The most common design in large-scale surveys is stratified multi-stage sampling. No new principles are involved when the object is to estimate the mean of a population divided into k strata and sampling within each stratum is independent. The population of fsu's is subdivided into k strata. Within each stratum, a sample of fsu's is selected and each of the selected fsu's is further sub-sampled.

Let the hth stratum contain N_h first-stage units, each with M_h second stage units. The corresponding sample numbers being n_h and m_h. The estimator of population mean per second-stage units is given by

$$\bar{y}_{st} = \frac{\sum\limits_{h}^{k} N_h M_h \bar{y}_h}{\sum\limits_{h}^{k} N_h M_h} = \sum\limits_{h}^{k} W_h \bar{y}_h \qquad (9.5.1)$$

where $\bar{y}_h$ is the sample mean in the hth stratum, and $W_h = \dfrac{N_h M_h}{\sum\limits_{h}^{k} N_h M_h}$

is the weight of the stratum in terms of the ssu's.

Applying Theorem 9.2.1 within each stratum, we have

$$V(\bar{y}_{st}) = \sum_h^k W_h^2 \left[\frac{(1 - f_{1h})}{n_n} S_{bh}^2 + \frac{(1 - f_{2h})}{n_h m_h} S_{wh}^2 \right] \qquad (9.5.2)$$

where $f_{1h} = \dfrac{n_h}{N_h}$, $f_{2h} = \dfrac{m_h}{M_h}$

Similarly, an unbiased estimator of sampling variance is given by

$$v(\bar{y}_{st}) = \sum_h^k W_h^2 \left[\frac{(1 - f_{1h})}{n_h} s_{bh}^2 + \frac{f_{1h}(1 - f_{2h})}{n_h m_h} s_{wh}^2 \right] \qquad (9.5.3)$$

To discuss the case of optimum allocation, we can write the cost function as

$$C = \sum_h^k c_{1h} n_h + \sum_h^k c_{2h} n_h m_h \qquad (9.5.4)$$

From relation (9.5.2), the variance may be written as

$$V(\bar{y}_{st}) = \sum_h^k W_h^2 \left[A_0 h + \frac{A_{1h}}{n_h} + \frac{A_{2h}}{n_h m_h} \right]$$

where $A_{0h} = -\dfrac{S_{bh}^2}{N_h}$, $A_{1h} = \dfrac{S_{bh}^2 - S_{wh}^2}{m_h}$, $A_{2h} = S_{wh}^2$

Hence, to minimize V for fixed C, or vice versa, we can apply Lagrange's method of undetermined multipliers. Differentiating partially w.r.t. n_h and m_h and equating to zero, we get

$$m_h = \frac{(A_{2h} c_{1h})^{1/2}}{(A_{1h} c_{2h})^{1/2}} = S_{wh} \frac{(c_{1h}/c_{2h})^{1/2}}{(S_{bh}^2 - S_{wh}^2/M_h)^{1/2}} \qquad (9.5.5)$$

and

$$n_h = n W_h \frac{(A_{1h}/c_{1h})^{1/2}}{\sum W_h (A_{1h}/c_{1h})^{1/2}}$$

$$= \frac{n W_h [(S_{bh}^2 - S_{wh}^2/M_h)/c_{1h}]^{1/2}}{\sum W_h [(S_{bh}^2 - S_{wh}^2/M_h)/c_{1h}]^{1/2}} \qquad (9.5.6)$$

Since $W_h \propto N_h M_h$, we may express

$$n_h \propto \frac{N_h M_h [S_{bh}^2 - S_{wh}^2 / M_h]^{1/2}}{\sqrt{c_{1h}}}$$

$$\propto N_h M_h \frac{S_h'}{\sqrt{c_{1h}}}$$

where

$$S_h' = \left[S_{bh}^2 - \frac{S_{wh}^2}{M_h} \right]^{1/2}$$

It can be seen easily from the above relations that the formula for optimum m_h is exactly the same as in unstratified sampling. Similarly, the optimum value of n_h takes the same form as for uni-stage stratified sampling.

9.6 TWO-STAGE SAMPLING WITH UNEQUAL FIRST STAGE UNITS: ESTIMATORS OF MEAN AND THEIR VARIANCES

Durbin (1953), Des Raj (1966) and Rao (1975) have discussed various estimators of multi-stage sampling at length. This section is devoted to a description of some estimators that are in common use. Let the population under consideration consist of N first-stage units. The ith fsu consists M_i second-stage units. Further, units are selected without replacement, with equal or unequal probabilities. A sample of n fsu's is selected and from the ith selected fsu, a sample of m_i ssu's is selected. Let us denote

$M_i = $ the number of ssu's in the ith fsu, $(i = 1, 2, \ldots, N)$

$M_0 = \sum_i^N M_i = $ the total number of ssu's in the population

$m_i = $ the number of ssu's to be selected from the ith fsu included in the sample

$m_0 = \sum_i^n m_i = $ the total number of ssu's in the sample

$\overline{Y}_i = \sum_j^{M_i} y_{ij} / M_i = $ the ith fsu mean

$$\overline{Y}_N = \sum_i^N \overline{Y}_i / N = \text{the overall mean of fsu means}$$

$$\overline{Y} = \frac{\sum_i^N M_i \overline{Y}_i}{\sum_i^N M_i} = \sum_i^N W_i \overline{Y}_i = \text{the mean per ssu or the population mean per element}$$

There are several estimators of the population mean $\overline{Y}$ but we propose only to study some of the practical methods which are,

$$\bar{y} = \frac{1}{n} \sum_i^n u_i \bar{y}_{i\cdot} = \frac{\sum_i^n M_i \bar{y}_{i\cdot}}{n\overline{M}} \tag{9.6.1}$$

$$\bar{y}_1 = \frac{\sum_i^n \bar{y}_{i\cdot}}{n} \tag{9.6.2}$$

$$\bar{y}_2 = \frac{\sum_i^n M_i \bar{y}_{i\cdot}}{\sum_i^n M_i} \tag{9.6.3}$$

where $\quad \bar{y}_{i\cdot} = \sum_j^{m_i} \frac{y_{ij}}{m_i}, \quad \overline{M} = \frac{M_0}{N}, \quad$ and $\quad u_i = \frac{M_i}{\overline{M}}$

THEOREM 9.6.1 Show that the estimator given by relation (9.6.1) is unbiased and its sampling variance is given by

$$V(\bar{y}) = (1 - f_1) \frac{S_b^2}{n} + \sum_i^N \frac{M_i^2}{nN\overline{M}^2} (1 - f_{2_i}) \frac{S_{w_i}^2}{m_i} \tag{9.6.4}$$

where $\quad S_b^2 = \dfrac{\sum_i^N (u_i \overline{Y}_i - \overline{Y})^2}{(N - 1)}$

and $\quad S_{w_i}^2 = \dfrac{\sum_j^{M_i} (y_{ij} - \overline{Y}_i)^2}{(M_i - 1)}$

Proof To prove that $\bar{y} = \frac{1}{n} \sum_i^n u_i \bar{y}_i.$ is an unbiased estimator, we can write

$$E(\bar{y}) = E_1 \left[\frac{1}{n} \sum_i^n E_2 (u_i \bar{y}_i.|i) \right]$$

$$= E_1 \left[\frac{1}{n} \sum_i^n u_i \bar{Y}_i \right] = \frac{1}{n} \sum_i^n E_1 (u_i \bar{Y}_i) = \bar{Y}$$

The sampling variance of the estimator is given by

$$V(\bar{y}) = V_1 E_2 (\bar{y}|n) + E_1 V_2 (\bar{y}|n)$$

$$= V_1 \left[\frac{1}{n} \sum_i^n u_i \bar{Y}_i \right] + E_1 \left[\frac{1}{n^2} \sum_i^n \frac{M_i^2}{\bar{M}^2} V (\bar{y}_i |n) \right]$$

$$= (1 - f_1) \frac{S_b^2}{n} + \sum_i^N \frac{M_i^2 (1 - f_{2i})}{n N \bar{M}^2} \frac{S_{w_i}^2}{m_i}$$

The units are chosen with equal probability in this method and the contribution made by fsu's to the components of this variance depends upon the variation between the fsu totals. If the units vary considerably in their sizes, this component will be large. The second component of variance is also to be large as there is likely to be positive correlation between M_i and S_{wi}^2. Frequently, this component is so large that this estimator is not preferred.

COROLLARY 1 Show that the estimator $\hat{Y} = \sum_i^n N M_i \frac{\bar{y}_i.}{n}$ estimates unbiasedly the population total Y and its sampling variance will be given by

$$V(\hat{Y}) = N^2 (1 - f_1) \frac{S_b^2}{n} + \sum_i^n (1 - f_{2i}) \frac{N M_i^2}{n \bar{M}^2} \frac{S_{w_i}^2}{m_i} \qquad (9.6.5)$$

COROLLARY 2 An unbiased estimator of variance in relation (9.6.4) is given by

$$v (\bar{y}) = (1 - f_1) \frac{s_b^2}{n} + \sum_i^n (1 - f_{2i}) \frac{u_i^2 s_{w_i}^2}{n N m_i} \qquad (9.6.6)$$

where s_b^2 and s_{wi}^2 are having their usual meanings.

THEOREM 9.6.2 Show that the estimator given by relation (9.6.2) is biased and its bias is given by

$$B = - \sum_i^N (M_i - \bar{M}) \bar{Y}_i / N\bar{M} \tag{9.6.7}$$

and sampling variance by

$$V(\bar{y}_1) = (1 - f_1) \frac{S_b'^2}{n} + \frac{1}{nN} \sum_i^N (1 - f_{2i}) \frac{S_{w_i}^2}{m_i} \tag{9.6.8}$$

where $$S_b'^2 = \frac{\sum_i^N (\bar{Y}_i - \bar{Y})^2}{N - 1}$$ and $S_{w_i}^2$ is as usual.

Proof To prove that $\bar{y}_1$ is a biased estimator, we can get

$$E(\bar{y}_1) = E\left(\sum^n \frac{\bar{y}_{i.}}{n}\right) = E_1 \left[\frac{1}{n} \sum_i^n E_2 (\bar{y}_{i.}|i)\right]$$

$$= E_1 \left[\frac{1}{n} \sum_i^n \bar{Y}_i\right] = \bar{Y}_N \neq \bar{Y}$$

which shows that $\bar{y}_1$ is a biased estimator.

Its bias can be obtained as

$$B = \bar{Y}_N - \bar{Y} = \sum_i^N \frac{\bar{Y}_i}{N} - \sum_i^N \frac{M_i \bar{Y}_i}{N\bar{M}}$$

$$= -\frac{1}{N\bar{M}} \left[\sum_i^N M_i \bar{Y}_i - \sum_i^N \bar{M} \bar{Y}_i\right]$$

$$= -\frac{1}{N\bar{M}} \left[\sum_i^N (M_i - \bar{M}) \bar{Y}_i\right]$$

The sampling variance of the estimator is given by

$$V(\bar{y}_1) = V_1 E_2 (\bar{y}_1|n) + E_1 V_2 (\bar{y}_1|n)$$

$$= V_1 \left[\frac{1}{n} \sum_i^n \bar{Y}_i\right] + E_1 \left[\frac{1}{n^2} \sum_i^n V_2 (\bar{y}_{i.}|i)\right]$$

$$= (1 - f_1) \frac{S_b'^2}{n} + E_1 \left[\frac{1}{n^2} \sum_i^n (1 - f_{2i}) \frac{S_{w_i}^2}{m_i}\right]$$

$$= (1 - f_1) \frac{S_b'^2}{n} + \frac{1}{nN} \sum_i^N (1 - f_{2i}) \frac{S_{w_i}^2}{m_i}$$

The bias in the estimator $\bar{y}_1$ appears due to the fact that the probabilities of selection of the ssu's vary from one unit to another, in the fsu's, due to their unequal sizes. If the M_i's do not vary considerably and the study variate is not correlated with M_i, the bias may not be large. Here, the MSE of $\bar{y}_1$ will consist of three components: one from the bias, one from variation within fsu's, and one arising from variation between the means of the fsu's. The values of m_i's are not specified and a proper choice of m_i can be helpful in controlling these components.

COROLLARY 1   An unbiased estimator of the bias is obtained by

$$\hat{B} = -\frac{(N-1)}{N\overline{M}(n-1)} \sum_i^n (M - \overline{M}')(\bar{y}_{i.} - \bar{y}_1) \qquad (9.6.9)$$

where

$$\overline{M}' = \frac{\sum_i^n M_i}{n}$$

COROLLARY 2   An unbiased estimator of the variance is given by

$$v(\bar{y}_1) = (1 - f_1)\frac{s_b'^2}{n} + \sum_i^n (1 - f_{2i})\frac{s_{w_i}^2}{nNm_i} \qquad (9.6.10)$$

where

$$s_b'^2 = \frac{\sum_i^n (\bar{y}_{i.} - \bar{y}_1)^2}{(n-1)}$$

and

$$s_{w_i}^2 = \frac{\sum_i^{m_i} (\bar{y}_{ij} - \bar{y}_{i.})^2}{(m_i - 1)}$$

THEOREM 9.6.3   Show that the estimator given by relation (9.6.3) is biased and its bias is given by

$$\text{Bias}(\bar{y}_2) = \left(\frac{1}{n} - \frac{1}{N}\right)\overline{Y}\left(S_M^2 - \frac{S_{My}}{\overline{Y}}\right) \qquad (9.6.11)$$

and its sampling variance by

$$V(\bar{y}_2) = \left(\frac{1}{n} - \frac{1}{N}\right)S_b''^2 + \frac{1}{nN}\sum_i^N \frac{M_i^2}{\overline{M}^2}(1 - f_{2i})\frac{S_{w_i}^2}{m_i} \qquad (9.6.12)$$

where
$$S_M^2 = \frac{\sum_i^N \left(\frac{M_i}{\overline{M}} - 1\right)^2}{(N-1)}$$

$$S_{My} = \frac{\sum_i^N \left(\frac{M_i}{\overline{M}} - 1\right)\left(\frac{M_i \overline{Y}_i}{\overline{M}} - \overline{Y}\right)}{(N-1)}$$

$$S_b''^2 = \frac{\sum_i^N M_i^2}{\overline{M}^2} \frac{(\overline{Y}_i - \overline{Y})^2}{(N-1)}$$

and $S_{w_i}^2$ is as usual.

Proof If we take $\sum_i^n \frac{M_i \bar{y}_i.}{\overline{M}} = n\bar{y}$ and $\sum_i^n M_i/\overline{M} = n\bar{u}$ then the estimator $\bar{y}_2$ can be written as the ratio estimator $\bar{y}/\bar{u}$. Applying relations (6.3.2) and (6.4.1), we can find the bias and variance.

Actually $\bar{y}_2$ is the ratio to size estimator, for which knowledge of $\overline{M}$ is not necessary. On similar lines, one may define the regression estimator and its bias, and sampling variance can be derived without any difficulty.

By comparing $V(\bar{y}_2)$ with $V(\bar{y})$, one may conclude that the second term in relation (9.6.12) is identical with the second term in relation (9.6.4). The first term however, is expected to be less than the corresponding term in the same if the fsu's sizes and their totals are positively correlated and the correlation coefficient is greater than half the ratio of their CV's. Similarly, if M_i and $(\overline{Y}_i - \overline{Y})$ are positively correlated and the bias in $\bar{y}_1$ is negligible, then $V(\bar{y}_2)$ will be larger than $V(\bar{y}_1)$. In general, if the M_i's vary considerably, the estimator $\bar{y}_2$ is likely to be more efficient than other estimators, provided n is sufficiently large and M_i is highly correlated with the study variate. In crop surveys in India, an empirical study on relative efficiencies of the three estimators $\bar{y}$, $\bar{y}_1$, and $\bar{y}_2$ was made when it was observed that the simple average $\bar{y}_1$ has the least standard error. Similar results are demonstrated in Example 9.2. The M_i's are found to vary considerably from village to village and the bias is found to be negligible. The estimator $\bar{y}$ was observed to be comparatively less efficient.

COROLLARY An unbiased estimator of the variance $V(\bar{y}_2)$ is given by

$$v(\bar{y}_2) = \left(\frac{1}{n} - \frac{1}{N}\right) \sum_i^n \frac{M_i^2}{\overline{M}^2} \frac{(\bar{y}_i. - \bar{y}_2)^2}{(n-1)}$$

$$+ \frac{1}{nN} \sum_i^n \frac{M_i^2}{\overline{M}^2} (1 - f_{2i}) \frac{s_{2i}^2}{m_i} \qquad (9.6.13)$$

EXAMPLE 9.2 For study of feeding and rearing practices of sheep and yield of wool in the Rajasthan State, during the year 1980-81, two stage sampling design with tehsils as first stage units and villages in the tehsil as second stage units was adopted. The data given below are the stationary sheep population in the selected villages in each of 4 tehsils selected from 12 tehsils of the Ajmer Division, as counted in the survey along with the number of villages in the tehsil.

Selected tehsil	Number of villages in the tehsil (M_i)	Stationary sheep population in the selected villages
Behrar	102	266, 890, 311, 46, 174, 31, 17, 186, 224, 31, 102, 46, 31, 109, 275, 128, 125, 267, 153, 152, 84, 21, 52, 10, 0, 48, 94, 123, 87, 89, 109, 0, 310, 3
Bairath	105	129, 57, 64, 11, 163, 77, 278, 50, 26, 127, 252, 194, 350, 0, 572, 149, 275, 114, 387, 53, 34, 150, 224, 185, 157, 244, 466, 203, 354, 816, 242, 140, 66, 590, 747, 147
Ajmer	200	247, 622, 225, 278, 181, 132, 659, 403, 281, 236, 595, 265, 431, 190, 348, 232, 88, 1165, 831, 120, 987, 938, 197, 614, 187, 896, 330, 485, 60, 60, 1051, 651, 552, 968, 987
Bansur	88	347, 362, 34, 11, 133, 36, 34, 61, 249, 170, 112, 42, 161, 75, 68. 0, 247, 186, 473, 0, 143, 198, 65, 0, 308, 122, 345, 0, 223, 302, 219, 120, 199, 35, 0, 0

Estimate the mean stationary sheep population in the Ajmer Division during the year 1980-81, together with its standard error when $\overline{M} = 124$.

Here we have

$$\bar{M} = 124, \quad M_1 = 102, \quad M_2 = 105, \quad M_3 = 200, \quad M_4 = 88$$
$$n = 4, \quad m_1 = 34, \quad m_2 = 36, \quad m_3 = 35, \quad m_4 = 36$$
$$\bar{y}_1 = 135, \quad \bar{y}_2 = 225, \quad \bar{y}_3 = 471, \quad \bar{y}_4 = 141$$

$$s_{w1}^2 = \frac{1}{33} (1427522 - 619650) = 24481$$

$$s_{w2}^2 = \frac{1}{35} (3219453 - 1822500) = 39912$$

$$s_{w3}^2 = \frac{1}{34} (11451530 - 7764433) = 105346$$

and

$$s_{w4}^2 = \frac{1}{35} (1274076 - 715716) = 15953$$

(i) *First Estimate* An unbiased estimate of the mean of the sheep population is given by

$$\bar{y} = \frac{1}{n\bar{M}} \sum_i^n M_i \bar{y}_i.$$

Thus

$$\bar{y} = \frac{(102 \times 135) + (105 \times 225) + (200 \times 471) + (88 \times 141)}{4 \times 124}$$

$$= \frac{144003}{496} \cong 290$$

The estimate of $V(\bar{y})$ is given by

$$v(\bar{y}) = \left(\frac{1}{n} - \frac{1}{N} \right) s_b^2 + \frac{1}{nN} \sum_i^n W_i^2 \left(\frac{1}{m_i} - \frac{1}{M_i} \right) s_{wi}^2$$

where

$$s_b^2 = \frac{1}{n-1} \sum_i^n (W_i \bar{y}_i. - \bar{y})^2$$

Calculating values,

$$s_b^2 = \frac{1}{4-1} [\{0.67 \times (135)^2 + 0.72 \times (225)^2$$

$$+ 2.60 \times (471)^2 + 0.50 \times (141)^2\} - 4 \times (290)^2]$$

$$= \frac{163054.79}{3} = 54351.60$$

and
$$\frac{1}{nN} \sum_i^n W_i^2 \left(\frac{1}{m_i} - \frac{1}{M_i} \right) s_{wi}^2$$

$$= \frac{1}{48} \left[\left(\frac{102}{124} \right)^2 \times 473 + \left(\frac{105}{124} \right)^2 \times 729 \right.$$

$$\left. + \left(\frac{200}{124} \right)^2 \times 2483 + \left(\frac{88}{124} \right)^2 \times 262 \right]$$

$$= \frac{7433.99}{48} = 154.87$$

Thus,

$$v(\bar{y}) = \frac{54351.60}{6} + 154.87 = 9213.47$$

$\therefore$ Standard error of $\bar{y} = \sqrt{9213.47} = 94.94$

(ii) *Second Estimate*   Another estimate of the mean stationary sheep population is

$$\bar{y}_1 = \frac{1}{n} \sum_i^n \bar{y}_i.$$

$$= \frac{1}{4} \left[\frac{4594}{34} + \frac{8093}{36} + \frac{16492}{35} + \frac{5080}{36} \right] = \frac{972}{4} = 243$$

An estimate of the variance of $\bar{y}_1$ is given by

$$v(\bar{y}_1) = \left(\frac{1}{n} - \frac{1}{N} \right) s_b^2 + \frac{1}{nN} \sum_i^n \left(\frac{1}{m_i} - \frac{1}{M_i} \right) s_{wi}^2$$

where
$$s_{wi}^2 = \frac{1}{m_i - 1} \sum_j^{m_i} (y_{ij} - \bar{y}_i.)^2$$

and
$$s_b^2 = \frac{1}{n-1} \sum_i^n (\bar{y}_i. - \bar{y}_1)^2$$

Here, we have

$$s_b^2 = \frac{1}{3} (310572 - 236196) = 24792$$

and s_{w1}^2, s_{w2}^2, s_{w3}^2, and s_{w4}^2, are as given before.

Thus,

$$v(\bar{y}_1) = \left(\frac{1}{4} - \frac{1}{12}\right) \times 24792 + \frac{1}{4 \times 12}\left[\left(\frac{1}{34} - \frac{1}{102}\right) \times 24481\right.$$

$$+ \left(\frac{1}{36} - \frac{1}{105}\right) \times 39912 + \left(\frac{1}{35} - \frac{1}{200}\right) \times 105346$$

$$\left. + \left(\frac{1}{36} - \frac{1}{88}\right) \times 15953\right]$$

$$= 4214$$

$\therefore$ Standard error of $\bar{y}_1 = \sqrt{4214} = 64.92$

(iii) *Third Estimate* We have $\bar{y}_2$, also an estimate of population mean, which is

$$\bar{y}_2 = \frac{\sum_i^n M_i\,\bar{y}_{i.}}{\sum_i^u M_i} = \frac{\frac{1}{n}\sum_i^n u_i\,\bar{y}_{i.}}{\frac{1}{n}\sum_i^n u_i} = \frac{\bar{y}}{\bar{u}}$$

We calculate, $\quad \bar{u} = \dfrac{\sum_i^n M_i}{n\bar{M}} = \dfrac{1}{4}(0.82 + 0.85 + 1.61 + 0.71)$

$$= 0.998$$

Hence $\quad \bar{y}_2 = \dfrac{\bar{y}}{\bar{u}} \cong 291$

Also, estimate of $V(\bar{y}_2)$ is

$$v(\bar{y}_2) = \left(\frac{1}{n} - \frac{1}{N}\right)s_b''^2 + \frac{1}{nN}\sum_i^n u_i^2\left(\frac{1}{m_i} - \frac{1}{M_i}\right)s_{wi}^2$$

where $\quad s_b''^2 = \displaystyle\sum_i^n \frac{u_i^2\,(\bar{y}_{i.} - \bar{y}_2)^2}{(n-1)}$

Here $\quad s_b''^2 = \dfrac{1}{(4-1)}[0.67\,(135-291)^2 + 0.72\,(225-291)^2$

$$+ 2.60\,(471 = 291)^2 + 0.05\,(141 - 291)^2]$$

$$= 38401.87$$

and the second part of the variance is as in (i).

Thus
$$v(\bar{y}_2) = \frac{38401.87}{6} + 154.87$$
$$= 6555.19$$

$\therefore$ Standard error $\bar{y}_2 = \sqrt{6555.19} = 80.86$

9.7 OPTIMUM ALLOCATION: UNEQUAL FIRST-STAGE UNITS

In section 9.3, we have discussed the optimum allocation when the first-stage units are of equal sizes. The cost and variance functions were seen to have opposite behaviours to an increase in n and m, and it is, therefore, necessary to consider these functions before the optimum allocations in the fsu's and the ssu's are discussed.

Cost Function In a two-stage sampling design, the cost of survey in a simple form can be written as

$$C = a + nc_1 + c_2 \sum_i^n m_i \qquad (9.7.1)$$

where a is the overhead cost, c_1 the average cost of selection per fsu, and c_2 the average cost of sampling per ssu.

In practice, c_1 is likely to be larger than c_2. Hence, a unit increase in n increases the cost as compared to a unit increase in m_i. Thus, the second component of cost function will vary from sample to sample for given n fsu's and, therefore, it becomes necessary to examine the question of optimum allocation of the total sample size in terms of the ssu's to the selected fsu's so the average number of sample ssu's per selected fsu is m. This problem has been discussed by Rangarajan (1957) and Rao (1961). They have given different methods by taking $m_i = r_i m$ with m as the average value, or $m_i = \lambda M_i$ with λ some positive constant, etc. Here we shall consider the average cost for the present discussion, which can be written as

$$C = a + nc_1 + \frac{nc_2}{N} \sum_i^N m_i \qquad (9.7.2)$$

Variance Function From the expressions for sampling variance of the estimators considered in the previous section, we can say that the total variance in two-stage sampling can be written in the form

$$V(\bar{y}) = A_0' + \frac{1}{n}\left(A_1' + \sum_i^N \frac{A_{2i}'}{m_i}\right) \qquad (9.7.3)$$

where A_0' is a constant term independent of n and m, and A_1' and A_{2i}' are functions of population parameters analogous to A_1 and A_2, independent of sample sizes n and m_i. If $m_i = r_i m$, where m is the average number of units selected per ssu, i.e. $m = E(m_i)$ and r_i depends on the method of determinating the set of values of m_i, then relation (9.7.3) takes the form

$$V(\bar{y}) = A_0' + \frac{1}{n}\left(A_1' + \frac{A_2'}{m}\right) \tag{9.7.4}$$

An increase in n and m plays a significant role for the variance decreases as n and m increases. Singh (1958) has discussed the behaviour of variance and efficiency when numbers of fsu's and ssu's are fixed. A guiding principle, for optimum allocation, is to minimize the variance for a fixed cost, i.e. the efficiency per unit of the cost is maximum, or to minimize the cost for a specified value of variance, i.e., the cost per unit is minimum for a specified value of efficiency. We shall discuss in brief both the cases.

(a) *When Cost is Fixed* Suppose the cost is fixed, say C_0 and the estimator $\bar{y}$ is used. Proceeding on the lines discussed in section 9.3, we have

$$m_{i(opt)} = \left(\frac{A_2' c_1}{A_1' c_2}\right)^{1/2} = \left(\frac{c_1 u_i^2 S_i^2}{c_2 \Delta}\right)^{1/2} \tag{9.7.5}$$

where

$$\Delta = S_b^2 - \sum_i^N \frac{u_i S_i^2}{N\overline{M}} \tag{9.7.6}$$

$$n_{opt} = \frac{(C_0 - a)(A_1'/c_1)^{1/2}}{(A_1' c_1)^{1/2} + (A_2' c_2)^{1/2}} \tag{9.7.7}$$

Substituting the values of $m_{i(opt)}$ and n_{opt} in the variance expression, we get the minimum variance as

$$V_{min}(\bar{y}) = A_0' + \frac{(\sqrt{A_1' c_1} + \sqrt{A_2' c_2})^2}{(C_0 - a)} \tag{9.7.8}$$

Suppose m_i is so determined as to be proportional to M_i, say $m_i = \lambda M_i$, where λ is some positive constant and can be obtained as

$$\lambda = \left(\frac{c_1}{c_2} \sum_i^N \frac{u_i S_i^2}{\Delta N\overline{M}^2}\right)^{1/2} \tag{9.7.9}$$

When the ssu's are of equal sizes, the optimum values of m and n are given by

$$m_{\mathrm{opt}} = \left[\frac{c_1}{c_2}\frac{S_w^2}{(S_b^2 - S_w^2/M)}\right]^{1/2} = \left[\frac{c_1}{c_2}\frac{A_2}{A_1}\right]^{1/2} \qquad (9.7.10)$$

$$n_{\mathrm{opt}} = \frac{(C_0 - a)}{(c_1 + c_2\, m_{\mathrm{opt}})} \qquad (9.7.11)$$

which are comparable with the values given in relations (9.3.7) and (9.3.5).

(b) *When Variance is Specified* Suppose the variance is specified at a given value, say V_0. By a similar procedure, we get the same value of $m_{i(\mathrm{opt})}$ as in relation (9.7.5). Further, we can have

$$n_{\mathrm{opt}} = \frac{(A_1'\, c_1)^{1/2} + (A_2'\, c_2)^{1/2}}{(V_0 - A_0')}\left(\frac{A_1'}{c_1}\right)^{1/2} \qquad (9.7.12)$$

Substituting the values of $m_{i(\mathrm{opt})}$ and n_{opt} in the cost function, we get the minimum cost as

$$C = a + \frac{(\sqrt{A_1'\, c_1} + \sqrt{A_2'\, c_2})^2}{V_0 - A_0'} \qquad (9.7.13)$$

In practice, the sampler will have to take into consideration other factors such as administrative and operational convenience in the field in addition to the variance–cost approach. In large scale surveys, field operations will always have a dominating role and the sampler will decide the choice himself.

9.8   TWO-STAGE pps SAMPLING

In the preceding sections, the theory for multi-stage sampling with equal probabilities of selection at each stage of sampling was discussed. When the fsu's are large and differ considerably in their sizes, it is desirable to select them with pps, sizes being M_i's. A system of sampling involving the use of varying probabilities has been used by Hansen and Hurwitz (1943, 1949). Singh (1954) compared two estimators in a two-stage sampling design where fsu's are selected with varying probability, wor, and ssu's with equal probability, wr, and wor, methods. An alternative estimate was suggested by Rao (1966). A simple method of variance estimation has been given by Durbin (1967). Brewer and Hanif (1970) have improved and extended the method to other estimators.

Suppose a sample of n fsu's is selected with pps, wr. From the selected ith fsu, a selection of m_i ssu's is made with simple random sampling, wor. If the ith fsu is selected more than once, then a fresh independent drawing of m_i ssu's is being made without replacement from the complete fsu each time. An unbiased estimator of Y is given by

$$\hat{Y}_{\text{pps}} = \frac{1}{n} \sum_i^n \frac{M_i y_i.}{p_i} \tag{9.8.1}$$

where p_i is the probability of selecting the ith fsu at each draw such that $\sum_i^N p_i = 1$, and $\bar{y}_i. = \dfrac{\sum_j^{m_i} y_{ij}}{m_i}$.

The sampling variance of the estimator is given by

$$V(\hat{Y}_{\text{pps}}) = \frac{1}{n} \sum_i^N p_i \left(\frac{Y_i}{p_i} - Y \right)^2 + \frac{1}{n} \sum_i^N \frac{M_i^2}{p_i} (1 - f_{2i}) \frac{S_{wi}^2}{m_i} \tag{9.8.2}$$

An unbiased estimator of $V(\hat{Y}_{\text{pps}})$ is

$$v(\hat{Y}_{\text{pps}}) = \frac{\sum_i^n \left(\dfrac{M_i \bar{y}_i.}{p_i} - \hat{Y} \right)^2}{n(n-1)} \tag{9.8.3}$$

which gives a good procedure of estimation, whatever the method of selection adopted at the second stage, provided the fsu's are selected with replacement. If one is interested in estimating the between and within components of variance, then the between component can be obtained by subtracting the within component from relation (9.8.3). The within fsu variance component can be estimated unbiasedly by

$$v_w(\hat{Y}_{\text{pps}}) = \sum_i^n \frac{M_i^2 (1 - f_{2i})}{n^2 p_i^2} \frac{s_{wi}^2}{m_i} \tag{9.8.4}$$

9.9 SELF-WEIGHTING DESIGNS

In large scale surveys, the estimator defined in relation (9.8.1) involves a term $M_i/np_i m_i$ which will vary from unit to unit. Due to the complexity of its varying nature, analysis of data will be quite cumbersome. In this situation, the technique of self-weighting samples described in section 3.7.2 will prove very useful. A design which provides a single common weight to all sample units is known as a *self-weighting design*, also called *equi-weighting design*. A design can be made self-weighting either at the field or estimation stage. If the selection of units is so done as to make all the weights equal to one another, the design is termed *self-weighting design at field stage*. For example, a multi-stage design can be made self-weighting at field stage by appropriate choice of the number of ultimate stage units to be chosen in the final stage of sampling. If self-weighting is achieved by adopting some devices at the estimation stage, the design is termed *self-weighting design at estimation stage*.

Usually a design should be made self-weighting at the field stage itself as it may be desirable from the point of view of both cost and operational convenience. For example, in crop-cutting surveys on paddy crop, it may be difficult to have the frame of all paddy-growing fields in a block, whereas the frame of all villages in the block may be easily available. Also, construction of the frame of the fields growing paddy crop in the entire block will involve a rise in cost, manpower and efforts; while these factors are minimum in selecting villages and then fields in the selected villages. However, it may not always be possible to arrive at such a design because, in practice, it is difficult to have only one common weight for all sample units. In such cases, two or more common weights are used, provided the number of such common weights is fairly small. Sometimes it may be necessary to make the design self-weighting at the estimation stage. Various methods are available for this purpose. In this section, both situations of making designs self-weighting at field and estimation stages, are considered.

The procedure of making the design self-weighting at the field stage has been considered by Hansen *et al* (1953) and Lahiri (1954). The problem of making the design self-weighting at the estimation stage has been discussed by Murthy and Sethi (1959, 1961). Som (1959) has given a procedure for making a stratified two-stage self-weighting design with equal numbers of ultimate units selected in each of the selected fsu's in the ultimate stage of sampling.

In stratified sampling, we have seen that proportional allocation is one of the class of designs which leads to self-weighting samples. In other words, if

$$n_i = \frac{n N_i}{N} \tag{9.9.1}$$

or $f_i = f$ the design will be self-weighting.

In the case of the two-stage pps sampling design considered in section 9.8, it can be seen from relation (9.8.1) that the design will be self-weighting if m_i's are taken so that

$$\frac{M_i}{np_i m_i} = k,$$

i.e.
$$m_i = \frac{M_i}{knp_i} \tag{9.9.2}$$

Further, if the average number of second stage units to be chosen from the selected fsu is m,

$$E(m_i) = \sum_i^N \frac{M_i}{knp_i} p_i = \sum_i^N \frac{M_i}{kn}$$

or
$$m = \frac{N\bar{M}}{kn}$$

giving
$$k = \frac{N\bar{M}}{nm} \tag{9.9.3}$$

which is the overall weight to be given to the ssu's in the sample.

Similarly, we can see that the estimator in relation (9.6.2) is self-weighting if $m_i = $ constant. When m_i and $\bar{Y}_i$ are uncorrelated, this estimator may be quite satisfactory. It should also be noted that its bias does not vanish even with large sample size. Also the estimator in relation (9.6.1) becomes self-weighting if

$$f_{2i} = \frac{m_i}{M_i} = \text{constant} = f_2 = \frac{\bar{m}}{\bar{M}} = \frac{N\bar{m}}{M_0} \tag{9.9.4}$$

In this event, the variance component within fsu's can be expressed in a more simplified form by putting

$$m_i = \frac{N\bar{m} M_i}{M_0} \tag{9.9.5}$$

Thus, we may conclude that the self-weighting system is not only very convenient for making computations but is also efficient from the operational point of view. In large scale surveys, it is always desirable that the weight is the same for every unit so that (i) tabulation becomes simpler, (ii) analysis becomes easier, and (iii) cost is minimized. Another advantage of the self-weighting design is that it gives a constant sample size from each selected fsu. Thus, the investigators will not be responsible for varying work loads in different fsu's. The only disadvantage of the system is that some of these methods may result in biased estimates though, in some cases, with smaller variance.

EXAMPLE 9.3 For estimating the total stationary sheep population in the Ajmer Division of Rajasthan during 1980-81, the design of survey was two-stage sampling with tehsils as first stage units and villages in the tehsil as second stage units. Tehsils were selected with replacement and with probability proportional to the sheep population as recorded in the 1976 livestock census, whereas the villages in a tehsil were selected with equal probability and without replacement. The data are given showing the stationary sheep population in selected villages of the Ajmer division, as counted in the second round of the survey.

Selected Tehsil	No. of villages in the tehsil (M_i)	Selection probability of the tehsil (p_i)	Stationary sheep population in the selected villages
Behrar	102	0.008568	266, 174, 224, 66, 109, 267, 21, 48, 87, 890, 31, 31, 102, 275, 153, 52, 94, 89, 311, 17, 108, 46, 128, 152, 10, 123, 109, 46, 186, 128, 39, 126, 84, 0
Bairath	105	0.015079	129, 163, 26, 350, 275, 34, 157, 354, 66, 57, 77, 127, 0, 114, 150, 244, 816, 590, 64, 278, 252, 572, 387, 224, 466, 242, 747, 11, 50, 194, 149, 53, 185, 203, 140, 174
Ajmer	200	0.073556	247, 181, 403, 265, 232, 130, 197, 330, 1051, 622, 987 281, 431, 88, 987, 614, 485, 651, 225, 132, 236, 190, 1165, 938, 187, 60, 552, 278, 650, 595, 348, 831, 968, 895, 60, 570
Bansur	88	0.012632	347, 133, 249, 161, 247, 143, 308, 223, 120, 362, 36, 170, 75, 186, 198, 122, 302, 199, 34, 34, 112, 68, 473, 65, 345, 219, 35, 11, 61, 42, 0, 0, 0, 0

Estimate the total sheep population in the Ajmer Division during the year 1980-81 along with its standard error.

The calculations were done as shown below:

Tehsil No.	M_i	p_i	m_i	$\sum_j y_{ij}$	$M_i \sum_j y_{ij}$	$m_i p_i$	$\sum_j \dfrac{M_i \, y_{ij}}{m_i \, p_i}$
1	102	0.008568	34	4592	468384	0.291312	1,607,843.1372
2	105	0.015079	36	8120	852600	0.542844	1,570,616.9728
3	200	0.073556	36	17062	3412400	2.648016	1,288,662.9084
4	88	0.012632	34	5080	447040	0.429488	1,040,867.2652
							5,507,290.2836

An unbiased estimate of the total sheep population is given by

$$\hat{Y}_{pps} = \frac{1}{n} \sum_i^n \sum_j^{m_i} \frac{M_i \, y_{ij}}{m_i \, p_i}$$

$$= \frac{1}{4} \times 5508290.2836$$

$$= 1377072.57$$

The estimate of $V(\hat{Y}_{pps})$ is given by

$$v\,(\hat{Y}_{pps}) = \frac{1}{n\,(n-1)} \left[\sum_i^n \left(\sum_j^{m_i} \frac{M_i \, y_{ij}}{m_i \, p_i} \right)^2 \right.$$

$$\left. - \frac{1}{n} \left(\sum_i^n \sum_j^{m_i} \frac{M_i \, y_{ij}}{m_i \, p_i} \right)^2 \right]$$

$$= \frac{1}{4 \times 3} (7796053512516 - 7584488460025)$$

$$= 17603404380.9167$$

$\therefore$ Standard error of $\hat{Y}_{pps} = \sqrt{17630404380.9167} = 132779.54$

and percentage relative standard error $= \dfrac{132779.54}{1377072.57} \times 100 = 9.61$

9.10 THREE-STAGE pps SAMPLING

We have considered the two-stage pps sampling design in the previous section which can be extended to three or more stages. Let a sample of nml units be selected in three stages by adopting pps, wr, at each stage. Suppose n fsu's are selected with p_i probabilities of selection for the ith fsu's $(i=1,\ldots,N)$. From each selected fsu, m, ssu's are selected with p_{ij}, probabilities of selection for the jth ssu $(j=1,\ldots,M_i)$ and from each selected ssu, l third stage units (tsu's) are selected with p_{ijk} probabilities of selection of the kth tsu of the jth ssu in the ith fsu $p(k=1,\ldots,L_{ij})$. Let y_{ijk} denote the value of the kth tsu in the jth ssu of the ith fsu $(i=1,\ldots,n; j=1,\ldots,m; k=1,\ldots,l)$ in the sample. An estimator of the population total Y can be defined as

$$\hat{Y} = \frac{1}{nml} \sum_i^n \frac{1}{p_i} \sum_j^m \frac{1}{p_{ij}} \sum_k^l \frac{y_{ijk}}{p_{ijk}} \tag{9.10.1}$$

It can easily be shown that the estimator is unbiased. i.e.

$$E(\hat{Y}) = E_1 E_2 E_3(\hat{Y}) = Y$$

and the variance of the estimator is obtained by

$$V(\hat{Y}) = V_1 E_2 E_3(\hat{Y}) + E_1 V_2 E_3(\hat{Y}) + E_1 E_2 V_3(\hat{Y})$$

Thus, $$V(\hat{Y}) = \frac{1}{n}\left(\sum_i^N \frac{Y_i^2}{p_i} - Y^2\right) + \frac{1}{nm}\sum_i^N \frac{1}{p_i}\left(\sum_i^{M_i} \frac{Y_{ij}^2}{p_{ij}} - Y_i^2\right)$$

$$+ \frac{1}{nml}\sum_i^N \frac{1}{p_i}\sum_i^{M_i}\frac{1}{p_{ij}}\left(\sum_k^{L_{ij}} \frac{y_{ijk}^2}{p_{ijk}} - Y_{ij}^2\right) \tag{9.10.2}$$

An unbiased estimator $V(\hat{Y})$ is given by

$$v(\hat{Y}) = \frac{1}{n(n-1)}\left(\sum_i^n y_i^2. - \frac{\hat{Y}^2}{n}\right) \tag{9.10.3}$$

where $$y_i. = \frac{1}{p_i}\sum_j^m \frac{1}{p_{ij}}\left(\sum_k^l \frac{y_{ijk}}{p_{ijk}}\right)$$

It should be noted that, like two-stage sampling, the sampling variance function in three-stage sampling can also be written as

$$V(\hat{Y}) = \frac{A_1}{n} + \frac{A_2}{nm} + \frac{A_3}{nml}$$

Similarly, the cost function can also be written in the form

$$C = a + nc_1 + nmc_2 + nmlc_3 \tag{9.10.4}$$

where a is the overhead cost and c_1, c_2 and c_3 have their usual meanings.

(i) If the cost of survey is fixed, say C_0, the optimum values of n, m and l are given by

$$
\left.
\begin{aligned}
l_{opt} &= \left(\frac{A_3c_2}{A_2c_3}\right)^{1/2} \\[2mm]
m_{opt} &= \left(\frac{A_2c_1}{A_1c_2}\right)^{1/2} \\[2mm]
n_{opt} &= \frac{(C_0 - a)\,(A_1/c_1)^{1/2}}{(A_1c_1)^{1/2} + (A_2c_1)^{1/2} + (A_3c_3)^{1/2}}
\end{aligned}
\right\}
\qquad (9.10.5)
$$

Substituting these values of l, m, and n, we get the minimum variance as

$$
V_{\min}(\hat{Y}) = \frac{\{(A_1c_1)^{1/2} + (A_2c_2)^{1/2} + (A_3c_3)^{1/2}\}^2}{(C_0 - a)}
\qquad (9.10.6)
$$

(ii) If the variance is specified, say V_0, the optimum values of n, m and l are given by

$$
\left.
\begin{aligned}
l_{opt} &= \left(\frac{A_3c_2}{A_2c_3}\right)^{1/2} \\[2mm]
m_{opt} &= \left(\frac{A_2c_1}{A_1c_2}\right)^{1/2} \\[2mm]
n_{opt} &= \frac{(A_1c_1)^{1/2} + (A_2c_2)^{1/2} + (A_3c_3)^{1/2}}{V_0(c_1/A_1)^{1/2}}
\end{aligned}
\right\}
\qquad (9.10.7)
$$

Substituting these optimum values of l, m and n, we get the minimum cost as

$$
C = a + \frac{1}{V_0}\left(\sqrt{A_1c_1} + \sqrt{A_2c_2} + \sqrt{A_3c_3}\right)^2
\qquad (9.10.8)
$$

SET OF PROBLEMS

9.1 Define multi-stage sampling and write its advantages over other sampling schemes. Write an unbiased estimator of the population total and derive its sampling variance.

9.2 What is meant by multi-stage sampling? Obtain an expression for the variance of the estimator of the population total for a suitable three-stage sampling design when units are of unequal size at each stage of sampling. Give the structure of the analysis of variance in three stage sampling and explain how the analysis can be used in planning similar surveys subsequently.

9.3   Suppose n fsu's are selected with pps, wr, and, from each selected fsu, m ssu's are selected with simple random sampling, wor. Give an unbiased estimator of the population total Y and derive an unbiased estimator of the sampling variance of the estimator.

9.4   If f_1 and f_2 are negligible in equal fsu two-stage sampling with equal size fsu and the cost function is linear, show that $m = 2$ gives a smaller value of $V(\bar{y})$ than $m = 1$ if

$$\frac{c_1}{c_2} > \frac{2S_b^-}{S_w^2}$$

9.5   If ρ is the correlation coefficient between ssu's in the same fsu, prove that

$$\frac{\rho}{1-\rho} = \frac{\left[(N-1)\dfrac{S_b^2}{N} - \dfrac{S_w^2}{M}\right]}{S_w^2}$$

9.6   A population consists of N fsu's, each containing M ssu's. To estimate the proportion P of units possessing a given attribute, n fsu's are selected by simple random sampling, wor, and from each selected fsu, m ssu's are selected by simple random sampling, wor. If p_i be the proportion of attribute in the ith selected fsu,

show that $\hat{P} = \overset{n}{\underset{i}{\Sigma}} p_i/n$ is an unbiased estimator of the population proportion P.

Obtain the variance of the estimator and show that an unbiased estimator of $V(\hat{P})$ is given by

$$v(\hat{P}) = (1 - f_1) \overset{n}{\underset{i}{\Sigma}} \frac{(p_i - \hat{P})^2}{n(n-1)} + (1 - f_2) \overset{n}{\underset{i}{\Sigma}} \frac{p_i(1-p_i)}{Nn(m-1)}$$

where f_1 and f_2 have their usual meanings.

9.7   In a pilot survey with two-stage sampling design, m' ssu's were chosen from each of n' fsu's. Estimate $V(\bar{y})$, when another sample of n fsu's is drawn from n' and, from each selected fsu, a selection of m ssu's is made. Show that an unbiased estimator of $V(\bar{y})$ is given by

$$v(\bar{y}) = (1 - f_1) \frac{s_b^2}{n} + \left[1 - \frac{m}{m'}(1-f_1) - f_1 f_2\right] \frac{s_w^2}{nm}$$

where s_b^2 and s_w^2 are obtained from the pilot sample.

9.8   Define a self-weighting design and discuss briefly its advantages and disadvantages. Show that a two-stage design, where n villages are selected with probability proportional to the number of households in them, in the first stage, and m households are selected with equal probability without replacement in the second stage, from each selected village, is self-weighting. Derive an unbiased estimator of variance of the self-weighting estimator.

9.9   A population is divided into k strata with M_i fsu's in the ith stratum ($i = 1, \ldots, k$). Each fsu contains N ssu's. A random sample of m fsu's is selected from each stratum and a random sample of n ssu's is taken for investigation in each of the selected fsu's. How will you obtain an unbiased estimate of the population total of a character from the sample? Derive a formula for estimation of the difference between the sampling variance of this estimated total and that of a linear unbiased estimate of the same population total which could have been obtained from an unstratified random sample of km fsu's with n ssu's taken up for investigation within each.

9.10 A sample of n fsu's is selected with simple random sampling, wor, and from each selected fsu a constant fraction f_2 of ssu's is taken. If r_i out of the m_i ssu's in the ith fsu possess an attribute, show that the estimator ratio to size $p(=\Sigma r_i/\Sigma m_i)$ estimates the population proportion of the attribute and an estimate of variance is given by

$$v(p) = \frac{(1 - f_1)}{n\overline{M}^2} \sum_i^n M_i^2 \frac{(p_i - p)^2}{n - 1} + \frac{f_1(1 - f_2)}{n^2 \, \overline{m} \, \overline{M}} \sum_i^n \frac{M_i m_i}{m_i - 1} \, p_i(1 - p_i)$$

where $p_i = \dfrac{r_i}{m_i}$

(Cochran, 1977)

9.11 In two-stage sampling, n fsu's are selected with pps, wr, and from the ith selected fsu having M_i units, m_i ssu's are selected with simple random sampling, wr. To estimate the population total Y, the sub-sampling numbers m_i are to be fixed such that (i) expected value of m_i is fixed at m, or (ii) total number of sample ssu's are fixed at m_0. Obtain the optimum values of m_i in both cases so that the variance of the estimator is minimum. Also compare their minimum variances. (Rangarajan, 1957)

9.12 In two-stage sampling, n fsu's are selected with pps, wr. If the ith fsu occurs r_i times in the sample, one of the following procedures may be adopted for second stage sampling:

(i) $r_i m_i$ ssu's are selected with simple random sample, wor;

(ii) r_i independent samples of m_i ssu's (drawn with simple random sample, wor) are taken; and

(iii) m_i units are selected without replacement and observations are weighted by r_i.

Obtain unbiased estimators of the population total Y and their sampling variances for all these cases. If V_1, V_2 and V_3 are the variances of the estimators, show that, for the same expected sample size,

$$V_1 \leqslant V_2 \leqslant V_3$$

(Rao, 1961)

9.13 In two-stage sampling, n fsu's are selected with pps, wor. From each sampled fsu, m ssu's are selected with simple random sampling, wor. Suggest an unbiased estimator of the population total. An estimator of the population total is $\hat{Y} = \sum_i^N \beta_i \hat{Y}_i$, where $\hat{Y}_i$ is an unbiased estimator of the ith fsu and β_i's are real numbers, predetermined for every sample with the restriction that β_i is zero if the ith fsu is not included in the sample. Show that the estimator is unbiased. Derive an unbiased estimator of variance. (Des Raj, 1966)

9.14   The following sampling schemes for estimating the population mean of a characteristic were considered:

(i) The population is divided into N clusters of M units each and two-stage sampling is adopted where n clusters and m units from each sampled cluster are selected with simple random sampling, wr, and

(ii) the population is divided into clusters of m' units each and a sample of n' such clusters is selected with simple random sampling, wr.

Show that, in both the cases, the sample mean is an unbiased estimator of the population mean and derive the variances in both cases. Derive the condition that the efficiencies of these two schemes be the same when $nm = n'm'$.

(Singh. D., 1956)

9.15   In a sample survey for estimating the number of standards of pepper in a Tehsil with 72 villages, a sample of 12 villages was selected with srs, wor, and from each selected village 5 clusters of 20 fields each were drawn with srs, wor. Data on the number of clusters in the sample villages and on the number of standards in the sample clusters are given below:

Sample village	No. of clusters	Number of standards in sample clusters				
		1	2	3	4	5
1	27	430	402	363	975	389
2	24	586	1234	100	368	344
3	14	1164	546	3060	1724	1274
4	116	693	218	836	1218	575
5	25	191	270	4502	4184	243
6	118	1036	1333	1179	728	1957
7	147	1555	254	950	382	355
8	36	910	452	129	122	243
9	91	340	0	92	28	340
10	171	57	59	0	0	21
11	86	159	45	242	1075	539
12	88	84	462	147	16	10

Estimate unbiasedly the total number of standards in the Tehsil and also obtain its standard error.

9.16   Raw wool contains varying amounts of grease, dirt and other impurities and its quality is measured by the percentage by weight of clean wool, termed clean content. To estimate the clean content, an electrical core-boring machine is used which takes cores of about 1/4 lb from a bale, which are then subjected to laboratory analysis. In an experiment, 6 bales were drawn from a large lot with equal probability and from each bale 4 cores were taken at random and the clean contents determined. The results of this experiment are given on next page.

Core	Sample bales					
	1	2	3	4	5	6
1	54.3	57.0	54.6	54.9	59.9	57.8
2	56.2	58.7	57.5	60.1	57.8	59.7
3	58.9	58.2	59.3	58.7	60.9	59.6
4	55.5	57.1	57.5	55.6	57.5	58.1

(i) Estimate the average clean content of wool for the lot and also obtain an estimate of its standard error.

(ii) Obtain the efficiency of sampling 12 bales and 2 cores from each bale as compared to that of the above scheme.

9.17 To estimate the total yield of paddy in a district, a stratified two-stage sampling design was adopted where 4 villages were selected from each stratum, with pps, wr, size being geographical area. Four plots were drawn from each sample village circular, systematically, for ascertaining the yield of paddy. The following data give the yield of paddy for the sample plots:

Stratum	Sample village	Inverse of probability	Total number of plots	Yield of paddy (kg)			
				1	2	3	4
1	1	440.21	28	104	182	148	87
	2	660.43	84	108	64	132	156
	3	31.50	240	100	115	50	172
	4	113.38	76	346	350	157	119
2	1	21.00	256	124	111	135	216
	2	16.80	288	123	177	106	138
	3	24.76	222	264	78	144	55
	4	49.99	69	300	114	68	111
3	1	67.68	189	110	281	120	114
	2	339.14	42	80	61	118	124
	3	100.00	134	121	212	174	106
	4	68.07	161	243	116	314	129

Estimate the total yield of paddy and obtain an estimate of its standard error.

9.18 A crop cutting survey by the method of stratified multi-stage random sampling was carried out in one district, on jute crop, for estimating the average yield of green weight of jute for the district, with its three administrative sub-divisions constituting the strata. In each administrative sub-division, a specified number of villages were selected at random within each selected village. Three fields under jute were chosen at random out of the total number of fields under jute in the village. In each field, a plot of 1/160th acre was located, harvested and the green weight of jute recorded in kg. The data obtained are shown below:

Sub-division	Total area under jute in acres	Yields of green weight of jute in kg per plot for villages and fields selected
1	5089	86, 85, 57, 81, 71, 92, 72, 37, 51, 81, 50, 43, 78, 71, 79
2	4133	86, 45, 81, 55, 56, 55, 91, 70, 64, 19, 62, 41
3	3007	81, 8, 43, 67, 48, 47, 35, 34, 37

Estimate the average yield of green weight of jute in kg per acre for the district and calculate its standard error.

REFERENCES

Brewer, K.W.R. and M. Hanif, "Durbins' new multi-stage variance estimator," *J.R. Statist. Soc.*, **32**B, 302-311, (1970).

Cochran, W.G., "The use of analysis of variance in enumeration by sampling," *J. Amer. Statist. Assoc.*, **34**, 492-510, (1939).

———*Sampling Techniques*, Third Edition, John Wiley and Sons, New York, (1977).

Des Raj, "Some remarks on a simple procedure of sampling without replacement," *J. Amer. Statist. Assoc.*, **61**, 391-397, (1966).

Durbin, J., "Some results in sampling theory when the units are selected with un-equal probabilities," *J.R. Statist. Soc.*, **15**B, 262-269, (1953).

———"Design of multi-stage surveys for the estimation of sampling errors," *Applied Statistics*, **16**, 152-164, (1967).

Ganguli, M., "A note on nested sampling," *Sankhya*, **5**, 449-452, (1941).

Hansen, M.H. and W.N. Hurwitz, "On the theory of sampling from finite popu-lations," *Ann. Math. Statist.*, **14**, 333-362, (1943).

———"On the determination of optimum probabilities in sampling," *Ann. Math. Statist.*, **20**, 426-432, (1949).

———and W.G. Madow, *Sample Survey Methods and Theory*, Vol. I, John Wiley and Sons, New York, (1953).

Lahiri, D.N., "Technical paper on some aspects of the development of the sample design," *Sankhya*, **14**, 332-362, (1954).

Mahalanobis, P.C., *Report on the Sample Census of Jute in Bengal*, Ind. Central Jute Committee, (1940).

Murthy, M.N. and V.K. Sethi, "Self-weighting design at tabulation stage," *National Sample Survey Working Paper*, No. 5 (1959); (also *Sankhya*, **27**B, 201-210, (1959).

————"Randomized rounded off multipliers," *J. Amer. Statist. Assoc.*, **56**, 328-334, (1961).

Rangarajan, R., "A note on two-stage sampling", *Sankhya*, **17**, 373-376, (1957).

Rao, J.N.K., "On sampling with varying probabilities in sub-sampling designs," *J. Ind. Soc. Agr. Statist.*, **13**, 211-217, (1961).

————"Alternative estimators in pps sampling for multiple characteristics," *Sankhya*, **23A**, 47-60, (1966).

————"Unbiased variance estimation for multi-stage designs," *Sankhya*, **32A**, (1975).

Roy, J., "A note on estimation of variance components in multi-stage sampling with varying probabilities, *Sankhya*, **17**, 367-372, (1957).

Singh, D., "The sampling with varying probabilities without replacement," *J. Ind. Soc. Agr. Statist.*, **6**, 48-57, (1954).

————"On efficiency of cluster sampling," *J. Ind. Soc. Agr. Statist.*, **8**, 44-55, (1956).

————"Estimates of variance components in finite populations," *J. Ind. Soc. Agr. Statist.*, **10**, 1-15, (1958).

Som, R.K., "Self-weighting sample design with an equal number of ultimate stage units in each of the selected penultimate stage units," *Bull. Cal. Statist. Assoc.*, **8**, 59-66, (1959).

Sukhatme, P.V., "Efficiency of sub-sampling designs in yield surveys," *J. Ind Soc. Agr. Statist.*, **2**, 212-228, (1950).

10

Multi-Phase Sampling

A man should never be ashamed to own he had been in the wrong, which is but saying, in other words, that he is wiser today than he was yesterday.

Jonathan Swift

10.1 INTRODUCTION

In sample surveys, the information on an auxiliary variate x is required many times, either for estimation or for selection or stratification to increase the efficiency-of the-estimator. When such information is lacking and it is relatively cheaper to obtain information on x, we can consider taking a large preliminary sample for estimating $\bar{X}$ or distribution of x as the case may be, and only a small sample (sometimes a sub-sample) for measuring the y variate, the character of interest for estimation. This could mean to devoting a part of the resources to this large preliminary sample and, therefore, reduction in sample size for measuring the study variate. This technique is known as *double sampling* or *two-phase sampling* and was proposed for the first time by Neyman (1938). When the sample for the main survey is selected in three or more phases, the sampling procedure is termed *multi-phase sampling*.

The difference between multi-phase sampling and multi-stage sampling procedures is that in multi-phase sampling it is necessary to have a complete sampling frame of the units whereas in multi-stage sampling, a sampling frame of the next stage units is necessary only for the sample units selected at the stage. This design is advantageous when the gain in precision is substantial as compared to the increase in cost due to collection of information on the auxiliary variate for large samples.

Neyman (1938) discussed the theory of double sampling for human populations. Robson (1952) and Robson and King (1953) extended the theory to multiple sampling. Srinath (1971) discussed the problem of optimum allocation, and Rao (1973) modified it and gave estimated variance under different conditions. We shall discuss, in the subsequent sections, the techniques for stratification, difference, ratio and regression methods of estimation and for pps selection.

10.2 DOUBLE SAMPLING FOR STRATIFICATION

In stratified sampling, the population is divided into k strata which are homogeneous within themselves and whose means are widely different. The strata weights are used in estimating unbiasedly the mean or the total of the character under study. If these weights are not known, the technique of double sampling can be used, which consists of selecting a preliminary sample of n' by simple random sampling, wor, to estimate the strata weights and then further selecting a sub-sample of n units with n_i units from the ith stratum, to collect information on the character under study, such that $\sum_i^k n_i = n$.

Let $W_i\ (= N_i/N)$ be proportion of units falling in the ith stratum, and $w_i\ (= n_i'/n')$ be proportion of first sample units falling in the ith stratum. As an estimator of the population mean $\bar{Y}$ we can take

$$\bar{y}_{std} = \sum_i^k w_i\, \bar{y}_i \qquad (10.2.1)$$

where $\bar{y}_i$ is the sample mean for the study variate in the ith stratum.

THEOREM 10.2.1 If the values of n_i do not depend on w_i, show that the estimator $\bar{y}_{std}$ is an unbiased estimator of the population mean and its sampling variance is given by

$$V(\bar{y}_{std}) = \sum_i^k \left[(1 - f_i) \frac{W_i^2 \, S_i^2}{n_i} + \frac{N - n'}{(N - 1)} \, \frac{W_i (1 - W_i)}{n'} \, (1 - f_i) \frac{S_i^2}{n_i} \right.$$

$$\left. + \frac{N - n'}{(N - 1) \, n'} \, W_i \, (\bar{Y}_i - \bar{Y})^2 \right]$$

$$= \sum_i^k \left[\left\{ W_i^2 + g \, \frac{W_i (1 - W_i)}{n'} \right\} (1 - f_i) \frac{S_i^2}{n_i} + g \, \frac{W_i \, (\bar{Y}_i - \bar{Y})^2}{n'} \right]$$

$$(10.2.2)$$

where $f_i = \dfrac{n_i}{N_i}$ and $g = \dfrac{(N - n')}{(N - 1)}$

Proof Whenever a new sample is drawn, it implies a fresh drawing of both first and second samples. Thus w_i and the sample mean $\bar{y}_i$ are both random variables. Since the first sample is a simple random sample, $E_1 (w_i) = W_i$. If we take expectation first, over samples in which w_i are fixed, $\bar{y}_i$ is the mean of a simple random sample from the stratum, $E_2 (\bar{y}_i|w_i) = \bar{Y}_i$. Hence, the expectation over different selections of the sample is given by

$$E(\bar{y}_{syd}) = E_1 \left[E_2 \sum_i^k w_i \bar{y}_i | w_i \right] = E_1 \left[\sum_i^k w_i \, \bar{Y}_i \right] = \sum_i^k W_i \, \bar{Y}_i = \bar{Y}$$

which shows that the estimator $\bar{y}_{std}$ is unbiased.

To calculate the sampling variance, we know that

$$V(\bar{y}_{std}) = E_1 V_2 (\bar{y}_{std}) + V_1 E_2 (\bar{y}_{std})$$

Here $\qquad E_1 V_2 (\bar{y}_{std}) = \sum_i^k E_1 V_2 (w_i \bar{y}_i | w_i) = \sum_i^k E_1 (w_i^2) \, V_2 (\bar{y}_i)$

Let us consider first the expectation over selections of the w_i, i.e. $E_1 (w_i) = W_i$ and $V_1 (w_i) = g \, W_i (1 - W_i)/n'$

Also, $\qquad\qquad V_1 (w_i) = E_1 (w_i^2) - [E_1 (w_i)]^2$

Therefore, $\qquad\quad E_1 (w_i^2) = V_1 (w_i) + [E_1 (w_i)]^2$

$$= \frac{g \, W_i \, (1 - W_i)}{n'} + W_i^2$$

Hence,
$$E_1 V_2 (\bar{y}_{std}) = \sum_i^k \left[g W_i \frac{(1 - W_i)}{n'} + W_i^2 \right] V (\bar{y}_i) \qquad \text{(i)}$$

Also
$$E_2 (\bar{y}_{std}) = \sum_i^k w_i \bar{Y}_i$$

$$\therefore \ V_1 E_2 (\bar{y}_{std}) = \sum_i^k \frac{g W_i (1 - W_i)}{n'} \bar{Y}_i^2 - \sum_i^k \sum_{i' \neq i}^k g \bar{Y}_i \bar{Y}_{i'} W_i W_{i'}$$

$$= \sum_i^k \frac{g}{n'} W_i (1 - W_i) \bar{Y}_i^2 - \frac{g}{n'} [(\sum_i^k W_i^2 \bar{Y}_i)^2$$

$$+ \sum_i^k W_i^2 \bar{Y}_i^2)]$$

$$= \frac{g}{n'} [\sum_i^k W_i \bar{Y}_i^2 - (\sum_i^k W_i \bar{Y}_i)^2]$$

$$= \frac{g}{n'} \sum_i^k W_i (\bar{Y}_i - \bar{Y})^2 \qquad \text{(ii)}$$

Combining (i) and (ii), we get

$$V (\bar{y}_{std}) = \sum_i^k \left[\left\{ W_i^2 + \frac{g W_i (1 - W_i)}{n'} \right\} (1 - f_i) \frac{S_i^2}{n_i} \right.$$
$$\left. + g \frac{W_i (\bar{Y}_i - \bar{Y})^2}{n'} \right]$$

COROLLARY 1 For large populations, prove that

$$V (\bar{y}_{std}) = \sum_i^k \left[W_i^2 + \frac{W_i (1 - W_i)}{n'} \right] \frac{S_i^2}{n_i} + \sum_i^k \frac{W_i (\bar{Y}_i - \bar{Y})^2}{n'}$$

$$(10.2.3)$$

For proportional allocation $n_i = n W_i$, the variance of the double sampling procedure is approximately given by

$$V_{prop} (\bar{y}_{std}) \cong \frac{1}{n} \sum_i^k W_i S_i^2 + \frac{1}{n'} \sum_i^k W_i (\bar{Y}_i - \bar{Y})^2 \qquad (10.2.4)$$

which shows that the between strata contribution to the variance would be much smaller with double sampling procedure.

COROLLARY 2 In stratified double sampling, an unbiased estimator of $V(\bar{y}_{std})$ is given by

$$v(\bar{y}_{std}) = \frac{n'}{(n'-1)} \sum_i^k \left[\left\{ w_i^2 - g\frac{w_i}{n'} \right\} \frac{s_i^2}{n_i} + g\frac{w_i(\bar{y}_i - \bar{y}_{std})^2}{n'} \right] \tag{10.2.5}$$

COROLLARY 3 If the second sample is drawn independently of the first so that n_i do not depend on w_i, and f_i are negligible, the variance in relation (10.2.3) is given by

$$V(\bar{y}_{std}) \cong \sum_i^k \frac{W_i S_i^2}{n_i} + \frac{g}{n'} \sum_i^k W_i (\bar{Y}_i - \bar{Y})^2 \tag{10.2.6}$$

COROLLARY 4 In the double sampling, $p_{std} = \sum_i^k w_i p_i$ is an unbiased estimator of the population proportion P, where w_i and p_i are the estimates of weight and proportion in the ith stratum.

Its sampling variance in large population is given by

$$V(p_{std}) \cong \sum_i^k \frac{W_i P_i (1 - P_i)}{n_i} + \frac{g}{n'} \sum_i^k W_i (P_i - P)^2 \tag{10.2.7}$$

where P_i is the proportion in the ith stratum.

10.3 OPTIMAL ALLOCATION

The cost function for double sampling can be written as

$$C = a + n' c' + nc \tag{10.3.1}$$

where a is the overhead cost, and c' and c are the costs per unit of measuring the auxiliary variate and study variate, respectively.

Usually, the problem is to obtain the values of n' and n_i (and ultimately n) so as to minimize the variance of the estimator for a given cost. The exact expressions for n_i and n' leading to minimum variance are rather complicated. However, Neyman (1938) suggested allocations of n_i proportional to $W_i S_i$. If these values are substituted in the variance expression after ignoring smaller terms, we have

$$V_{opt}(\bar{y}_{std}) = \frac{V_n}{n} + \frac{V_{n'}}{n'} \tag{10.3.2}$$

where

$$V_n = (\sum_i^k W_i S_i)^2$$

and

$$V_{n'} = \sum_i^k W_i (\bar{Y}_i - \bar{Y})^2$$

The optimum values of n and n' are given by

$$n = \frac{(C_0 - a) \sqrt{V_n c'}}{\sqrt{c \, c'} \, (\sqrt{c \, V_n} + \sqrt{c' \, V_{n'}})} \qquad (10.3.3)$$

and

$$n' = \frac{(C_0 - a) \sqrt{c \, V_{n'}}}{\sqrt{c \, c'} \, (\sqrt{c \, V_n} + \sqrt{c' \, V_{n'}})} \qquad (10.3.4)$$

For these values of n and n', the minimum variance of $\bar{y}_{std}$ is given by

$$V_{min} \, (\bar{y}_{std}) = \frac{(\sqrt{c V_n} + \sqrt{c' \, V_{n'}})^2}{(C_0 - a)} \qquad (10.3.5)$$

EXAMPLE 10.1   If the farm size is taken as x_i and area under wheat as y_i, the data from a census of all farms in a district divided into 2 strata are given below:

Stratum	N_i	$\bar{X}_i$	$\bar{Y}_i$	$S_{y_i}^2$
1	1580	82.56	19.40	312
2	430	244.85	51.63	922
Total	2010	117.28	26.30	620

Assuming that the cost of measuring the wheat area is 10 times that for farm size and the total amount available for the survey is Rs 100, draw a random sample of 100 farms by double sampling with optimum allocations.

We can obtain easily, from the given data,

$$V_n = (\sum_i^2 W_i \, S_{y_i})^2 = 417$$

$$V_{n'} = \sum_i^2 W_i \, (\bar{Y}_i - \bar{Y})^2 = 175$$

Hence, from relations (10.3.3) and (10.3.4), we have

$$\frac{n}{n'} = \left(\frac{417}{175} \times \frac{1}{10}\right)^{1/2} = 0.488$$

Also

$$100 = 0.1 \, n' + 0.488 \, n' = 0.588 \, n'$$

Therefore,

$$n' = \frac{100}{0.588} = 170$$

and

$$n = 170 \times 0.488 = 83$$

From relation (10.3.5), we have

$$V_{\text{opt}} = \frac{417}{83} + \frac{175}{170} = 6.05$$

For a random sample of size 100, with no double sampling, we have

$$V = \frac{620}{100} = 6.20$$

Hence, relative precision $= \dfrac{6.20}{6.05} \times 100 = 102.47$ percent, i.e. a gain

by 2.47% is obtained from double sampling.

The main objective in optimum allocation is to chose n' and n_i, which minimize $V(\bar{y}_{\text{std}})$ for a given cost. The actual cost function for double sampling can be written as

$$C = a + n'\, c' + \sum_i^k n_i\, c_i \tag{10.3.6}$$

where c_i is the cost of measuring a unit in the ith stratum and the rest are as usual.

The n_i's are random variates and, by application of the Schwartz inequality, we can get optimum allocations. This can easily be checked with the treatment given by Cochran (1977).

Since the variance and cost functions involve population parameters which are unknown, we propose discussing the outlines as done by Rao (1973), which can be used to get optimum allocation in situations when only sample estimates are given. From relation (10.2.4), we can derive an estimate of the variance of $\bar{y}_{\text{std}}$ as

$$v(\bar{y}_{\text{std}}) = \frac{1}{n'} \sum_i^k w_i (\bar{y}_. - \bar{y}_{\text{std}})^2 + \sum_i^k \frac{w_i^2\, s_i^2}{n_i} \tag{10.3.7}$$

Thus the optimum values of n' and n_i are given by

$$n'_{\text{opt}} = \frac{(C - a)[\sum_i w_i (\bar{y}_i - \bar{y}_{\text{std}})^2]^{1/2}}{c'^{1/2}\,[\{c \sum_i w_i (\bar{y}_i - \bar{y}_{\text{std}})^2\}^{1/2} + c'^{1/2} \sum_i w_i\, s_i]} \tag{10.3.8}$$

and

$$n_{i\text{opt}} = \frac{(C - a)\, w_i s_i}{c^{1/2}\,[\{c \sum_i w_i (\bar{y}_i - \bar{y}_{\text{std}})^2\}^{1/2} + c'^{1/2} \sum_i w_i\, s_i]} \tag{10.3.9}$$

For these values of n' and n_i, the minimum variance $v(\bar{y}_{\text{std}})$ is given as

$$v_{\text{min}}(\bar{y}_{\text{std}}) = \frac{[\{c \sum_i w_i(\bar{y}_i - \bar{y}_{\text{std}})^2\}^{1/2} + c'^{1/2} \sum_i w_i\, s_i]^2}{(C - a)} \tag{10.3.10}$$

EXAMPLE 10.2   In a survey conducted by an agency in Saraibal block of Santhal Parganas in Bihar (India) during the Rabi season of 1967-68, 499 fields growing wheat were selected at random and the yield of wheat (kg/ha) was recorded by eye estimation. The data showing the distribution of the number of fields falling in each of the 6 strata are given below. A sub-sample of the fields was selected from the fields falling in each of 6 strata and crop-cutting experiments were conducted. The yield rates observed are given in column (4) for each of the selected fields.

Strata	No. of eye estimates	No. of crop-cutting experiments	Yield rate in kg/ha based on crop cutting in the fields selected in each stratum
(1)	(2)	(3)	(4)
101-200	154	40	200, 208, 152, 224, 104, 168, 160, 152, 247, 178, 84, 360, 340, 380, 340, 340, 360, 184, 420, 172, 216, 128, 136, 128, 114, 139, 160, 112, 104, 140, 108, 136, 120, 81, 82, 70, 380, 440, 400, 400.
201-300	189	46	104, 152, 148, 256, 280, 260, 320, 288, 288, 140, 144, 124, 496, 450, 492, 256, 256, 252, 332, 384, 276, 330, 344, 243, 296, 292, 314, 424, 360, 416, 200, 192, 195, 326, 248, 322, 104, 144, 61, 38, 112, 137, 144, 206, 496, 535.
301-400	91	22	280, 192, 192, 280, 200, 304, 440, 320, 192, 448, 448, 345, 326, 420, 212, 326, 163, 326, 325, 496, 304, 243.
401-500	40	15	288, 156, 280, 136, 384, 472, 345, 333, 300, 326, 324, 720, 672, 568, 520.
501-600	13	6	428, 368, 506, 824, 624, 768.
Above 600	12	2	344, 712.
Total	499	131	

(i) Obtain an estimate of the average yield of wheat in kg/ha for the block and also the standard error of the estimate.

(ii) Assuming the cost function

$$150 = n + 0.05n'$$

and the optimum allocation to be used for allocation of n to various strata, estimate the values of n and n' so that the variance of the double sampling estimate is minimum. Also estimate the variance of the estimate with these values of n and n'.

(i) Given that $n' = 499$, $n = 131$

n_i	n'	w_i	$\bar{y}_i$	Σy_i^2	$\bar{y}_i^2$	$n_i \bar{y}_i^2$
40	154	0.3086	209.15	2253606	43743.7225	1749748.00
46	189	0.3788	263.00	3822072	69169.0000	3181774.00
22	91	0.1824	308.27	2286088	95030.3929	2090668.64
15	40	0.0802	588.26	2668590	150753.5929	2261303.88
6	13	0.0261	586.33	2232820	343782.8689	2062691.21
2	12	0.0241	528.00	625280	278784.0000	557568.00

$w_i \bar{y}_i$	s_i^2	$(\bar{y}_i - \bar{y}_{std})^2$	$n_i' (\bar{y}_i - \bar{y}_{std})^2$
65.5440	12912.413	4445.5775	761618.9350
82.4968	14228.844	271.4190	51298.1910
56.2280	9305.684	829.3190	78468.0290
47.1780	29091.865	11835.6780	473427.1200
15.3030	34024.558	94162.4450	1224111.7967
12.7250	67712.000	61764.6756	741176.1072

$$\sum_i^6 w_i \bar{y}_i = 279.4748$$

An unbiased estimate of the population mean is given by

$$\bar{y}_{std} = \sum_i^6 w_i \bar{y}_i = 279.4748$$

An estimate of the variance of $\bar{y}_{std}$ is given as

$$v(\bar{y}_{std}) = \frac{1}{n'} \sum_i^6 w_i (\bar{y}_i - \bar{y}_{std})^2 + \sum_i^6 \frac{w_i^2 s_i^2}{n_i}$$

$$= 13.368 + 113.8781 = 127.2405$$

and standard error of $\bar{y}_{std} = \sqrt{127.2405} = 11.28$

(ii) The given cost function is

$$150 = n + 0.05 \, n'$$

Here, $\quad C = 150, \quad\quad a = 0, \quad\quad c' = 0.05 \quad$ and $\quad\quad c = 1$

On substituting these values in relations (10.3.8) and (10.3.9), we get

$$n'_{opt} = \frac{80.6 \times 150}{0.224 \, (0.224 \times 80.6 + 196.88)} = 249$$

$$n_{1 \, opt} = 74, \quad\quad n_{2 \, opt} = 29, \quad\quad n_{3 \, opt} = 11,$$

$$n_{4 \, opt} = 14, \quad\quad n_{5 \, opt} = 6 \quad\quad n_{6 \, opt} = 3$$

Using relation (10.3.10), we get

$$v_{opt} \, (\bar{y}_{std}) = \frac{(215.09)^2}{150} = 368$$

10.4 DOUBLE SAMPLING FOR DIFFERENCE ESTIMATOR

A difference estimator for estimating the population mean $\bar{Y}$ when information on x is not available in advance and it is considered important to use the auxiliary variate to derive more precise estimates, is discussed here. A preliminary random sample, wor, of size n' is taken and the information on x is collected. A sub-sample of size n is drawn, wor, from the preliminary sample and information on y is measured. The difference estimator of $\bar{Y}$ may be defined by

$$\bar{y}_{dd} = \bar{y} + \beta \, (\bar{x}' - \bar{x}) \tag{10.4.1}$$

where β is taken as known in the population; $\bar{y}$, $\bar{x}$ are the sub-sample means for y and x, respectively; and $\bar{x}'$ is the preliminary sample mean of x.

THEOREM 10.4.1 Show that $\bar{y}_{dd}$ is an unbiased estimator of the population mean, its sampling variance is given by

$$V(\bar{y}_{dd}) = \left(\frac{1}{n} - \frac{1}{N} \right) S_y^2 + \left(\frac{1}{n} - \frac{1}{n'} \right) (\beta \, S_x^2 - 2\rho S_y \, S_x) \tag{10.4.2}$$

Proof Given the first sample, let $\bar{x}'$ be the mean value.

$$E_2 \, (\bar{y}_{dd} \mid \bar{x}') = E_2 \, \{[\, \bar{y} + \beta \, (\bar{x}' - \bar{x})] \mid \bar{x}'\} = \bar{y}'$$

Hence $\quad\quad\quad E_1 \, (\bar{y}') = \bar{Y}$

or $\quad\quad\quad\quad E \, \bar{y}_{dd}) = \bar{Y}$

This shows that the estimator is unbiased. For the sampling variance of $\bar{y}_{dd}$, we have

$$V(\bar{y}_{dd}) = V_1 E_2 (\bar{y}_{dd}|\bar{y}') + E_1 V_2 (\bar{y}_{dd}|\bar{y}')$$

Here

$$V_1 E_2 (\bar{y}_{dd}|\bar{y}') = \left(\frac{1}{n'} - \frac{1}{N}\right) S_y^2$$

and

$$E_1 V_2 (\bar{y}_{dd}|\bar{y}') = E_1 \left(\frac{1}{n} - \frac{1}{n'}\right) \sum_i^{n'} \frac{(y_i - \beta x_i - \bar{y}' + \beta \bar{x}')^2}{(n' - 1)}$$

$$= \left(\frac{1}{n} - \frac{1}{n'}\right) \sum_i^{N} \frac{(y_i - \beta x_i - \bar{Y} + \beta \bar{X})^2}{(N - 1)}$$

$$= \left(\frac{1}{n} - \frac{1}{n'}\right)(S_y^2 + \beta^2 S_x^2 - 2\rho\beta S_y S_x)$$

Combining both results, we prove the theorem.

COROLLARY 1 An unbiased estimator of the sampling variance given in relation (10.4.2) can be written as

$$v(\bar{y}_{dd}) = \left(\frac{1}{n} - \frac{1}{N}\right) s_y^2 + \left(\frac{1}{n} - \frac{1}{n'}\right) s_d^2 \qquad (10.4.3)$$

where

$$s_y^2 = \frac{\sum_i^{n}(y_i - \bar{y})^2}{(n - 1)}$$

and

$$s_d^2 = \frac{\sum_i^{n}[y_i - \bar{y} - \beta(x_i - \bar{x})]^2}{(n - 1)}$$

COROLLARY 2 If a direct random sample is taken without using the double sampling procedure, the sample size for the same cost

$$C = a + nc + n'c'$$

will be obtained by

$$n_0 = \frac{C}{c} = n + \frac{a + n'c'}{c} \qquad (10.4.4)$$

and the sampling variance of the sample mean will be

$$V(\bar{y}_d) = \left(\frac{1}{n_0} - \frac{1}{N}\right) S_y^2 \qquad (10.4.5)$$

COROLLARY 3 Taking $\beta = k S_y/S_x$, the condition that double sampling is more precise than a direct random sampling will be obtained by

$$2\rho > \left[k\left(1 - \frac{n}{n'}\right)\left(1 + \frac{nc}{n'c'}\right) \right]^{-1} \qquad (10.4.6)$$

A method using multi-auxiliary information in the first sample has been discussed by Des Raj (1965), showing how this information may be used for achieving higher precision by applyng the double sampling technique.

10.5 DOUBLE SAMPLING FOR RATIO ESTIMATOR

In a number of situations, it happens that the population mean $\bar{X}$ is not known and the ratio estimator cannot be used to estimate the population mean $\bar{Y}$. The usual procedure in such situations is to use the method of double sampling. The sampling procedure is similar to that given in Section 10.4, i.e. consists of drawing a first sample of size n' to estimate the population mean $\bar{X}$ and then a sub-sample of size n from n' to estimate the mean of the main variate under study. Several estimators of the population mean $\bar{Y}$ can be framed and the simplest one is the usual biased estimator

$$\bar{y}_{Rd} = \frac{\bar{y}}{\bar{x}} \bar{x}' = \hat{R} \bar{x}' \tag{10.5.1}$$

where $\bar{y}$, $\bar{x}$ are sub-sample means for y and x, respectively, and $\bar{x}'$ is the mean of x in the first sample.

THEOREM 10.5.1 Show that the estimator $\bar{y}_{Rd}$ is biased and its relative bias is given by

$$\text{Rel Bias} = \left(\frac{1}{n} - \frac{1}{n'}\right)(C_x^2 - \rho C_x C_y) \tag{10.5.2}$$

Its sampling variance is given by

$$V(\bar{y}_{Rd}) = \left(\frac{1}{n'} - \frac{1}{N}\right) S_y^2 + \left(\frac{1}{n} - \frac{1}{n'}\right)(S_y^2 + R^2 S_x^2 - 2 R S_{yx}) \tag{10.5.3}$$

Proof By substituting,

$$\bar{y} = \bar{Y}(1 + e)$$
$$\bar{x} = \bar{X}(1 + e_1)$$
$$\bar{x}' = \bar{X}(1 + e_2)$$

where $E(e) = E(e_1) = E(e_2) = 0$, we have

$$\bar{y}_{Rd} = \bar{Y}[(1 + e)(1 + e_1)^{-1}(1 + e_2)]$$
$$= \bar{Y}[1 + (e - e_1 + e_2) + \ldots]$$

Taking expectation, we find that the bias is zero to the first order of approximation, and in this case the variance is just similar to that of the ratio estimator in simple random sampling. If the terms up to the second order of approximation are retained, we have

$$\bar{y}_{Rd} = \bar{Y} \left[1 + (e - e_1 + e_2) + (-ee_1 + ee_2 - e_1 e_2 + e_1^2) + \ldots \right]$$

Now $\qquad E(ee_1) = \text{cov}(\bar{y}, \bar{x}) = \left(\dfrac{1}{n} - \dfrac{1}{N} \right) S_{yx}$

$$E(ee_2) = \text{cov}(\bar{y}, \bar{x}')$$
$$= \text{cov}\left[E(\bar{y}|n'), E(\bar{x}'|n') \right], + E\left[\text{cov}(\bar{y}, \bar{x}'|n') \right]$$
$$= \text{cov}(\bar{y}', \bar{x}') = \left(\dfrac{1}{n'} - \dfrac{1}{N} \right) S_{xy}$$

$$E(e_1 e_2) = V(\bar{x}') = \left(\dfrac{1}{n'} - \dfrac{1}{N} \right) S_x^2$$

and $\qquad E(e_1^2) = V(\bar{x}) = \left(\dfrac{1}{n'} - \dfrac{1}{N} \right) S_x^2$

Hence, the expectation of $\bar{y}_{Rd}$, to the second order of approximation, is obtained as

$$E(\bar{y}_{Rd}) \cong \bar{Y} \left[1 + \left(\dfrac{1}{n} - \dfrac{1}{n'} \right) (C_x^2 - \rho\, C_x\, C_y) \right]$$

Thus, its relative bias is given by

$$\text{Rel Bias} = \left(\dfrac{1}{n} - \dfrac{1}{n'} \right) (C_x^2 - \rho\, C_x\, C_y)$$

Again, to find the sampling variance, we have, to the first order of approximation,

$$V(\bar{y}_{Rd}) = E(\bar{y}_{Rd} - \bar{Y})^2 = \bar{Y}^2\, E(ee_1 + e_2)^2$$

Expanding and taking expectation term by term, we obtain

$$V(\bar{y}_{Rd}) = \left(\dfrac{1}{n'} - \dfrac{1}{N} \right) S_y^2 + \left(\dfrac{1}{n} - \dfrac{1}{n'} \right) (S_y^2 + R^2 S_x^2 - 2R\, S_{yx})$$

COROLLARY 1 The estimator of $V(\bar{y}_{Rd})$ is given by

$$v(\bar{y}_{Rd}) = \left(\dfrac{1}{n'} - \dfrac{1}{N} \right) s_y^2 + \left(\dfrac{1}{n} - \dfrac{1}{n'} \right) (s_y^2 + \hat{R}^2 s_x^2 - 2\hat{R}\, s_{yx})$$

$$(10.5.4)$$

where $\qquad s_y^2 = \sum_i^n \dfrac{(y_i - \bar{y})^2}{(n - 1)}$

$$s_x^2 = \sum_i^n \dfrac{(x_i - \bar{x})^2}{(n - 1)}$$

$$s_{yx} = \sum_i^n \dfrac{(y_i - \bar{y})(x_i - \bar{x})}{(n - 1)}$$

COROLLARY 2   Show that the estimator $\bar{y}_{Rd}$ is more efficient than the estimator based on simple random sampling when no auxiliary variate is used, if

$$\rho > \frac{C_x}{2C_y} \tag{10.5.5}$$

10.5.1   Optimum Allocation

Here, we shall discuss the case where, for a fixed cost, the variance of the estimator is minimized. Assuming N large, we can write

$$V(\bar{y}_{Rd}) = \frac{S_y^2}{n'} + \left(\frac{1}{n} - \frac{1}{n'}\right)(S_y^2 + R^2 S_x^2 - 2R S_{yx})$$

$$= \frac{V}{n} + \frac{1}{n'}(S_y^2 - V) \tag{10.5.6}$$

where $\qquad V = \dfrac{\displaystyle\sum_i^N (y_i - R x_i)^2}{(N-1)}$

The cost function for double sampling method can be written as

$$C = a + cn + c'n' \tag{10.5.7}$$

where a is the overhead cost, c and c' are the costs per unit of measuring the main variate and the auxiliary variate, respectively. To obtain the optimum values of n and n' which can minimize $V(\bar{y}_{Rd})$, we may write

$$\phi = V(\bar{y}_{Rd}) + \lambda(C - C_0)$$

where C_0 is fixed cost.

Differentiating w.r.t. n and n' and equating to zero, we have

$$\frac{n\sqrt{c}}{\sqrt{V}} = \frac{n'\sqrt{c'}}{\sqrt{S_y^2 - V}} = \frac{(C_0 - a)}{\sqrt{cV} + \sqrt{c'(S_y^2 - V)}} \tag{10.5.8}$$

Substituting the values of n and n' in relation (10.5.6), we find that the minimum variance of $\bar{y}_{Rd}$ is given by

$$V_{min}(\bar{y}_{Rd}) = \frac{[\sqrt{cV} + \sqrt{c'(S_y^2 - V)}]^2}{(C_0 - a)} \tag{10.5.9}$$

Also, the minimum variance in case of simple random sampling without use of the auxiliary variate is given by

$$V_{min}(\bar{y}) = \frac{cS_y^2}{(C_0 - a)} \tag{10.5.10}$$

Hence, the relative precision of the double sampling estimator $\bar{y}_{Rd}$, as compared to the estimator $\bar{y}$, is obtained as

$$\text{RP} = \frac{cS_y^2}{[\sqrt{cV} + \sqrt{c'(S_y^2 - V)}]^2} = [\psi + \sqrt{u(1 - \psi^2)}\,]^2 \quad (10.5.11)$$

where $$u = \frac{c'}{c} \quad \text{and} \quad \psi = \sqrt{\frac{V}{S_y^2}}$$

10.6 DOUBLE SAMPLING FOR REGRESSION ESTIMATOR

In this section, we shall discuss the regression estimator in those situations where the population mean of x is not known in advance. In such cases, a two-phase sampling scheme can be used. Assume that the sampling procedure as given in section 10.5, is followed. A linear regression estimator of the population mean $\bar{Y}$ may be defined as

$$\bar{y}_{ld} = \bar{y} + b\,(\bar{x}' - \bar{x}) \quad (10.6.1)$$

where $\bar{y}$ and $\bar{x}$ are the sub-sample means for y and x, respectively, $\bar{x}'$ is the mean of x in the first sample and b is the least square estimate of the regression coefficient, β in the population.

THEOREM 10.6.1 Show that the estimator $\bar{y}_{ld}$ is not unbiased and its bias is approximated by

$$B(\bar{y}_{ld}) = -\beta\left(\frac{1}{n} - \frac{1}{n'}\right)\left[\frac{\mu_{21}}{S_{yx}} - \frac{\mu_{30}}{S_x^2}\right] \quad (10.6.2)$$

and its sampling variance is given by

$$V(\bar{y}_{ld}) \cong \frac{(1 - \rho^2)}{n} S_y^2 + \frac{\rho^2 S_y^2}{n'} - \frac{S_y^2}{N} \quad (10.6.3)$$

Proof It has been shown in Theorem 10.4.1 that the difference estimator is unbiased. In the present case, β is estimated from the sample, hence we have

$$E(\bar{y}_{ld}) = E_1 E_2(\bar{y}_{ld}|\bar{x}') = E_1[\bar{y}' - E_2 b(\bar{x} - \bar{x}')]$$

$$= \bar{Y} - E_1 E_2\{b(\bar{x} - \bar{x}')|\bar{x}'\}$$

Therefore, $$B(\bar{y}_{ld}) = - E_1 E_2 \{b(\bar{x} - \bar{x}')|\bar{x}'\}$$

Let
$$\bar{x} = \bar{X}(1 + e),$$

$$\bar{x}' = \bar{X}(1 + e'),$$

$$s_{xy} = S_{xy}(1 + e_1),$$

$$s_x^2 = S_x^2(1 + e_2)$$

Then, we have

$$E(e) = E(e') = E(e_1) = E(e_2) = 0$$

$$B(\bar{y}_{ld}) = - E_1 E_2 \left[\beta(e - e')\left(1 + \frac{e_1}{S_{xy}}\right)\left(1 + \frac{e_2}{S_x^2}\right)^{-1} \right]$$

$$= - E_1 E_2 \{\beta(e - e')(e_1 - e_2)\}$$

$$= - \beta \left[\frac{\text{cov}(S_{xy}, \bar{x}') - \text{cov}(S_{xy}, \bar{x})}{S_{xy}} \right.$$

$$\left. - \frac{\text{cov}(s_x^2, \bar{x}') - \text{cov}(s_x^2, \bar{x})}{S_x^2} \right]$$

$$= - \beta \left[\frac{\mu_{21}}{S_{xy}} - \frac{\mu_{30}}{S_x^2} \right]$$

This expression of bias shows that it is just similar to that given in relation (7.3.2) and reduces to this value when $n' = N$. Hence, it is seen that the bias will be negligible if the sample size is large enough.

The variance of $\bar{y}_{ld}$ is to be obtained over the two phases and is given by

$$V(\bar{y}_{ld}) = V_1 E_2(\bar{y}_{ld}) + E_1 V_2(\bar{y}_{ld})$$

Let $u_i = y_i - \beta x_i$

Since the sub-sample was drawn at random from the first sample,

$$E_2(\bar{y}_{ld}) = \bar{y}', \quad V_2(\bar{y}_{ld}) = \left(\frac{1}{n} - \frac{1}{n'}\right) s_u'^2$$

where $s_u'^2$ is the variance of u in the first sample.

Thus, $$V_1 E_2(\bar{y}_{ld}) = E_1 \left(\frac{1}{n'} - \frac{1}{N}\right) S_y^2$$

and
$$E_1 V_2(\bar{y}_{ld}) = E_1 \left(\frac{1}{n} - \frac{1}{n'}\right) s_u'^2 = \left(\frac{1}{n} - \frac{1}{n'}\right) S_u'^2$$

$$= \left(\frac{1}{n} - \frac{1}{n'}\right)(1 - \rho^2) S_y^2$$

Adding both, we get the result.

This expression for the variance shows that the variance in two-phase sampling is larger than that in uni-phase sampling. It should however, be noted that collection of information on the main variate for all the n' units in the first phase may be expensive. Therefore, the cost of two-phase sampling is expected to be less than that of uni-phase sampling.

COROLLARY   The estimator of $V(\bar{y}_{ld})$ is given by

$$v(\bar{y}_{ld}) = \left(\frac{1}{n} - \frac{1}{N}\right)(1 - r^2) s_y^2 + \left(\frac{1}{n'} - \frac{1}{N}\right) r^2 s_y^2 \qquad (10.6.4)$$

where
$$s_y^2 = \frac{\sum_i^n (y_i - \bar{y})^2}{(n - 1)}, \quad s_x^2 = \frac{\sum_i^n (x_i - \bar{x})^2}{(n - 1)}$$

and
$$r = \frac{\sum_i^n (y_i - \bar{x})(x_i - \bar{x})}{(n - 1) s_x s_y}$$

10.6.1 Optimum Allocation

The cost function for the double sampling scheme can be written as given in relation (10.5.7). Assuming N to be large, we can write the variance

$$V(\bar{y}_{ld}) = (1 - \rho^2) \frac{S_y^2}{n} + \frac{\rho^2 S_y^2}{n'} \qquad (10.6.5)$$

The optimum values of n and n' may be obtained by minimizing the variance for a fixed cost, or minimizing the cost for a fixed variance.

(i) *When cost is fixed*

To minimize the function $\phi = V + \lambda(C - C_0)$, we may differentiate it w.r.t. n and n', and equating to zero, we have

$$\frac{\sqrt{cn}}{\sqrt{S_y^2 (1 - \rho^2)}} = \frac{\sqrt{c'n'}}{\rho S_y} = \frac{(C_0 - a)}{[\sqrt{c (1-\rho^2)S_y^2} + \sqrt{c'\rho^2 S_y^2}]} \quad (10.6.6)$$

Substituting the values of n and n' in relation (10.6.5), we obtain the minimum variance of $\bar{y}_{ld}$, which is given by

$$V_{\min}(\bar{y}_{ld}) = \frac{S_y^2 [\sqrt{c (1 - \rho^2)} + \rho \sqrt{c'}]^2}{(C_0 - a)} \quad (10.6.7)$$

(ii) *When variance is fixed*   To minimize the cost for a fixed variance V_0, we get the minimum values of n and n' as

$$n' = \frac{\rho^2 S_y^2}{V_0} \left[1 + \left(\frac{(1 - \rho^2)}{\rho^2} \frac{c'}{c} \right)^{1/2} \right] \quad (10.6.8)$$

and
$$n = n' \left(\frac{1 - \rho^2}{\rho^2} \frac{c}{c'} \right)^{1/2} \quad (10.6.9)$$

The minimum cost is given by

$$C = a + \frac{S_y^2}{V_0} [\sqrt{c' \rho^2} + \sqrt{c(1 - \rho^2)}]^2 \quad (10.6.10)$$

COROLLARY   Show that the condition in which double sampling method is more precise than taking a simple random sample for the same cost, is obtained as

$$\rho^2 > \frac{4cc'}{(c + c')^2} \quad (10.6.11)$$

By applying relation (10.6.7) and comparing with the variance of simple random sample, the condition may be derived.

For discussion of the case when the second sample is drawn independently of the first, one may consider the ideas proposed by Bose (1943). Further, Khan and Tripathi (1967) have extended the idea of double sampling with regression to the case where p auxiliary variates are measured in the second sample.

EXAMPLE 10.2   The following data, relating to the yield of paddy, have been taken from crop cutting surveys conducted in certain parts of Uttar Pradesh (India). In all, 40 random cuts of size 1/100th of an acre were taken and the yield of paddy observed on the day of harvest was noted. On a sub-sample of 20 cuts out of 40, the dry yield was also noted. The results are given below:

S.N. of cut	Harvest yield (in kg)	Dry yield (in kg)	S.N. of cut	Harvest yield (in kg)
1	16.8	15.2	21	8.7
2	12.7	11.8	22	11.6
3	18.8	17.5	23	11.5
4	13.9	12.5	24	14.4
5	11.3	10.4	25	17.8
6	10.9	10.1	26	8.4
7	12.5	11.2	27	8.7
8	17.4	15.8	28	14.6
9	14.1	13.0	29	12.1
10	11.9	10.8	30	7.9
11	13.4	12.3	31	8.9
12	13.5	12.4	32	11.1
13	8.3	7.6	33	13.0
14	13.7	12.5	34	10.5
15	14.6	13.3	35	14.2
16	14.5	13.5	36	12.7
17	17.1	16.2	37	11.9
18	14.5	13.5	38	15.5
19	11.4	10.3	39	17.1
20	14.0	12.8	40	10.9

(i) Estimate the average yield per acre of dry paddy, along with its sampling error, by utilizing whatever information is available both on harvest as well as on dry yield.

What would have been loss in precision, had we neglected the additional information on harvest yield from the sub-sample of only 20 cuts for which dry yield is available.

(ii) If the cost of obtaining dry weight from the cut is $1\frac{1}{2}$ times that of obtaining the harvest weight of the cut, determine the optimum proportion of cuts for which dry weight is to be observed.

(a) Ratio method of estimation

(i) The double sampling ratio estimate of the average yield of dry paddy per acre is given by

$$\bar{y}_{Rd} = \frac{\bar{y}}{\bar{x}} \; \bar{x}'$$

and its estimated variance for large N, is given by

$$v(\bar{y}_{Rd}) = \left(\frac{1}{n} - \frac{1}{n'}\right)(s_y^2 + \hat{R}^2 s_x^2 - 2\hat{R} s_{xy}) + \frac{1}{n'} s_y^2)$$

where

$$\bar{y} = \frac{252.7}{20} = 12.635$$

$$\bar{x} = \frac{276.3}{20} = 13.765$$

and

$$\bar{x}' = \frac{513.8}{40} = 12.845$$

Thus,

$$\bar{y}_{Rd} = \frac{12.635 \times 12.845}{13.765} = 11.78 \times 100 \text{ kg/acre}$$

For estimating the variance, we calculate

$$\hat{R} = \frac{\bar{y}}{\bar{x}} = \frac{12.635}{13.765} = 0.9178$$

$$s_y^2 = \frac{1}{19}[3296.49 - 20(12.635)^2] = 5.45$$

$$s_x^2 = \frac{1}{19}[3906.73 - 20(13.765)^2] = 6.15$$

and

$$s_{xy} = \frac{1}{19}[3588.19 - 20(12.635)(13.765)] = 5.77$$

Thus,

$$v(\bar{y}_{Rd}) = \left(\frac{1}{20} - \frac{1}{40}\right)\left[5.45 + (0.9178)^2 \times 6.15\right.$$

$$\left. - 2 \times 0.9178 \times 5.77 + \frac{1}{40} \times 5.55\right]$$

$$= 0.1363 \times 10^4$$

and

standard error of $(\bar{y}_{Rd}) = \sqrt{0.1363 \times 10^4} = 36.9 \text{ kg/acre}$

Now, if additional information on the green yield is neglected, the estimate of the average yield of dry paddy and its estimated variance are given by

$$\bar{y} = 12.635 \times 100 \text{ kg/acre}$$

and

$$v(\bar{y}) = \frac{s_y^2}{n} = \frac{5.45}{23} = 0.27 \times 10^4,$$

respectively.

Percentage gain in efficiency due to additional information is

$$= \left(\frac{v(\bar{y})}{v(\bar{y}_{\text{Rd}})} - 1 \right) \times 100$$

$$= \left(\frac{0.27 \times 10^4}{0.1363 \times 10^4} - 1 \right) \times 100 = 98.91$$

(ii) For the given cost function, we have

$$\frac{c}{c'} = \frac{3}{2}$$

The optimum proportion of cuts is given by

$$\frac{(n)_{\text{opt}}}{n'} = \sqrt{\frac{vc}{v'c'}}$$

We have
$$V = s_y^2 + \hat{R}^2 s_x^2 - 2\hat{R}s_{xy}$$
$$= 5.45 + (0.9178)^2 \times 6.15 - 2 \times 0.9178 \times 5.77$$
$$= 0.0465$$

and $V' = 2\hat{R}s_{xy} - \hat{R}^2 s_x^2 = 2 \times 0.9178 \times 5.77 - (0.9178)^2 \times 6.15$

$$= 5.4033$$

Thus $\left(\frac{n}{n'} \right)_{\text{opt}} = \sqrt{\frac{2}{3} \times \frac{0.0465}{5.4033}} = \sqrt{57.4 \times 10^{-4}} = 0.0757$

(b) *Regression method of estimation*

(i) The double sampling regression estimate of the average dry yield of paddy is given by

$$\bar{y}_{ld} = \bar{y} + b(\bar{x}' - \bar{x})$$

and an estimate of the variance by

$$v\left(\bar{y}_{ld}\right) = \frac{s_y^2\left(1-r^2\right)}{n} + \frac{s_y^2\,r^2}{n'}$$

where

$$b = \frac{s_{xy}}{s_x^2}$$

and

$$r = \frac{s_{xy}}{s_x s_y}$$

On substituting values, we get

$$\bar{y}_{ld} = 12.635 + 0.9381\,(12.845 - 13.765)$$
$$= 11.77 \times 100 \text{ kg/acre}$$

We calculate,

$$r = \frac{5.77}{\sqrt{6.15}\ \sqrt{5.45}} = 0.9966$$

$$r^2 = 0.9933 \qquad \text{and} \qquad 1 - r^2 = 0.0067$$

Thus,

$$v(\bar{y}_{ld}) = \frac{5.45 \times 0.0067}{20} + \frac{5.45 \times 0.9933}{40}$$
$$= 0.1372 \times 10^4$$

Standard error of

$$\bar{y}_{ld} = \sqrt{0.1372 \times 10^4} = 37.04$$

Percentage gain due to additional information

$$= \left(\frac{v(\bar{y})}{v(\bar{y}_{ld})} - 1\right) \times 100$$

$$= \left(\frac{0.27 \times 10^4}{0.1372 \times 10^4} - 1\right) \times 100 = 96.8$$

(ii) We are given $c = 3/2\,c'$. Optimum proportion of cuts is given by

$$\left(\frac{n}{n'}\right)_{\text{opt}} = \sqrt{\frac{vc}{v'c'}}$$

$$v = s_y^2\,(1 - r^2) = 5.45 \times 0.0067 = 0.0366$$

and

$$v' = s_y^2\,r^2 = 5.45 \times 0.9933 = 5.4134.$$

Thus
$$\left(\frac{n}{n'}\right)_{opt} = \sqrt{\frac{2}{3} \times \frac{0.0366}{5.4134}} = \sqrt{45.07 \times 10^{-4}} = 0.0671$$

10.7 DOUBLE SAMPLING FOR pps ESTIMATOR

The information on auxiliary variate x can also be utilized for selection to increase the precision of the estimator. When information on the auxiliary variate is lacking, the same can be collected by the double sampling techniques proposed by Des Raj (1964) and Singh and Singh (1965). A preliminary sample of size n' is selected from the given population, having N units, with simple random sampling, wor, and the auxiliary variate is measured on these n' units. From this sample, a sub-sample of n units is selected by pps, wr, and the study variate y is measured on these units. An estimator of the population mean $\overline{Y}$ may be written as

$$\bar{y}_{dpps} = \sum_{i}^{n} \frac{z_i}{nn'} \qquad (10.7.1)$$

where
$$z_i = \frac{y_i}{p_i}, \quad p_i = \frac{x_i}{x'}, \quad x' = \sum_{i}^{n'} x_i$$

and y_i and x_i are the values of the study variate for the ith unit in the sub-sample and the ith unit for the auxiliary variate in the preliminary sample, respectively.

THEOREM 10.7.1 Show that $\bar{y}_{dpps}$ is an unbiased estimator of the population mean and its sampling variance is given by

$$V(\bar{y}_{dpps}) = \left(\frac{1}{n'} - \frac{1}{N}\right)S_y^2 + \frac{(n' - 1)}{n\,n'\,N\,(N - 1)}\,S_z^2 \qquad (10.7.2)$$

where
$$S_y^2 = \frac{\sum_{i}^{N} (y_i - \overline{Y})^2}{(N - 1)}$$

$$S_z^2 = \sum_{i}^{N} \frac{p_i (z_i - \overline{Y})^2}{(N - 1)}$$

Proof Given the first sample, $E(\bar{y}_{\text{dpps}}|x') = \bar{y}'$,

Therefore, $E(\bar{y}_{\text{dpps}}) = E_1 E_2(\bar{y}_{\text{dpps}}|x') = E_1(\bar{y}') = \bar{Y}$

This shows that $\bar{y}_{\text{dpps}}$ is an unbiased estimator.

The variance of $\bar{y}_{\text{dpps}}$ can be obtained by noting that

$$V(\bar{y}_{\text{dpps}}) = V_1 E_2(\bar{y}_{\text{dpps}}) + E_1 V_2(\bar{y}_{\text{dpps}})$$

Here $\quad V_1 E_2(\bar{y}_{\text{dpps}}) = V_1(\bar{y}') = \left(\dfrac{1}{n'} - \dfrac{1}{N}\right) S_y^2$

and $\quad E_1 V_2(\bar{y}_{\text{dpps}}) = E_1 \left[\dfrac{1}{n'^2} \cdot \dfrac{1}{n} \sum_i^{n'} p_i \left(\dfrac{y_i}{p_i} - \bar{y}'\right)^2\right]$

$$= \dfrac{n'(n'-1)}{n\, n'^2\, N(N-1)} \sum_i^N \sum_{j>i} p_i p_j \left(\dfrac{y_i}{p_i} - \dfrac{y_j}{p_j}\right)^2$$

$$= \dfrac{(n'-1)}{n\, n'\, N(N-1)} \sum_i^N p_i (z_i - \bar{Y})^2$$

Combining both, we get the result.

COROLLARY Show that an unbiased estimator of $V(\bar{y}_{\text{dpps}})$ is given by

$$v(\bar{y}_{\text{dpps}}) = \dfrac{1}{n'^2(n-1)} \left[\sum_i^n z_i^2 - (\sum_i^n z_i)^2\right] + \left(\dfrac{1}{n'} - \dfrac{1}{N}\right) \dfrac{1}{n(n'-1)}$$

$$\times \left[\sum_i^n p_i z_i^2 - \dfrac{1}{n'(n-1)} \left\{\sum_i^n z_i^2 - (\sum_i^n z_i)^2\right\}\right] \qquad (10.7.3)$$

10.7.1 Optimum Allocation

If n' is large, $(n'-1)/n'$ may be taken to be approximately equal to unity and hence the variance in relation (10.7.2) may be written as

$$V(\bar{y}_{\text{dpps}}) = A_0 + \dfrac{A'}{n'} + \dfrac{A}{n} \qquad (10.7.4)$$

where, A_0, A and A' are constants.

The cost function in this case has the form given in relation (10.5.7).

The optimum values of n and n' may be obtained by minimizing the variance for a fixed cost or minimizing the cost for a fixed variance.

(i) *When cost is fixed* It can be easily shown that the optimum values of n and n', which minimize the variance for a fixed cost are

$$n' = \frac{(C_0 - a)\,(A/c)^{1/2}}{(Ac)^{1/2} + (A'c')^{1/2}} \tag{10.7.5}$$

and

$$n = \frac{(C_0 - a)\,(A'/c')^{1/2}}{(Ac)^{1/2} + (A'c')^{1/2}} \tag{10.7.6}$$

Substituting the values of n and n' in relation (10.7.4), we obtain the minimum variance of $\bar{y}_{dpps}$.

(ii) *When variance is fixed* To minimize the cost for a fixed variance V_0, we get the minimum values of n and n' as

$$n' = \frac{(Ac)^{1/2} + (A'c')^{1/2}}{V_0 - A_0} \left(\frac{A'}{c'}\right)^{1/2} \tag{10.7.7}$$

and

$$n = \frac{(Ac)^{1/2} + (A'c')^{1/2}}{V_0 - A_0} \cdot \left(\frac{A}{c}\right)^{1/2} \tag{10.7.8}$$

Substituting these values, we get the minimum value of the cost.

COROLLARY For the cost function given in relation (10.5.7), the pps estimator based on double sampling will be more precise than that based on one sample with simple random sampling, if

$$S_z^2 < \frac{(n' - n_0)\,n\,S_y^2}{(n' - 1)\,n_0} \tag{10.7.9}$$

where n_0 is the sample size for one sample.

SET OF PROBLEMS

10.1 If (y_i, x_i) follows a bivariate normal distribution with means (μ_y, μ_x), variances (σ_y^2, σ_x^2) and correlation coefficient ρ, then a double sampling estimator $\bar{y}_{1j} = \bar{y}_n + b\,(\bar{x}'_n - \bar{x}_n)$ is an unbiased estimator of μ_y and its sampling variance is given by

$$V(\bar{y}_{1d}) = \sigma_y^2\,(1 - \rho^2) \left(\frac{1}{n} - \frac{1}{n'}\right) \left[1 + \frac{1}{n-3}\right] + \frac{\sigma_y^2}{n'}$$

Also show that an unbiased estimator of the variance is obtained by

$$v(\bar{y}_{1d}) = \left(\frac{1}{n} - \frac{1}{n'}\right) \frac{Q}{n-3} + \frac{s_y^2}{n'}$$

where n and n' are sample sizes at the first and second phases, respectively,

and

$$Q = \sum_i^n [(y_i - \bar{y}_n) - b\,(x_i - \bar{x}_n)]^2$$

(Tikkiwal, 1960)

10.2 In order to estimate population mean, a random sample of n' units is selected to measure the auxiliary variate x and a sub-sample of size n is taken to observe the study variate y. Show that estimators

$$t_1 = \bar{x}'\,\frac{\bar{y}}{\bar{x}} \quad \text{and} \quad t_2 = \bar{r}\,\bar{x}' + \frac{n\,(n'-1)}{n'\,(n-1)}\,(\bar{y} - \bar{r}\bar{x})$$

where $\bar{r} = \dfrac{1}{n} \sum\limits_i^n \left(\dfrac{y_i}{x_i} \right)$, are two good estimators. Derive their sampling variances and compare their efficiencies.

10.3 A population of N units has been grouped into k unequal strata. A sample of size n is drawn from the whole population by simple random sampling, wr, such that n_i' units fall in the ith stratum, $i = 1, 2, \ldots, k$. By the method of proportional allocation n_i units are drawn from the ith stratum by a similar method $(i = 1, 2, \ldots, k)$. If P_{ds} denotes the relative precision of double sampling with stratification for estimating the population mean as compared to simple random sampling and if P_{st} denotes the relative precision of stratified sampling with proportional allocation when the sizes of strata are known, show that

$$P_{ds} = \frac{P_{st}}{[1 + (n'/n)(P_{st}-1)]} \qquad \text{(Cochran, 1963)}$$

10.4 In a population, initial random sample of size n' is drawn for forming strata and a sub-sample of size n is taken from it to study the main variate y. Show that the estimator $t = \sum\limits_i w_i' \, \bar{y}_i$ is an unbiased estimator and its sampling variance can be estimated unbiasedly by

$$v(t) = \frac{n'}{n-1} \sum_i \left[\left(w_i'^2 - \frac{w_i'}{n'} \frac{N-n'}{N-1} \right) \frac{s_i^2}{n_i} + \frac{(N-n') \, w_i'}{n' \, (N-1)} (\bar{y}_i - \sum_i^n w_i' \, \bar{y}_i)^2 \right]$$

10.5 A random sample of 32 cuts, each of size 1/200th of hectare, was taken in a crop-cutting survey for estimating the average yield per hectare of dry paddy. The green weight of paddy was recorded for all 32 cuts. For a sub-sample of 16 cuts out of 32, the dry weight of paddy was also recorded. The data on yield of paddy in kilograms per hectare were as follows:

S. N. of cut	Green wt (kg)	Dry wt (kg)	S. N. of cut	Green wt (kg)
1	7.6	6.9	17	3.9
2	5.8	5.3	18	5.3
3	8.5	7.9	19	6.5
4	6.3	5.7	20	8.1
5	5.1	4.7	21	3.8
6	4.9	4.6	22	6.6
7	5.7	5.1	23	5.5
8	7.9	7.2	24	3.6
9	6.4	5.9	25	4.0
10	5.4	4.9	26	5.0
11	6.1	5.6	27	5.9
12	3.8	3.4	28	4.8
13	6.2	5.7	29	6.4
14	6.6	6.0	30	5.8
15	7.8	7.3	31	5.4
16	6.4	5.8	32	7.0

Estimate the average yield per hectare of dry paddy and its sampling error by the following two methods.

(i) Taking the sub-sample of 16 cuts for which dry weight is available as simple random sample and

(ii) using the green weight as an ancillary variate estimate the relative efficiency of the two procedures.

10.6 A survey was conducted in a tehsil, in two consecutive years, for estimating the total bovine population. The tehsil has 77 village in all. A sample of 24 villages was selected, with equal probability, without replacement. In the second year, the survey was confined to a sub-sample of 12 villages selected from the 24 villages. The number of bovines in the selected villages are given below:

S. N. of village	Bovine population	
	First year	Second year
1	220	189
2	1337	758
3	654	583
4	1186	681
5	1787	1195
6	399	143
7	2212	1677
8	228	111
9	1416	643
10	813	616
11	528	267
12	666	176
13	237	—
14	681	—
15	1063	—
16	2743	—
17	502	—
18	405	—
19	1228	—
20	1085	—
21	472	—
22	466	—
23	675	—
24	542	—

Estimate the total number of bovines in the second year of the survey with and without using the figures in the first year and compare their efficiencies.

REFERENCES

Bose, Chameli, "Note on the sampling error in the method of double sampling," *Sankhya*, **6**, 330, (1943).

Cochran, W.G., *Sampling Techniques*, third edition, John Wiley & Sons (Second Edition, 1963), (1977).

Des Raj, "On double sampling for pps estimation," *Ann. Math. Statist.*, **35** 900-902. (1964).

_____ "On a method of using multiauxiliary information in sample surveys," *J. Amer. Statist. Assoc.*, **60**, (1965).

Khan, S. and T.P. Tripathi, "The use of multivariate auxiliary information in double sampling," *J. Ind. Statist. Assoc.*, **5**, 42-48, (1967).

Neyman, J., "Contribution to the theory of sampling human populations", *J. Amer. Statist. Assoc.*, **33**, 101-116, (1938).

Rao, J.N.K., "On double sampling for stratification and analytical surveys," *Biometrika*, **60**, 125-133, (1973).

Robson, D.S., "Multiple sampling of attributes," *J. Amer. Statist. Assoc.*, **47**, 203-215, (1952).

Robson, D.S., and A.J. King, "Double sampling and the curtis impact surveys," *Cornell Uni. Agr. Exp. St. Mem.*, 231, (1953).

Singh, D. and B.D. Singh, "Some contributions to two phase sampling," *Aus. J. Statist.*, **2**, 45-67, (1965).

_____ "Double sampling for stratification on successive occasions," *J. Amer. Statist. Assoc.*, (1966).

Srinath, K.P., "Multiphase sampling in non-response problems," *J. Amer. Statist. Assoc.*, **16**, 583-586, (1971).

Tikkiwal, B.D., "On theory of classical regression and double sampling estimation," *J.R. Statist. Soc.*, **22**, 131-138, (1960).

11

Successive Sampling

Success depends on three things: who says it, what he says, how he says; and of these three things, what he says is the least important.

John Morley

11.1 REPETITIVE SURVEYS

Surveys often have to be repeated on many occasions (over years or seasons) for estimating the same characteristic at different points of time. The information collected on previous occasions can be used to study the change or the total value over occasion for the character, in addition to studying the average value for the most recent occasion. For example, in milk yield surveys, one may be interested in estimating (i) the average milk yield for the current season, (ii) the change in average milk yield for two different seasons, and (iii) the total milk production for the year. The successive method cf sampling consists of selecting sample units on different occasions such that some units are common with samples selected on previous occasions. If sampling on successive occasions is done according to a specific rule, with partial replacement of sampling units, it is known as *rotation sampling*. Hansen *et al* (1955) and Rao and Graham (1964) have discussed rotation designs for successive sampling. Singh and Singh (1965), Singh (1968), and Singh and Kathuria (1969) have extended successive sampling for many other sampling designs.

There are many considerations for successive sampling, such as the nature of matching in the sample at different occasions, the fraction of sample to be retained, cost, efficiency and administrative convenience. The main point of interest is whether the same sample should be used on every occasion, or an entirely new or mixture of old and new samples be used. Generally, the main objective of successive surveys is to estimate the change with a view to studying the effects of the forces acting upon the population. For this, it is better to retain the same sample from occasion to occasion. For populations where the basic objective is to study the overall average or the total, it is better to select a fresh sample for every occasion. If the objective is to estimate the average value for the most recent occasion, the retention of a part of the sample over occasions provides efficient estimates as compared to other alternatives.

There are a few more points which should also be borne in mind in connection with sampling on successive occasions. First, repeated survey of the same sampling units may invite some personal biases and the desired information may not be ascertained accurately. Second, repeated survey may bring about some modification in these sampling units as compared to the rest of the population. A solution to this problem was proposed by Yates (1949) for two occasions. An extension for more than two occasions has been discussed by Patterson (1950). For details, the reader is referred to these references. We shall restrict ourselves to discussing the theory as to how the sample on the first occasion could possibly be used to make estimates on the second occasion.

11.2 SAMPLING ON TWO OCCASIONS

Consider a population of N units. Assume that the number of units remains the same but the value of the unit changes over occasions. Further, let the variability of the population remain the same over occasions. For simplicity, assume that a sample of size n is drawn by the simple random sampling method, wr, from the population, on the first occasion. Out of this sample of size n, a portion np ($= m$), of the sample is retained (*matched*) on the second occasion and supplemented by an additional sample of size (*unmatched*) nq ($= u$) selected from the remaining $(N - n)$ units, where q is the replacement fraction such that $p + q = 1$. Thus, the second sample is also of size n.

For developing the relevant theory, the notation x and y will be used for the values of the characteristic under study on the previous and current occasions, respectively. Further, let

$\bar{x}_n$ = mean per unit for the first occasion based on all n units of the sample

$\bar{x}_m$ = mean per unit for the first occasion, for np units that are common to both the occasions

$\bar{x}_u$ = mean per unit for the first occasion, for nq units that are selected on the second occasion only

$\bar{y}_m$ = mean per unit for the second occasion for np units that are common to the sample on both occasions

$\bar{y}_u$ = mean per unit for the second occasion for nq units that are selected on the second occasion only.

11.2.1 Estimator of Mean on the Current Occasion

Suppose we wish to estimate the mean for the second occasion. We can have two independent estimators. A difference estimator of the mean, on the second occasion, based on matched sample, is given by

$$\bar{y}'_m = \bar{y}_m - \bar{x}_m + \bar{x}_n \qquad (11.2.1)$$

Another estimator of the population mean on the second occasion, independent of $\bar{y}'_m$, can be made on the basis of u units selected on the second occasion, i.e.

$$\bar{y}_u = \sum_i^u \frac{y_i}{u} \qquad (11.2.2)$$

It can be easily shown that both estimators are unbiased.

For large N relative to sizes of samples to be drawn, after ignoring fpc, variances can be written as

$$V(\bar{y}_u) = \frac{V_y}{nq} \qquad (11.2.3)$$

and $$V(\bar{y}'_m) = [1 + (1 - 2\rho) q] V_y/np \qquad (11.2.4)$$

These estimators can be improved by modifying them by the procedure discussed in the following theorem.

THEOREM 11.2.1 Show that the best linear combined estimator of the average of the study variate for the current occassion is given by

$$\bar{y} = \phi_2 \bar{y}_u + (1 - \phi_2) \bar{y}'_m \qquad (11.2.5)$$

where ϕ_2 is a positive function of variances, less than 1. Derive its sampling variance and the optimum value of q.

Proof To obtain the best overall estimator, we choose ϕ_2 which minimizes the variance of relation (11.2.5), i.e. we minimize the expression

$$(1 - \phi_2)^2 \, V(\bar{y}'_m) + \phi_2^2 \, V(\bar{y}_u).$$

It can be easily shown that for a minimum of the above expression, we have

$$\phi_2 \, V(\bar{y}_u) = (1 - \phi_2) \, V(\bar{y}'_m)$$

giving
$$\phi_2 = \frac{V(\bar{y}'_m)}{V(\bar{y}_u) + V(\bar{y}'_m)} = \frac{1/V(\bar{y}_u)}{1/V(\bar{y}_u) + 1/V(\bar{y}'_m)}$$

$$= \frac{W_u}{(W_u + W_m)} \tag{11.2.6}$$

where W_u and W_m are the inverse variances of $\bar{y}_u$ and $\bar{y}'_m$, respectively.

By substituting the value in relation (11.2.5), we get

$$\bar{y} = \frac{(W_u \bar{y}_u + W_m \bar{y}_m)}{(W_u + W_m)} \tag{11.2.7}$$

By the principle of least squares, the variance of $\bar{y}$ can be derived and is given by

$$V(\bar{y}) = \frac{V_y}{n} \, [1 + (1 - 2\rho) \, pq]^{-1} \tag{11.2.8}$$

It should be noted that this variance reduces to V_y/n in both cases of complete matching ($q = 0$) and no matching ($q = 1$). The optimum value of q which minimizes the above variance expression can be obtained by

$$q_{\text{opt}} = [1 + \sqrt{2(1 - \rho)}]^{-1} \tag{11.2.9}$$

By substituting q_{opt}, the minimum variance can be worked out as

$$V_{\text{min}}(\bar{y}) = \frac{V_y}{2n} \, [1 + \sqrt{2(1 - \rho)}] \tag{11.2.10}$$

$$= \frac{V_y}{n} \, [\tfrac{1}{2} + \sqrt{(1 - \rho)/2}] \tag{11.2.11}$$

If a completely independent sample is taken on the second occasion, the estimate, $\bar{y}_u = \sum_{i}^{n} y_i/n$ has sampling variance V_y/n, which will be larger than $V_{\min}(\bar{y})$ for $\rho > \frac{1}{2}$. If the same sample is taken on the second occasion, the variance will remain the same. Hence, for making current estimates, it is always better to replace the sample partially.

COROLLARY If the difference estimator in relation (11.2.5) is replaced by a regression estimator

$$\bar{y}''_m = \bar{y}_m + b(\bar{x}_n - \bar{x}_m) \qquad (11.2.12)$$

where b is the regression coefficient estimate from the common part, the linear estimator $\bar{y}$ will have the sampling variance

$$V(\bar{y}) = \frac{(1 - \rho^2 q)}{(1 - \rho^2 q^2)} \frac{V_y}{n} \qquad (11.2.13)$$

and the optimum value of the replacement fraction is obtained by

$$q_{\text{opt}} = [1 + \sqrt{(1 - \rho^2)}]^{-1} \qquad (11.2.14)$$

The results given in the next theorem are based on the works of Yates (1949) and Patterson (1950). A more generalized treatment is given in Hansen, Hurwitz and Madow (1953). We shall discuss the outlines here, in brief, with the following theorem which leads to the same results as in Theorem 11.2.1.

THEOREM 11.2.2 A best linear estimator of the mean for the second occasion is given by the weighted estimate

$$\bar{y}_w = \frac{p}{1 - q^2 \rho^2} [\bar{y}_m + q\rho (\bar{x}_u - \bar{x}_m)] + \frac{q(1 - q \rho^2)}{1 - q^2 \rho^2} \bar{y}_u \qquad (11.2.15)$$

with its sampling variance

$$V(\bar{y}_w) = \frac{V_y}{n} \frac{(1 - q \rho^2)}{(1 - q^2 \rho^2)} \qquad (11.2.16)$$

Derive its optimum q and the minimum variance.

Proof Let us consider a linear estimator of the form

$$\bar{y}_w = a\bar{x}_u + b\bar{x}_m + c\bar{y}_m + d\bar{y}_u$$

Since $\qquad\qquad E(\bar{x}_u) = E(\bar{x}_m) = \bar{X}$

and $\qquad\qquad E(\bar{y}_u) = E(\bar{y}_m) = \bar{Y}$

we have $\qquad\quad E(\bar{y}_w) = (a + b) \bar{X} + (c + d) \bar{Y}$

If it is desired that $\bar{y}_w$ be an unbiased estimator of $\bar{Y}$, we must have

$$a + b = 0, \qquad c + d = 1$$

Hence $\qquad \bar{y}_w = a(\bar{x}_u - \bar{x}_m) + c\bar{y}_m + (1 - c) \bar{y}_u \qquad (11.2.17)$

Now, let us proceed to work out its sampling variance. If the estimator is a minimum variance unbiased estimator, it must be uncorrelated with every zero function (Rao, 1952). Thus, we have

$$\text{cov}(\bar{y}_m, \bar{y}_w) = \text{cov}(\bar{y}_u, \bar{y}_w) \qquad \text{(A)}$$

$$\text{cov}(\bar{x}_m, \bar{y}_w) = \text{cov}(\bar{x}_u, \bar{y}_w) \qquad \text{(B)}$$

Putting

$$\text{cov}(\bar{y}_m, \bar{x}_u) = \text{cov}(\bar{y}_m, \bar{y}_u) = 0,$$

$$\text{cov}(\bar{y}_m, \bar{x}_m) = \frac{\rho S_y^2}{np}$$

$$\text{cov}(\bar{y}_m, \bar{y}_m) = \frac{S_y^2}{np}$$

$$\text{cov}(\bar{y}_u, \bar{x}_u) = \text{cov}(\bar{y}_u, \bar{x}_m) = 0$$

and

$$\text{cov}(\bar{y}_u, \bar{y}_u) = \frac{\sigma_y^2}{nq}$$

From (A) and (B), and putting $S_y^2 = \sigma^2$, we get

$$\frac{-a\,\rho\sigma^2}{np} + \frac{c\sigma^2}{np} = \frac{(1-c)\,\sigma^2}{nq}$$

and

$$\frac{-a\sigma^2}{np} + \frac{c\rho\sigma^2}{np} = \frac{a\sigma^2}{nq}$$

Solving for a and c, we have

$$a = \frac{pq\rho}{(1 - q^2\,\rho^2)}$$

$$c = \frac{p}{(1 - q^2\,\rho^2)}$$

Substituting these values in relation (11.2.17), we get $\bar{y}_w$ in the desired form. Hence, its variance is

$$V(\bar{y}_w) = \frac{V_y}{n} \frac{(1 - q\,\rho^2)}{(1 - q^2\,\rho^2)}$$

Equating to zero the derivative of $V(\bar{y}_w)$ w.r.t. to q, we find that, for a given sample size, $V(\bar{y}_w)$ will have its minimum value if we choose

$$q_{\text{opt}} = [1 + (1 - \rho^2)^{1/2}]^{-1} \qquad (11.2.18)$$

Therefore, $\quad V_{\min}(\bar{y}_w) = \frac{V_y}{2n}[1 + (1 - \rho^2)^{1/2}] = \frac{S_y^2}{2nq} \qquad (11.2.19)$

COROLLARY 1   If information on the second occasion alone is used, the estimator will be

$$\bar{y}' = p\,\bar{y}_m + q\,\bar{y}_u$$

with sampling variance

$$V(\bar{y}') = \frac{S_y^2}{n}$$

Hence, relative precision $= \dfrac{(1 - q\,\rho)}{(1 - q^2\,\rho^2)}$ (11.2.20)

It may be noted that the above variance reduces to σ^2/n in the case of complete matching ($q = 0$) and no matching ($q = 1$). Also, if $\rho = 0$, we have $p = \frac{1}{2}$ and if $\rho = \pm 1$, we have $p = 0$. This shows that the matching proportion should not exceed $\frac{1}{2}$. It can also be seen that the gain in relative precision decreases as ρ increases. Hence, for current estimates, there should not be more than 50 per cent matching in the samples.

COROLLARY 2   If W_u and W_m are estimated from the samples and their estimates are

$$\frac{1}{w_u} = \frac{s_y^2}{nq}$$

and

$$\frac{1}{w_m} = \frac{(1 - r^2)}{np}\,s_y^2 + r^2\,s_y^2$$

where r is the estimate of correlation coefficient ρ.
Then, an estimate of $V(\bar{y}_w)$ is obtained by

$$v(\bar{y}_w) = \frac{s_y^2\,(1 - r^2 q)}{n(1 - r^2 q^2)}$$ (11.2.21)

11.2.2 Estimation of Change

In the previous section we have studied how and under what circumstances, the estimator of mean for the second occasion can be improved by utilizing the information collected on the first occasion. Another important problem in sampling on two occasions is to estimating the change in the total value of the study variate during the period. The estimation of change presents a somewhat different problem of applying information provided by the samples. In this section, we shall discuss the problem of estimating the change in the population mean over a period, based on samples taken on two occasions.

One obvious estimator of the change $\bar{Y} - \bar{X}$ is given by

$$\bar{d} = p(\bar{y}_m - \bar{x}_m) + q(\bar{y}_u - \bar{x}_u) \qquad (11.2.22)$$

with its sampling variance

$$V(\bar{d}) = 2(1 - p\,\rho)\,\frac{S_y^2}{n} \qquad (11.2.23)$$

However, the other unbiased estimator of the change is

$$\bar{d}' = p\,\frac{(\bar{y}_m - \bar{x}_m)}{(1 - \rho q)} + \frac{q(1 - \rho)\,(\bar{y}_u - \bar{x}_u)}{(1 - q\rho)} \qquad (11.2.24)$$

and its sampling variance is given by

$$V(\bar{d}') = \frac{2(1 - \rho)\,S_y^2}{n(1 - q\rho)} \qquad (11.2.25)$$

By replacing S_y^2 by s_y^2 and ρ by r, we can get an estimate of $V(\bar{d}')$, which may be written as

$$v(\bar{d}') = \frac{2(1 - r)\,s_y^2}{n(1 - rq)} \qquad (11.2.26)$$

If we consider a more general linear estimator of the change in the form

$$\bar{d}'' = a\bar{x}_u + b\bar{x}_m + c\bar{y}_m + d\bar{y}_u \qquad (11.2.27)$$

With the condition that it provides an unbiased estimate of the change $\bar{Y} - \bar{X}$, we get $a + b = -1$ and $c + d = 1$.

Proceeding on similar lines as for the estimator of the mean on the second occasion, we find an estimate that minimizes the variance. It is given by

$$\bar{d}'' = \frac{q(1 - q\rho^2)}{(1 - q^2\rho^2)}\,(\bar{y}_u - \bar{x}_u) + \frac{p}{(1 - q^2\rho^2)}\,(\bar{y}_m - \bar{x}_m)$$

$$+ \frac{pq\rho}{(1 - q^2\rho^2)}\left[(\bar{x}_u - \bar{x}_m)\,\frac{S_y}{S_x} - (\bar{y}_u - \bar{y}_m)\,\frac{S_x}{S_y}\right] \qquad (11.2.28)$$

In case $S_x = S_y$, the estimator $\bar{d}''$ reduces to $\bar{d}'$. Further, if the ratio of $V(\bar{d})$ to $V(\bar{d}')$ is considered, we get the relative precision

$$\text{RP} = \frac{(1 - p\rho)(1 - q\rho)}{(1 - \rho)} = \frac{(1 + pq\rho^2 - \rho)}{(1 - \rho)} \qquad (11.2.29)$$

Hence, the gain in the relative precision is given by

$$\text{Gain} = \frac{pq\rho^2}{(1 - \rho)} \qquad (11.2.30)$$

It can be seen that a higher gain in precision can be achieved by using the estimator $\bar{d}'$ if the correlation coefficient is sufficiently large. For any positive value of ρ, the optimum value of q is 0, i.e. $V(\bar{d}')$ is minimum if there is a complete matching of samples on both occasions. Thus, we find that the optimum choice of q for estimating the change from one occasion to another is quite contrary to the optimum for estimating the mean for the second occasion. In practice, we can make a compromise between these situations by assigning due weights to the importance of the estimates of change and mean. For details, the reader is referred to Hansen, Hurwitz and Madow (1953).

11.2.3 Estimation of the Sum of Means

The next problem is to estimate the sum of the means over two occasions. The obvious estimator of this sum is the sum of the estimators, which can be written as

$$\hat{S} = p(\bar{y}_m + \bar{x}_m) + q(\bar{y}_u + \bar{x}_u) \qquad (11.2.31)$$

with its sampling variance

$$V(\hat{S}) = 2(1 + p\rho)\frac{S_y^2}{n} \qquad (11.2.32)$$

A better unbiased estimator of the sum is

$$\hat{S}' = \frac{p}{1 + q\rho}(\bar{y}_m + \bar{x}_m) + \frac{q(1 + \rho)}{1 + q\rho}(\bar{y}_u + \bar{x}_u) \qquad (11.2.33)$$

and its sampling variance is

$$V(\hat{S}') = \frac{2(1 + \rho)S_y^2}{(1 + q\rho)n} \qquad (11.2.34)$$

which can be estimated by

$$v(\hat{S}') = \frac{2(1 + r)s_y^2}{(1 + qr)n} \qquad (11.2.35)$$

If we consider a more general estimator of the sum in the form given by relation (11.2.27) we can derive it by the same approach discussed above. In case $S_x = S_y$, the estimator conforms to relation (11.2.32). Further, if the ratio of $V(\hat{S})$ to $V(\hat{S}')$ is considered, we get the relative precision

$$RP = \frac{(1 + \rho + pq\rho^2)}{(1 + \rho)} \qquad (11.2.36)$$

Hence the gain in relative precision is given by

$$\text{Gain} = \frac{pq\rho^2}{(1 + \rho)} \qquad (11.2.37)$$

If ρ is positive, the best matching policy for estimating the sum on occasions is to retain $p = 0$ or $q = 1$, which means that an independent sample should be drawn at the second occasion. It can also be seen that the gain in precision is not substantial even for high values of ρ and the estimator using the simple average of each occasion can equally be used.

The reader should note that there is generally lack of agreement between estimates of change, the estimate of the mean on the second occasion, and the estimate of the sum of means for both occasions. If the suggested estimates are used, the estimate of the mean at the current occasion will not be the sum of the estimate of the mean at the first occasion and the estimate of the change. To overcome this difficulty, two approaches are suggested. One is to revise the estimate for the first occasion in the light of the information obtained from the sample at the second occasion. Patterson (1950) suggested an estimator to this alternative, which could be delivered on symmetry basis from relation (11.2.15) as

$$\bar{x}_w = \frac{p}{1 - q^2\rho^2} [x_m + q\rho (\bar{y}_u - \bar{y}_m)] + \frac{q(1 - q\rho^2)}{(1 - q^2\rho^2)} \bar{x}_u \qquad (11.2.38)$$

Its variance will be the same as the variance of the estimated mean on the second occasion. It can be easily seen that the estimated change will be the difference between the estimate of the mean on the second occasion obtained by relation (11.2.15) and the estimate of the mean on the first occasion obtained by relation (11.2.30). A second alternative is to estimate the mean of the second occasion and the change in such a way that the estimated change will be the difference between the estimate of the mean on the second occasion and the estimate of the mean on the first occasion. But, in this procedure, neither of these estimates will have smaller variance as given by the estimates discussed above.

Another problem of interest is that all these estimates depend upon the values of ρ, p and q which are generally not known in advance. If the value of ρ is assumed on the basis of past experience, the estimates will remain unbiased but their variances will be increased. If ρ is estimated from the sample, the estimates are no longer unbiased and all the above treatments will not be rigorously valid.

The theory of regression and double sampling estimation for finite population, when the regression coefficient is estimated from the sample, has been derived by Tikkiwal (1960). This approach has been extended by Prabhu Ajgaonkar (1968) when different constants are estimated from the sample. Kathuria and Singh (1971) have discussed at length some alternative procedures in two-stage sampling on successive occasions. The minimum variance unbiased linear estimators for multi-purpose and repeat surveys in two-stage designs have been discussed by Singh and Singh (1973) and Singh, Singh and Srivastava (1976), respectively. For double sampling for stratification on successive occasions, the reader is referred to Singh and Singh (1965).

EXAMPLE 11.1 In an Arecanut survey, the total number of arecanut trees in 30 randomly selected villages in a sub-division were counted for two consecutive years, 1980-81 and 1981-82, to study the change in the number of trees. The first 10 villages were common to the two years. The total number of villages in a sub-division was 1978.

S.N. of villages	Number of arecanut trees		S.N. of villages	Number of arecanut trees	
	1980-81	1981-82		1980-81	1981-82
1	1954	1933	26	12	
2	400	406	27	4410	
3	1536	1677	28	320	
4	1073	1096	29	4531	
5	364	357	30	1862	
6	205	246	31		0
7	41	16	32		93
8	1893	2132	33		1983
9	607	680	34		994
10	75	86	35		145
11	834		36		10,648
12	5840		37		0
13	0		38		5604
14	0		39		211
15	2371		40		5641
16	0		41		2251
17	4187		42		862
18	20		43		584
19	0		44		729
20	517		45		500
21	0		46		607
22	0		47		47
23	0		48		3575
24	5536		49		0
25	2541		50		0

(i) Estimate the change in the average number of arecanut trees in two years and the average number of arecanut trees in the subdivision in the second year, using available information for the previous year.

(ii) Give the standard errors of these estimates.

With notations having their usual meanings, described earlier, we have

$$N = 1878, \; n = 30, \; p = \frac{1}{3} \text{ and } q = \frac{2}{3}$$

$$\bar{x}_u = \frac{32981}{20} = 1649.05$$

$$\bar{x}_m = \frac{8848}{10} = 884.80$$

$$\bar{y}_u = \frac{34420}{20} = 1721.00$$

$$\bar{y}_m = \frac{8017}{10} = 801.70$$

$$\bar{x}_n = \frac{41829}{30} = 1394.30$$

$$s_{xu}^2 = \frac{1}{19} [138873081 - 54387318.05] = 4446619.1$$

$$s_{yu}^2 = \frac{1}{19} [201592612 - 59236820.0] = 7492410.1$$

$$s_{xm}^2 = \frac{1}{9} [13614666 - 7828710.4] = 642884.0$$

$$s_{ym}^2 = \frac{1}{9} [12660535 - 6427228.9] = 692589.5$$

$$s_{xm,ym} = \frac{1}{9} [12723198 - 7093441.6] = 625528.5$$

$$b = \frac{s_{xm,ym}}{s_{xm}^2} = 0.9730$$

$$r = \frac{625528.5}{(642884.0)^{1/2} \times (692589.5)^{1/2}} = 0.9374$$

$$s_y^2 = \frac{1}{29} [214253147 - (1414.57)^2 \times 30] = 5318030.9$$

$$W_u = \frac{1}{\frac{s_y^2}{nq}} = 0.0000037$$

and $$W_m = \frac{1}{(1 - r^2) \frac{s_y^2}{np} + \frac{r^2 s_y^2}{n}} = 0.0000045$$

An estimate of the average number of arecanut trees on the current occasion, using the information from the previous occasion, is given by

$$\bar{y} = \phi \, \bar{y}_u + (1 - \phi) \, \bar{y}'_m$$

where $$\bar{y}'_m = \bar{y}_m + b (\bar{x}_n - \bar{x}_m)$$

and $$\phi = \frac{W_u}{W_u + W_m}$$

Here we can calculate

$$\bar{y}'_m = 801.70 + 0.9730 (1394.30 - 884.80) = 1297.44$$

and $$\phi = \frac{0.0000037}{0.0000045 + 0.0000037} = 0.45$$

Thus, the estimate is given by

$$\bar{y} = 0.45 \times 1721.50 + (1 - 0.45) \times 1297.44 = 1488.09$$

The estimate of variance of $\bar{y}$ is given by

$$v(\bar{y}) = \frac{s_y^2 (1 - r^2 q)}{n (1 - r^2 q^2)}$$

On substituting the values, we get

$$v(\bar{y}) = \frac{5318030.98 (1 - 0.9374^2 \times 2/3)}{30 [1 - (0.9374)^2 \times (2/3)^2]} = 120466.03$$

Standard error of $\bar{y} = \sqrt{120466.03} = 347.08$

The estimate of change in the average number of arecanut trees on two years is given by

$$\bar{d} = \frac{p}{1 - rq} (\bar{y}_m - \bar{x}_m) + \frac{q (1 - r)}{(1 - rq)} (\bar{y}_u - \bar{x}_u)$$

On substituting the values of the terms involved in the above expression, we get

$$\bar{d} = - 65.82$$

The estimate of the variance of $\bar{d}$ is given by

$$v(\bar{d}) = \frac{2(1 - r) s_y^2}{n(1 - rq)} = 59167.99$$

Therefore standard error of $\bar{d} = \sqrt{59167.99} = 243.24$

11.3 SAMPLING ON MORE THAN TWO OCCASIONS

When information is available for more than two occasions, the esti mate on the current occasions can be worked out through a recurrence formula. The method is slightly modified from that for two occasions. Let the characteristic be observed on n units on the first occasion. On the second occasion, $m_2 + u_2$ units are observed of which m_2 are common (matched) and u_2 are uncommon (unmatched) to the first occasion. On the hth occasion let $m_h + u_h$ be observed, m_h and u_h being the numbers of units matched and unmatched, respectively, with the $(h - 1)$th occasion. Let us denote

$\bar{y}_{mh} = $ the mean of matched portions on the hth, occasion

$\bar{y}_{uh} = $ the mean of unmatched portions on the hth occasion

$\bar{x}_{mh} = $ the mean of matched portions on the preceding occasion

$\bar{y}_{h-1} = $ the mean of the whole sample on the $(h - 1)$th occassion

Now we may form an estimator of the population mean on the hth occasion as

$$\bar{y}'_{mh} = \bar{y}_{mh} + \beta_{h, h-1}(\bar{y}_{h-1} - \bar{x}_{mh}) \qquad (11.3.1)$$

where $\beta_{h, h-1}$ is some specified value of the regression coefficient.

Hence, the best combined estimator of the average of the character under study for the current occasion is given by

$$\hat{\bar{Y}}_h = \phi_h \bar{y}_{uh} + (1 - \phi_h) \bar{y}_{mh} \qquad (11.3.2)$$

where ϕ_h is some positive number less than 1. Weighting inversely as the variance, we have

$$\phi_h = \frac{W_{uh}}{(W_{uh} + W_{mh})} \qquad (11.3.3)$$

where W_{uh} and W_{mh} have meanings similar to that given in relation (11.2.6)

A general theory has been investigated by Patterson (1950) and Tikkiwal (1951) when the regressions $\beta_{h,h-1}$ are assumed to be known. With increasing h, ϕ_h rapidly tends to a limiting value which depends on ρ and q_h. Its limiting value is obtained by

$$\phi = \lim_{h \to \infty} \phi_h = -\frac{(1 - \rho^2) + [(1 - \rho^2)\{(1 - \rho^2) + 4\,\rho^2 pq\}]^{1/2}}{2\,p\rho^2}$$

Hence, the limiting value of $V(\hat{\bar{y}}_h)$ is

$$V(\hat{\bar{Y}}_h) = \frac{\phi S_y^2}{nq_{\text{opt}}} \qquad (11.3.4)$$

The details of the limiting values of $V(\hat{\bar{y}}_h)$, when q_{opt} is used, have been given by Cochran (1977). We are discussing these outlines in brief:

Let
$$V(\hat{\bar{Y}}_h) = (W_{uh} + W_{mh})^{-1} = \frac{G_h S_y^2}{n} \qquad (11.3.5)$$

where G_h is the ratio which can be derived from ϕ_h and can be written as

$$G_h^{-1} = 1 - p_h + \left[\rho^2 G_{h-1} + \frac{(1 - \rho^2)}{p_h}\right]^{-1} \qquad (11.3.6)$$

The variance of $\hat{\bar{Y}}_h$ will be minimum when the term in relation (11.3.6) is maximum. Applying calculus methods, we get

$$p_h = \frac{(1 - \rho^2)^{1/2}}{G_{h-1}} \cdot [1 + (1 - \rho^2)^{1/2}] \qquad (11.3.7)$$

which gives

$$G_h^{-1} = 1 + \frac{[1 - (1 - \rho^2)^{1/2}]^2}{\rho^2 G_{h-1}} \qquad (11.3.8)$$

From relations (11.3.7) and (11.3.8), the limiting values of p_h and G_h are obtained, and we have

$$p = \lim p_h = \tfrac{1}{2},$$

which shows that the matching fraction to be used is $\tfrac{1}{2}$, at the most, whatever the value of ρ may be. Similarly,

$$G = \lim_{h \to \infty} G_h = 2\,\frac{[(1 - \rho^2)^{1/2} - (1 - \rho^2)]}{\rho^2} \qquad (11.3.9)$$

Thus, the limiting value of $V(\hat{\bar{Y}}_h)$ can be obtained.

Similarly, an appropriate estimator of the change between h and $h-1$ occasions can be obtained by $\hat{\bar{Y}}_h - \hat{\bar{Y}}_{h-1}$. Its sampling variance is given by

$$V(\hat{\bar{Y}}_h - \hat{\bar{Y}}_{h-1}) = 2\phi[1 - \rho(1 - \phi)]\frac{S_y^2}{nq} \qquad (11.3.10)$$

11.4 SAMPLING FOR A TIME SERIES

The theory of sampling on successive occasions can also be applied, with slight modifications, to time series data for estimating the mean value at the current occasion or change, etc. Eckler (1955) has discussed rotation sampling designs where, every month, an independent random sample is taken each of size n. During the survey, each member of the sample provides information for the current and previous periods. After each count an unbiased estimate $\bar{y}_h$ is made for the mean value of the current period. From the same count, an unbiased estimate $\bar{x}_h$ is also made for the previous period. Assuming that ρ is the correlation coefficient between consecutive periods and σ^2 remains unchanged for each period, the best linear estimator $\hat{\bar{Y}}_h$ of the hth period is given by

$$\hat{\bar{Y}}_h = \bar{y}_h + c_h(\hat{\bar{Y}}_{h-1} - \bar{x}_{h-1}) \qquad (11.4.1)$$

and

$$\hat{\bar{Y}}_{h-1} = \bar{y}_{h-1} + c_{h-1}(\hat{\bar{Y}}_{h-2} - \bar{x}_{h-2}) \qquad (11.4.2)$$

given,

$$c_1 = 0$$

In the special case where $c_h = 1$, an estimator analogous to the changed estimator is obtained. Applying the conditions for a minimum variance estimator discussed in the previous section, we have sampling variance

$$V(\hat{\bar{Y}}_h) = (1 - \rho c_h)\frac{\sigma^2}{n} \qquad (11.4.3)$$

A detailed discussion of the procedure is given by Hansen, Hurwitz and Madow (1953). Eckler (1955) and Woodruff (1971) have discussed estimation of changes in higher level rotation sampling designs. In another study by Rao and Graham (1964), a different approach has been investigated. Scott and Smith (1974) have also examined the performance of composite estimators in time series data in repeated surveys. The reader is referred to these for further study.

SET OF PROBLEMS

11.1 A survey is planned on two occasions to estimate $\theta = a\bar{Y}_1 + b\bar{Y}_2$, where a, b are known constants and $\bar{Y}_1$, $\bar{Y}_2$ are the means on the two occasions. A sample of n units is taken on the first occasion. On the second occasion, a sub-sample of m units from the first occasion is retained and an independent sample of $u = n - m$ units is taken. If the total cost of the survey is represented by the function

$$C = C_0 + cn + C_1 m + C_2 u,$$

obtain the optimum values of n and m so that, for a given cost, the variance of the estimator is minimized.

11.2 From a population of size N, a large sample of size n is drawn by simple random sampling method, wr, in the first phase, to obtain information on the auxiliary variate x. In the second phase, a sub-sample of n_1 units is drawn from the first phase sample and a sample of n_2 units is taken independently from the whole population by simple random sampling method, wr, to study the main variate y. Considering two estimators of the population mean $\bar{y}$, (i) regression estimator based on the two-phase sample of (n_1, n_2) units and (ii) unbiased estimator based on direct sample of n_2 units, obtain the best combined linear estimator and derive its sampling variance.

(Patterson, 1950)

11.3 In sampling on two occasions, n units are selected with simple random sampling, wr, on the first occasion. On the second occasion, a sub-sample of np of these units selected by a similar method, is retained for the second occasion and supplemented by a sample of $n(1-p)$ units taken independently by a similar method. To estimate the population mean on the second occasion $\bar{Y}_2$, an unbiased estimator is taken,

$$\bar{y} = \lambda_1\bar{y}_1 + \lambda_2\bar{y}_2 + \mu_1 y_1' + \mu_2\bar{y}_2'$$

where λ_i, μ_i $(i = 1, 2)$ are constants and $\bar{y}_i\,\bar{y}_i'$ $(i = 1, 2)$ are the sample means for the ith occasion, based on the common np units and $n-np$ units, respectively.

Determine the optimum values of λ_i, μ_i and p which would minimize the sampling variance and find an expression for the minimum variance.

(Hansen, Hurwitz and Madow, 1953)

11.4 In simple random sampling on two occasions, the estimator

$$\bar{y}_m' = \phi\bar{y}_u + (1 - \phi)\,\bar{y}_m'$$

where

$$\bar{y}_m' = \bar{y}_m + \hat{\beta}(\bar{x}_n - \bar{x}_m)$$

is the best combined linear estimator. Its variance is given by

$$V(\bar{y}_m') = \frac{S^2}{n}\left[\frac{\phi^2}{\lambda} + \frac{(1 - \phi^2)}{\mu}(1 + \lambda)(1 - 2\rho)\right]$$

for large N,

where

$$\mu = m/h, \lambda = u/n$$

Show that the optimum value of ϕ lies between λ and $\lambda/(1 + \lambda)$ if correlation coefficient $\rho \geqslant 1/2$.

(Cochran, 1963)

11.5 A simple random sample of 28 cities was selected from 196 cities in 1980-81 for estimating the total number of inhabitants in 196 cities. Of these, 10 cities were retained, which constituted the matched sample for observation in 1981-82, and a fresh sample (unmatched sample) or 18 cities was selected from the remaining population independently. The data on the total number of inhabitants (in 000's) obtained from the sampled cities in 1980-81, as also from matched and unmatched samples in 1981-82, are given below:

S. N. of cities	No. of inhabitants (000's)		S. N. of cities	No. of inhabitants (000's)	
	1980-81	1981-82		1980-81	1981-82
1	40	64	24	66	
2	40	60	25	60	
3	56	142	26	46	
4	64	77	27	2	
5	64	63	28	507	
6	77	89	29		80
7	44	53	30		143
8	50	64	31		67
9	121	113	32		50
10	179	260	33		464
11	76		34		48
12	138		35		63
13	67		36		115
14	29		37		69
15	381		38		459
16	23		39		104
17	37		40		183
18	120		41		106
19	61		42		86
20	387		43		57
21	93		44		65
22	172		45		50
23	78		46		634

Estimate, along with the standard error, the change in number of inhabitants in 1981-82 as compared to, 1980-81, the total number of inhabitants in 1981-82 and also the revised estimate of the total number of inhabitants in 1980-81, using the information obtained for 1981-82.

11.6   For estimating the area and production of arecanut and coconut in Assam, a pilot sample survey was jointly conducted by the Indian Central Arecanut and Indian Central Coconut Committees during the year 1958-59. The design of the survey was one of stratified two stage sampling with subdivisions in a district as strata, villages as first stage units and trees in a village as second stage units. The number of arecanut trees in the 33 randomly selected villages from the Nowgong subdivision of Nowgong district for the two consecutive years 1960-61 and 1961-62 are given below:

Sampled villages kept fixed over both years			Sampled villages not kept fixed over both years		
S. N. of selected villages	No. of arecanut trees		S. N. of selected villages	No. of arecanut trees	
	1960-61	1961-62		1960-61	1961-62
1	10	6	18	2371	146
2	1584	1933	19	0	10648
3	400	406	20	4187	0
4	1536	1677	21	5604	20
5	1073	1097	22	0	211
6	364	357	23	517	5641
7	205	246	24	0	2251
8	41	16	25	0	862
9	1893	2132	26	0	534
10	607	680	27	5536	729
11	75	86	28	2541	500
12	5291	5344	29	607	12
13	834	0	30	4410	47
14	5840	93	31	320	3575
15	0	1983	32	4531	0
16	0	994	33	1862	0
17	7400	2491			

The first 17 villages selected during 1960-61 were retained for the next year for studying the change in the number of trees.

(a) Estimate the change in number of arecanut trees from 1960-61 to 1961-62 along with its standard error.

(b) Estimate the total number of arecanut trees in the Nowgong subdivision during 1961-62, using the available information for the previous year and give its standard error.

REFERENCES

Cochran, W.G., *Sampling Techniques*, John Wiley & Sons, New York, (1977).

Eckler, A.R., "Rotation sampling," *Ann. Math. Statist.*, **26**, 664-685, (1955).

Graham, J.B., "Composite estimation in two cycle rotation sampling designs," *Comm. in Statist.*, **1**, 419-413, (1973).

Hansen, M.H., W.N. Hurwitz and W.G. Madow, *Sampling Survey Methods and Theory*, John Wiley & Sons, New York, Vols. I & II, (1953).

Hansen, M.H., W.N. Hurwitz, H. Nisselson, and J. Steinberg, "The redesign of the consus current population survey," *J. Amer. Statist. Assoc.*, **50**, 701-719, (1955).

Kathuria, O.P. and D. Singh, "Relative efficiencies of some alternative procedures in two-stage sampling on successive occasions," *J. Ind. Soc. Agr. Statist.*, **23**, 101-144, (1971).

Prabhu, Ajgaonkar, S.G., "The theory of univariate sampling on successive occasions under the general correlation pattern, *Aust. J. Statist.*, **10**, (1968).

Patterson, H.D. "Sampling on successive occasions with partial replacement of units," *J. R. Statist. Soc.*, **12B**, 241-255, (1950).

Rao, C.R., "Some theorems on minimum-variance unbiased estimation," *Sankhya*, **12**, 27-42, (1952).

Rao, J.N.K. and J.B. Graham, "Rotation designs for sampling on repeated occasions," *J. Amer. Statist., Assoc.*, **59**, 492-509, (1964).

Scott, A.J. and T.M.F. Smith, "Analysis of repeated surveys using time series methods," *J. Amer. Statist. Assoc.*, **69**, 674-678, (1974).

Singh, D., "Estimated in successive sampling using multi-stage design," *J. Amer. Statist. Assoc.*, **63**, 99-112, (1968).

―――and O.P. Kathuria, "On two-stage successive sampling," *Aust. J. Statist.*, **11**, 59-66, (1969).

―――and B.D. Singh, "Double sampling for stratification on successive occasions," *J. Amer. Statist. Assoc.*, **60**, 784-792, (1965).

―――and R. Singh, "Multi-purpose surveys on successive occasions," *J. Ind. Soc. Agr. Statist.*, **25**, 81-90, (1973).

Singh, D., Shivtar Singh, and A.K. Srivastava, On repeat surveys in two-stage sampling design, *J. Ind. Soc. Agr. Statist.*, **18**, 55-70, (1976).

Tikkiwal, B.D., "Theory of successive sampling," *Unpublished thesis for Diploma*, I.C.A.R., New Delhi, (1951).

―――"On the theory of classical regression and double sampling estimation," *J.R. Statist. Soc.*, **22**, 131-138, (1960).

Woodruff, R.S., "A simple method for approximating the variance of complicated estimates," *J. Amer. Statist. Assoc.*, **66**, 411-414, (1971).

Yates, F., *Sampling Methods for Censuses and Surveys*, Charles Griffin & Co., London, (1949).

12

Sequential Sampling

*Give your decisions, never your reasons; your decisions
may be right, but your reasons are sure to be wrong.*
Willian Murray

12.1 INTRODUCTION

A random sample is, by definition, one which has a fixed size and a
determinate probability of selection. As yet, nobody has tried to
explain why the sample size is fixed well in advance. If the validity of
the results is unchallengeable at sample size n, it must be equally
good for sample size $(n + 1)$ or $(n - 1)$, and so on. This led Wald
(1943) to device a sampling method which had no prior fixing of the
number of observations but was based on a definite rule related to the
observations themselves. He gave it the nomenclature *Sequential
sampling*.

An essential feature of the sequential procedure, as distinguished
from classical fixed-size procedure, is that the number of observations
required for the sequential method depends on the outcome of the
observations and is, therefore, a random variate. In scientific research,
there always exists the possibility that the researcher may change his
mind in the light of further information gathered by him in the process of investigation. But if a fixed-sample size method is adopted, the
researcher will not have this advantage. Let us suppose that a new
fertilizer treatment has been suggested and one is confronted with the

question as to whether its use should be recommended or not. Obviously in a situation like this, an experiment may be undertaken which will estimate the effect of the suggested treatment and the expected loss, if any, in the event of it's being an incorrect recommendation. Further series of the experiment may reduce the expected loss and the results become more precise. But the experiment will need money, and continuance of the experiment will invariably increase the cost. Hence, a decision rule will have to be framed, which is based on minimizing the expected loss or the cost of the experiment.

The idea of sequential sampling was first given by Sukhatme (1935) when the problem of optimal sampling was studied under conditions where stratum sizes were known but standard deviations were not. Another method of sampling with variable sample size was derived by Tweedie (1945) and was termed *inverse sampling* as the procedure was the inverse of the classical fixed-size procedure. Haldane (1945), Wald (1947, 1951) and Anscombe (1949, 1952, 1953, 1954) have discussed the sequential methods of estimation. Singh (1977) has discussed it at length and has described it as the choice of rethinking in terms of flexible sampling procedures which may be modified, if necessary, in the course of an investigation. The sequential approach, in a broader sense, is a field where one may come across the unusual phenomenon of theory lagging behind practice. Another interesting but unexplored subject is the multi-sample sequential approach to sampling theory and the formulation of an estimation procedure. This may be a line of worthwhile investigation for scientific workers. In this chapter, we shall discuss some methods of estimation of population size and population mean, and generalized sequential estimators.

12.2 ESTIMATION OF POPULATION SIZE

(i) *Classical or Direct Method* Pettersen (1896) and Lincoln (1930) discussed the problem of estimation of population size by the fixed-size (classical) method. The procedure used is simple and may be summarized here.

Notations

N = the size of the population being studied

n = the size of the sample being studied

t = the number of marked individuals in the population

$s =$ the number of marked individuals recovered in the sample

$p \ (=t/N) =$ the proportion of marked individuals in the population

THEOREM 12.1.1 If sampling is with replacement, the probability of having to sample n individuals to obtain s marked ones is given by the binomial distribution

$$P_r(s) = \binom{n}{s} p^s (1 - p)^{n-s}$$

where n is fixed and s is a random variate. The estimator of the population size N is given by

$$\hat{N}_{c_1} = \frac{nt}{s} \tag{12.1.1}$$

which is unbiased, but its variance is not measurable.
The proof is obvious.

THEOREM 12.1.2 If sampling is without replacement, the probability of having to sample n individuals to obtain s marked ones is given by hypergeometric distribution

$$P_r(s) = \frac{\binom{t}{s}\binom{N-t}{n-s}}{\binom{N}{n}}$$

where n is fixed and s is a random variate. The estimator of the population size N is given by

$$\hat{N}_{c_2} = \frac{nt}{s} \tag{12.1.2}$$

which is unbiased, but its variance is not measurable.
The proof is obvious.

(ii) *Sequential or Inverse Sampling Method* There are situations where complete enumeration or census is well-nigh impossible and well known sampling methods do not suffice for the purpose, where sequential estimators are proved to be useful. Bailey (1951), Chapman (1952) and Leslie and Chittey (1952) used sequential methods for the estimation of biological populations. Anscombe (1953) and Singh (1977) have given a detailed discussion and a new line of approach has been sketched. Singh and Singh (1979), Chaudhary and Khatri (1980) and Chaudhary and Gosawmi (1981) have presented different techniques for the estimation of population size when it is finite but unknown.

A single modification of the step is to invert the sampling procedure, i.e. to fix the number of tagged individuals to be recovered by sampling rather than fixing the sample size. This sampling procedure is known as *inverse sampling*. The selection of items is made one at a time and estimates of the population size are taken on the basis of the number of tagged individuals recaptured in the sample. Thus, sample size n is treated as a random variate and s is a fixed parameter.

THEOREM 12.1.3 If sampling is with replacement, the probability of having to sample n individuals to obtain s marked ones is given by the negative binomial distribution.

$$P_r(n) = \binom{n-1}{s-1} p^s (1-p)^{n-s}$$

where s is fixed and n is a random variate. The estimator of the population size N is given by

$$\hat{N}_{1n_1} = \frac{nt}{s} \qquad (12.1.3)$$

which is unbiased. Its sampling variance is obtained by

$$V(\hat{N}_{1n_1}) = \frac{N(N-t)}{s} \qquad (12.1.4)$$

The proof is obvious.

Since relation (12.1.4) involves the unknown parameter N it will have to be replaced by its estimate to obtain the estimated variance. Thus, an unbiased estimate of the variance of $\hat{N}_{1n_1}$ is given by

$$v(\hat{N}_{1n_1}) = \frac{nt^2(n-s)}{s^2(s+1)} \qquad (12.1.5)$$

THEOREM 12.1.4 If sampling is without replacement, the probability of having to sample n individuals to obtain s marked ones is given by the negative hypergeometric distribution, i.e.

$$P_r(n) = \frac{\binom{t-1}{s-1}\binom{N-t}{n-s}\frac{t}{N}}{\binom{N-1}{n-1}}$$

where s is fixed and n is a random variate. The estimate of the population size N is given by

$$\hat{N}_{1n_2} = \frac{nt}{s} - 1. \qquad (12.1.6)$$

which is unbiased. Its sampling variance is obtained by

$$V(\hat{N}_{1n_2}) = \frac{(N+1)(N-t)(t-s+1)}{s(t+2)} \qquad (12.1.7)$$

The proof is obvious.

EXAMPLE 12.1   During the year 1982-83, a subsidy grant for the construction of 3000 biogas plants was issued to users in a given zone of Haryana State in India. The list of users to whom it was issued, along with the site of each plant, was available. It was reported by an agency that a large part of the subsidy was misused. To estimate the extent of misuse, a sample survey study was conducted by taking a sample from the list and the names of users who had constructed the plant were checked. If the list of 300 users who have biogas plant up to 1981-82 is also available, how will you estimate the total number of users in 1982-83 (i) when a sample of size 48 had 9 names from the list up to 1981-82, (ii) when the size of the sample was 55 until it included 10 names from the list up to 1981-82? Estimate its standard error.

(i) *Sample size is fixed*   When a sample of size 48 had 9 names from the list up to 1981-82, we have

$$n = 48, \qquad s = 9, \qquad t = 300$$

Therefore, $$\hat{N} = \frac{48 \times 300}{9} = 1600$$

No simple formula is available for its standard error.

(ii) *Sample size is a variate*   When the sample included 9 names from the list up to 1981-82 and the sample size was 55, we have

$$n = 55, \qquad s = 10, \qquad t = 300$$

Therefore, $$\hat{N}_1 = \frac{55 \times 300}{10} = 1650$$

The procedure is sequential and the sampling variance of the estimate is obtained by

$$v(\hat{N}_1) = \frac{55(55 - 10) \times 300^2}{10 \times 10 \times 11} = 193500$$

$$\sigma_{\hat{N}_1} = 440$$

Hence, the upper 3-sigma limit is 2970 and it shows that there were at least 30 bogus names in the 1982-83 list and that a thorough inquiry should be conducted to derive the exact number.

12.3 COMPARATIVE STUDY

It will be shown here that, on an average, the sequential sampling procedure is better than the classical one. Let n_1 and s_1 be the fixed sample size and number of tagged individuals recaptured respectively, in the classical method. Let n and s be the sample size and number, respectively, of tagged individuals captured in sequential sampling. We know that

$$E(n_{seq}) = \frac{(N+1)s}{(t+1)} \tag{12.3.1}$$

If s of the sequential method is kept equal to s_1 of the classical method, it can be easily seen that

$$s = s_1 = \frac{n_1 t}{N} \tag{12.3.2}$$

Putting the value from relation (12.3.2) in relation (12.3.1), we have

$$E(n_{seq}) = \frac{(N+1)}{(t+1)} \frac{n_1 t}{N} = \frac{(1+1/N)}{(1+1/t)} n_1 < n_1 \tag{12.3.3}$$

Hence, we may conclude very safely that, with comparatively less average effort, sequential sampling will give better results. On the other hand, if the experimenter knows absolutely nothing about the possible population size, then, by an improper choice of t and s, $E(n)$ may be extremely large. It is also derived that

$$V(n_{seq}) = \frac{N^2 s}{t^2} \tag{12.3.4}$$

which is very large and may be taken to be undesirable property of this procedure.

Further, it may be seen that there is a slight difference in the estimates given by both sequential methods and their values are not in agreement. Their absolute difference is given by

$$| \hat{N}_{in_2} - \hat{N}_{in_1} | = \frac{n}{s} - 1 = \frac{(n-s)}{s} \tag{12.3.5}$$

EXAMPLE 12.2 When sampling is with replacement, the sequential estimator of population size N is given by $\hat{N} = nt/s$. On the other hand, when sampling is without replacement, the sequential estimator of N is given by $\hat{N} + 1 = n(t + 1)/s$. If we assume $N + 1$ as N, $(t + 1)$ as t, $(n + 1)$ as n and $(s + 1)$ as s, we may define a new estimator as $\hat{N} = (n - 1) t/(s - 1)$. On this basis, some more estimators are suggested by changing n, t and s. Compare them by calculating their bias, sampling variance, estimated bias and sampling variance for sequential scheme with replacement.

When sampling is with replacement sequential scheme, by changing n, t and s we may define some sequential estimators of the population size N as under:

$$\hat{N}_1 = \frac{(n - 1)}{(s - 1)} t, \qquad \hat{N}_2 = \frac{(n - 1)}{s} t, \qquad \hat{N}_3 = \frac{nt}{(s - 1)}$$

$$\hat{N}_4 = \frac{(n + 1)}{(s + 1)} t, \qquad \hat{N}_5 = \frac{n(t + 1)}{s}$$

Assuming that the distribution for a random sample size is a negative binomial distribution, we can calculate bias and its estimate and the variance and its estimate from the sample for these estimators. A tabular presentation of the results thus obtained may be written and a comparative study is done as shown below.

Table 12.1

Estimator	Bias	Estimated bias	Variance	Estimated variance	Precision
$\hat{N}$	0	0	$\dfrac{N(N-t)}{s}$	$\dfrac{n(n-s)t^2}{s^2(s+1)}$	—
$\hat{N}_1$	$\dfrac{N-t}{s-1}$	$\dfrac{(n-s)t}{(s-1)^2}$	$\dfrac{N(N-t)s}{(s-1)^2}$	$\dfrac{n(n-s)t^2}{(s-1)^2(s+1)}$	$\left(1 - \dfrac{1}{s}\right)^2$
$\hat{N}_2$	$\dfrac{-t}{s}$	$\dfrac{-t}{s}$	$\dfrac{N(N-t)}{s}$	$\dfrac{n(n-s)t^2}{s^2(s+1)}$	1
$\hat{N}_3$	$\dfrac{N}{(s-1)}$	$\dfrac{nt}{(s-1)^2}$	$\dfrac{N(N-t)s}{(s-1)^2}$	$\dfrac{n(n-s)t^2}{(s-1)^2(s+1)}$	$\left(1 - \dfrac{1}{s}\right)^2$
$\hat{N}_4$	$-(N-t)$	$\dfrac{-(n-s)t}{(s+1)^3}$	$\dfrac{N(N-t)s}{(s+1)^2}$	$\dfrac{n(n-s)t^2}{(s+1)^3}$	$\left(1 - \dfrac{1}{s}\right)^2$
$\hat{N}_5$	$\dfrac{N}{t}$	$\dfrac{n(t+1)}{ts}$	$\dfrac{N(N-t)(t+1)^2}{st^2}$	$\dfrac{n(n-s)(t+1)^2}{(s+1)s^2}$	$\left(t + \dfrac{t}{1}\right)^2$

It may be concluded ˉthat the term t/s controls the variance of the estimator and that a change in sample size brings a bias. Here $\hat{N}_2$ is negatively biased but it is equally acceptable as the unbiased estimator $\hat{N}$. On the other hand, $\hat{N}_1$ and $\hat{N}_3$ are positively biased estimators but they are of equal precision. Bias changes with sample size. $\hat{N}_2$ is another estimator which is negatively biased but its precision is relatively higher than that of others.

12.4 ESTIMATION OF POPULATION MEAN

Let a finite population $\mathcal{U}$ consist of N distinguishable units U_i associated with a real variate y_i ($i = 1, 2, \ldots, N$). A parameter θ ($= \theta(y)$) is a point in the class of point sets (for brevity $\mathcal{R}$) and it can be expressed as a sum of single-valued set functions defined over the class $\mathcal{R}$, i.e.,

$$\theta = \sum_{a_i \in \mathcal{R}} \lambda_i f(a_i) \qquad (12.4.1)$$

where λ_i is some adjustment constant, $f(a_i)$ is a single-valued set function defined over the class $\mathcal{R}$, and $\sum_{a_i \in \mathcal{R}}$ is the summation over all sets a_i belonging to the class $\mathcal{R}$.

Elaborating the idea further, a more generalized method to express a parametric function may be written as

$$\theta = \sum_{a_i \in \mathcal{R}} \lambda_i \, \pi_{a_i} f(a_i) \qquad (12.4.2)$$

where π_{a_i} is a probability measure defined over a_i in the class $\mathcal{R}$.

To illustrate, the population mean $\overline{Y}$ for a characteristic y can be expressed as θ in relation (12.5.2), with a_i as a point set $\{U_i\}$, $f(a_i)$ as y_i, along with probability measure π_{a_i} as $1/N$ and λ_i as 1. Similarly, the population variance σ^2 can be expressed as relation (12.5.2), with a point set of two units $\{U_i, U_j\}$, with $f(a_i)$ and π_{a_i} as $(y_i - y_j)^2$ and $1/N^2$, respectively, with λ_i as 1.

This is one of the important reasons why probability samples are preferred over non-probability samples. Whatever the sampling scheme or design may be, the search for a method which provides a technique to evaluate the probability measure in a form acceptable to all occasions continues.

A statistic t defined over the probability field is a function over the samples. A statistic used to estimate a parametric function θ is called an estimator to θ. The most general form of a linear estimator may be

$$t = \frac{\sum\limits_{a_i \in s} p\,(a_i,\, s)\, f\,(a_i)}{\sum\limits_{a \in s} p\,(a_i,\, s)} \tag{12.4.3}$$

where $p\,(a_i,\, s)$ is the probability measure defined over point a_i in the sample s.

In case, $\sum\limits_{a_i \in \mathscr{R}} p\,(a_i,\, s) = 1$, then t is called an unbiased estimator.

The estimator in relation (12.4.3) may also be written as

$$t = \frac{\sum\limits_{a_i \in s} f\,(a_i)\, p\,(a_i,\, s/E_{a_i})}{\sum\limits_{a \in s} p\,(a_i,\, s/E_{a_i})} \tag{12.4.4}$$

where E_{a_i} is an event depending on the occurrence of set a_i in the sample and $p\,(a_i,\, s/E_{a_i})$ is the probability measure for a given E_{a_i}

In other words, the event E_{a_i} may be termed a *terminal event* because it decides the number of observations in the sample, and sampling terminates accordingly, i.e. the stopping rules are defined in terms of this event. In a non-sequential process, the minimum value of the loss in estimating the mean value with known σ^2 is $2\,(c\,\sigma^2)^{1/2}$, where c is the cost per observation. Therefore, in a sequential process, sampling will terminate as soon as it touches this value. Likewise, there may be a number of ways to define the terminal event. Thus, a class of sequential estimators may be obtained by defining the event E_{a_i} in a number of ways. The estimator given by relation (12.4.4) may be termed a *generating estimator* for sequential sampling schemes.

To illustrate, let us consider the sequential mean

$$\bar{y}_n = \tfrac{1}{2}\,\bar{y}_{n-1} + \tfrac{1}{2}\,y_n$$

which can also be written as

$$\bar{y}_n = \frac{y_1}{2^{n-1}} + \frac{y_2}{2^{n-1}} + \frac{y_3}{2^{n-2}} + \ldots + \frac{y_{n-1}}{2^2} + \frac{y_n}{2}$$

That is, it gives rise to a random sample with unequal probability for different items. Similarly, the sequential mean can be written as

$$\bar{y}_{\text{seq}} = \frac{(n-1)}{n}\,\bar{y}_{n-1} + \frac{1}{n}\,y_n$$

The sequence of these means in terms of previous ones can be expressed as below.

$$\bar{y}_2 = \tfrac{1}{2} y_1 + \tfrac{1}{2} y_2$$
$$\bar{y}_3 = \tfrac{2}{3} \bar{y}_2 + \tfrac{1}{3} y_3$$
$$\bar{y}_4 = \tfrac{3}{4} \bar{y}_3 + \tfrac{1}{4} y_4$$
$$\cdot$$
$$\cdots \quad \cdots \quad \cdots$$

Thus, a generalized form of a sequential estimator can be

$$\bar{y}_{\text{seq}} = \pi_{n-1} \, \bar{y}_{n-1} + \pi_n \, y_n \tag{12.4.5}$$

where $\qquad \pi_{n-1} + \pi_n = 1,$

and π_n is the probability of introducing the nth unit in the sample.

THEOREM 12.4.1 Show that the estimator defined in relation (12.4.5) is unbiased and its sampling variance is given by

$$V(t) = \sigma^2 \sum p^2 \, (a_i, s | E_{a_i}) \tag{12.4.6}$$

Proof With the application of n-dimensional Euclidean geometry, one can provide the proof.

12.5 ACCEPTABLE SEQUENTIAL ESTIMATORS

The concept of admissible decisions was used by Wald (1947). Godambe (1960), Roy and Chakravorthy (1960), Pathak (1962, 1964), Hanurav (1968), Joshi (1968), Prabhu Ajgaonkar (1969) and others contributed on the best admissible estimators in sampling for finite populations. Singh (1977) applied the concept to sequential sampling and a class of acceptable estimators was discussed. In this section, we shall present a procedure similar to that given by Pathak (1964) and Singh and Singh (1981). Before discussion let us define the terms used in the text.

Probability Field Consider a non-negative function p defined for every combination (s_j) of s_j. A probability measure may be constructed in which the combination (s_j) will be taken with probability proportionate to p (s_j) over the combination such that $\sum p$ $(s_j) = 1$ and will be referred to as probability field (Ω).

Sampling Design It is any function P on $\mathscr{R}$, the set of all possible subsets of s of $\mathscr{U}$ such that $p(s) \geqslant 0$ and $\sum p(s) = 1$, $s \in \mathscr{R}$.

Sampling System It may be considered the specification of all possible samples along with their probability field over the combination of units in the sample with reference to θ, i.e., it is a combination of estimators of ordered sequence of samples s from $\mathscr{U}$ with probability field (Ω). Symbolically,

$$F \equiv F(t, \Omega)$$

Sampling Structure A sampling system F along with its risk function d defined over Ω is called a sampling structure for estimation of θ. Symbolically,

$$D \equiv D(F, d) \equiv D(t, \Omega, d).$$

A sampling structure is said to be *unbiased* if t is an unbiased estimator of θ. A sampling structure is said to be *ultimate acceptable* if t is consistent and a minimum risk unbiased estimator (MRUE). If the risk of D_1 is smaller than that of D_2, i.e., $d_1 < d_2$, then t is said to be an *acceptable estimator*.

Risk Function. The risk function of a sequential rule (ϕ, δ) is the expected value of the loss in t when θ is the true value of the parameter. Symbolically,

$$d = E[\theta, (\phi, \delta)].$$

Sequential Rule. A sequential rule is a pair (ϕ, δ) in which ϕ is a stopping rule and δ is a terminal rule.

Stopping Rule. A stopping rule is a sequence of function

$$\phi(t) = \{\phi_0, \phi_1,(t_1), \phi_2(t_2), \ldots\}$$

where $\phi_j(t_j)$ is such that $0 \leqslant p_j \leqslant 1$ for all j.

It means that ϕ stands for conditional probability that the experimenter will cease sampling when he has taken j observations.

Terminal Rule. A terminal rule is a sequence of function

$$\delta(t) = \{\delta_0, \delta_1(t), \delta_2(t_2), \ldots\}$$

for all j that in the probability distribution of σ-field for which expected loss $E[\theta, (\phi, \delta)]$ is finite, δ_j is a sequential terminal rule for a statistical problem.

Acceptable Estimators

Given a sequential sampling scheme that the sample units are arranged in ascending order of their unit indices from an *order statistic* which can be written as

$$t = [y_{(1)}, y_{(2)}, \ldots, y_{(m)}]$$

Another order statistic, if all the units are distinct, may be formed

$$t' = [y_{(1)}, y_{(2)}, \ldots, y_{(v)}]$$

Such estimators are sufficient to form a class of estimators which belong to some defined probability fields. Let us consider sequential sampling with replacement method in which two types of estimators of population mean are defined as

$\bar{y}_n = y/n \equiv$ average of all sample units

$\bar{y}_v = y/v \equiv$ average of all v distinct units in the sample

In a sequential sample with n as variate sample size, the number of distinct units v is also a random variate. So a generalized form of sequential estimators of the population mean may be written as

$$\bar{y}_{nseq} = \phi_1(v) + \phi_2(v)\,\bar{y}_v \qquad (12.5.1)$$

where $\phi_1(v)$ and $\phi_2(v)$ are two functions.

If the number of distinct units v in the sample is given, we have

$$E(\bar{y}|v) = E[\{\phi_1(v) + \phi_2(v)\,\bar{y}_v\}|v]$$
$$= \phi_1(v) + \phi_2(v)\,\overline{Y}$$

Therefore, $\qquad E(\bar{y}_{nseq}) = E\phi_2(v) + E\phi_2(v)\,\overline{Y}$

Obviously, the necessary and sufficient conditions for $\bar{y}_{nseq}$ being an unbiased estimator of $\overline{Y}$ are

and $\qquad \left. \begin{array}{l} E(\phi_1(v)) = 0 \\[2mm] E(\phi_2(v)) = 1 \end{array} \right\} \qquad (12.5.2)$

In order to choose an acceptable estimator from the field, the risk function of sampling structure D should be minimized. Thus, the problem is to choose unbiased estimators in Ω, which have a minimum variance and satisfy the conditions in relation (12 6.2). Hence, we can minimize the variance $V(\bar{y}_{nseq})$. We know that

$$V(\bar{y}_{nseq}) = E_1\,V_2(\bar{y}_{nseq}|v) + V_1\,E_2(\bar{y}_{nseq}|v)$$

Thus we get

$$V(\bar{y}_{nseq}) = E\left[\phi_2^2(\nu)\left(\frac{1}{\nu} - \frac{1}{N}\right)S^2\right] + E\left[\phi_1(\nu) + \phi_2(\nu)\bar{Y}\right.$$

$$\left. - E\{\phi_1(\nu) + \phi_2(\nu)\bar{Y}\}\right]^2 \qquad (12.5.3)$$

To choose a sampling structure which gives an unbiased estimator having uniformly minimum risk, a proper choice of $\phi_1(\nu)$ and $\phi_2(\nu)$ should be made in conformity with conditions in relation (12.5.2). Minimizing the first term in relation (12.5.3) which involves $\phi_2(\nu)$ only, we have, by using Schwartz's inequality,

$$\phi_2(\nu) = \frac{\left(\frac{1}{\nu} - \frac{1}{N}\right)^{-1}}{E\left(\frac{1}{\nu} - \frac{1}{N}\right)^{-1}} = \frac{\dfrac{\nu N}{(N-\nu)}}{E\left(\dfrac{\nu N}{(N-\nu)}\right)} \qquad (12.5.4)$$

Minimizing the second term in relation (12.5.3) in terms of $\phi_2(\nu)$, we have

$$\phi_1(\nu) = \bar{Y}\left[1 - \phi_2(\nu)\right] \qquad (12.5.5)$$

Since in relation (12.5.5) $\phi_1(\nu)$ is expressed in terms of unknown parameter $\bar{Y}$, the value of $\phi_1(\nu)$ cannot be determined except in the case where $\phi_2(\nu) = 1$. Thus $\bar{y}_{nseq}$ reduces to $\bar{y}_\nu$ only. If some prior information about $\bar{Y}$ is available, say $\bar{X}$ (information on previous occasion), then relation (12.5.5) can be written as

$$\phi_1(\nu) = \bar{X}\left[1 - \phi_2(\nu)\right] \qquad (12.5.6)$$

Hence the estimator in relation (12.5.1) becomes

$$\bar{y}_{nseq} = \bar{X}\left[1 - \frac{\nu N/(N-\nu)}{E(\nu N/(N-\nu))}\right] + \frac{\dfrac{\nu N}{(N-\nu)}}{E\left(\dfrac{\nu N}{(N-\nu)}\right)}\bar{y}_\nu \qquad (12.5.7)$$

Further, if ν/N can be ignored, relation (12.5.7) becomes

$$\bar{y}'_{nseq} = \bar{X}\left[1 - \frac{\nu}{E(\nu)}\right] + \frac{\nu\bar{y}_\nu}{E(\nu)} \qquad (12.5.8)$$

On the other hand if no information about $\bar{X}$ is given, then $\bar{X} = 0$ and we can write relations (12.5.7) and (12.5.8) as

$$\bar{y}'_{nseq} = \frac{\dfrac{\nu N}{(N - \nu)}}{E\left(\dfrac{\nu N}{(N - \nu)}\right)} \bar{y}_\nu \qquad (12.5.9)$$

and $\qquad \bar{y}^{**}_{nseq} = \dfrac{\nu}{E(\nu)} \bar{y}_\nu \qquad$ respectively. $\qquad (12.5.10)$

In relations (12.5.7), (12.5.8), (12.5.9) and (12.5.10), the sequential estimators of the mean have been derived and can generate commonly used estimators if some restrictions are imposed. In order to define a sampling structure D completely, the risk function should also be derived. Let us assume that the risk function is taken as the sampling variance. Since the estimators in relations (12.5.7) and (12.5.9) involve the term $E[\nu N/(N - \nu)]$ which requires a ready reference, discussing them at length has been postponed. At present we consider the estimators given in relations (12.5.8) and (12.5.10) and their sampling variances. If ν is taken as given, we have

$$V(\bar{y}_\nu | \nu) = \left(\frac{1}{\nu} - \frac{1}{N} \right) S^2$$

Therefore, $\qquad V(\bar{y}_\nu) = \left[E\left(\frac{1}{\nu} \right) - \frac{1}{N} \right] S^2 \qquad (12.5.11)$

Similarly, we find that

$$V(\bar{y}^{**}_{nseq}) = \frac{\bar{Y}^2 V(\nu)}{[E(\nu)]^2} + \frac{S^2 E(\nu^2)}{[E(\nu)]^2} \left[E\left(\frac{1}{\nu} \right) - \frac{1}{N} \right] \qquad (12.5.12)$$

Comparing relations (12.5.11) and (12.5.12) we get

$$V(\bar{y}_\nu) - V(\bar{y}^{*}_{nseq}) = k_1^2 S^2 - k_2^2 \bar{Y}^2 \qquad (12.5.13)$$

where $\qquad k_1^2 = \dfrac{E(\nu) E\left(\dfrac{1}{\nu} \right)}{E(\nu^2)} + \dfrac{V(\nu)}{N \{E(\nu)\}^2}$

and $\qquad k_2^2 = \dfrac{V(\nu)}{\{E(\nu)\}^2}$

Since $E(\nu) E(1/\nu) > 1$ and, therefore, coefficients of both $\bar{Y}^2$ and S^2 are positive, we conclude that $\bar{y}_\nu$ is better than $\bar{y}^{**}_{nseq}$ if

$$cv \leqslant \left| \frac{k_2}{k_1} \right| \qquad (12.5.14)$$

and worse otherwise.

Similarly, the variance of the estimator given by relation (12.5.9) may be obtained by

$$V(\bar{y}') = V\left[\frac{\nu}{E(\nu)}(\bar{y}_\nu - \bar{X})\right]$$

If we have prior information $\bar{X}$, we find only a change, viz. $\bar{y}_\nu$ has been replaced by $(\bar{y}_\nu - \bar{X})$. Thus, proceeding on similar lines as above, it can be shown that $\bar{y}_\nu$ is better than $\bar{y}_{nseq}^{**}$ if

$$\frac{S}{(\bar{Y} - \bar{X})} \leqslant \left|\frac{k_2}{k_1}\right| \qquad (12.5.15)$$

and worse otherwise.

This result shows that if $\bar{X}$ provides a close approximation of $\bar{Y}$, it is always better to use $\bar{y}_{nseq}$ rather than $\bar{y}_\nu$.

SET OF PROBLEMS

12.1 Let p the proportion of units possessing a rare attribute be estimated by sequential (inverse) sampling without replacement. Show that $\hat{p} = (m-1)/(n-1)$, where m is a fixed number of units having the rare attribute in a sample of variable size n, is an unbiased estimator with its sampling variance

$$V(\hat{p}) = \frac{(m-1)p + Np(1-p)}{(N-1)}$$

with its estimate

$$v(\hat{p}) = \frac{(m-1)^2}{(n-1)^2} - \frac{(N-1)}{N}\frac{(m-1)(m-2)}{(n-1)(n-2)} - \frac{(m-1)}{N(n-1)}$$

12.2 In Problem 12.1, if the sampling procedure is with replacement, show that $\hat{p}$ is still an unbiased estimator of the population proportion p. Derive its sampling variance and show that it is estimated by

$$v(\hat{p}) = \frac{\hat{p}(1-\hat{p})}{(n-2)}$$

12.3 To estimate the mean $\bar{Y}$ of the population with size N, a sample of size n is drawn with replacement till ν (a fixed number) of distinct units are included in the sample. Show that

(i) $E(n) = N[1/N + 1/(N-1) + \ldots + 1/(N-n+1)]$

(ii) $E(1/n) > 1/E(n) > (N-n)/n(N-1)]$

(iii) $\bar{y}_n = \overset{n}{\underset{i}{\Sigma}} y_i/n$ and $\bar{y}_\nu = \overset{\nu}{\underset{i}{\Sigma}} y_i/\nu$ are unbiased

Hence, or otherwise, prove that $V(\bar{y}_n) > V(\bar{y}_v)$.

12.4 Suppose that 500 fish captured in a lake are tagged and released. After a day, another catch of 500 fish is made and it is found that 50 among them are tagged. Estimate the total number of fish in the lake and calculate the standard error of the estimate.

12.5 N is the unknown number of individuals from a minority community. Let t be the number of card-holders from the community allowed to mix freely in the society. A sample of n individuals is taken, in which s are card-holders. $p_s(N)$ denotes the probability that the sample contains a card-holder. Show that

$$p_s(N): p_s(N-1) = (N-t)(N-n): N(N-n-t+s)$$

12.6 In a well-defined area A, a random sample of n points is marked. The points are connected by line segments with a random start from any point. Show that the length of the shortest path P, is never less than

$$E(P) \geqslant \frac{1}{2}\left[\frac{A(1-m)^2}{m}\right]^{1/2}$$

REFERENCES

Anscombe, F.J., "Large sample theory of sequential estimation," *Biometrika*, **36**, 445-58, (1949).

———"Large sample theory of sequential estimation," *Proc. Camb. Phil. Soc.* **48**, 600-607, (1952).

———"Sequential estimation," *J.R. Statist. Soc.*, **15**, 1-25, (1953).

———"Fixed sample size analysis of sequential observations. *Biometrics*, **10**, 89-100, (1954).

Bailey, N.T.J., "On estimating the size of mobile populations from recapture data, "*Biometrika*, **38**, 292-306, (1951).

Chaudhary, F.S. and J.B. Singh, "On generalized sequential estimator," *J. Ind. Soc. Agr. Statist.*, **31**, 72, (1979).

———and R.S. Khatri, "A sequential estimation of population size," *J. Ind. Soc. Agr. Statist.*, **32**, 136, (1980).

———and R.P. Goswami, "Sequential estimation of biological population in the field," *J. Ind. Soc. Agr. Statist.*, **33**, 103, (1981).

Chapman, D.G., "Inverse, multiple and sequential sample censuses," *Biometrics*, **8**, 286-306, (1952).

Godambe, V.P., "An admissible estimate for any sampling design," *Sankhya*, **22**, 285-288, (1960).

Haldane, J.B.S., "On a method of estimating frequencies," *Biometrika*, **33**, 222-225, (1945).

Hanurav, T. V., "Hyper-admissibility and optimum estimators for sampling finite populations," *Ann. Math. Statist.*, **39**, 621-642, (1968).

Joshi, V.M., "Admissibility of the sample mean as estimate of the mean of a finite population," *Ann. Math. Statist.*, **39**, 606-620, (1968).

Leslie, P.H. and D. Chittey, "The estimation of population parameters from data obtained by capture-recapture method I," "The minimum likelihood equations for estimating the death rate," *Biometrika*, **38**, 269-292, (1951).

Lincoln, F.C., "Calculating waterfowl abundance on the basis of banding returns," *Circ. V.S. Dep. Agric.* 118, (1930).

Murthy, M.N. and M.P. Singh, "On the concept of best admissible estimators in sampling theory," *Sankhya*, **31**, 343-354, (1969).

Pathak, P.K., "On estimating the size of a population and its inverse by capture mark method," *Sankhya*, **26A**, 75-80, (1964).

————"On simple random sampling with replacement," *Sankhya*, 24, 287-302, (1962).

Peterson, G.G.J., "The yearly immigration of young plaice into limfjord from the German sea, etc.," *Danish Biol. Stat. Report for 1895*, **61**, 1-48, (1896).

Prabhu Ajgaonkar, S.G., "A best estimator for the entire class of linear estimators," *Sankhya*, **31A**, 455-462, (1969).

Roy, J. and I M. Chakravorthy, "Estimating mean of a finite population," *Ann. Math. Statist.*, **31**, 392-398, (1960).

*Singh, F., "Sequential Approach to Sample Surveys," *Ph. D. Thesis*, (1977).

*Singh, F. and D. Singh, "Sequential estimation of population size—a graphic approach," *Unpublished paper*, (1979).

*————"On generalized sequential estimators," *J. Ind. Soc. Agr. Statist.*, **32**, 1-5, (1980).

*————"Acceptable sequential estimators of the population mean," *J. Ind. Soc. Agr. Statist.*, **33**, 29-37, (1981).

Sukhatme, P.V., "Contribution to the theory of the representative method," *Supple. J.R. Statist. Soc.*, **2**, 263-268, (1935).

Tweedie, M.C.K. "Inverse statistical variates," *Nature*, **155**, 453, (1945).

Wald, A., "Sequential analysis of statistical data, theory," *A report by statistical research group*, Columbia University, (1943).

————*Sequential Analysis*, Wiley & Sons, New York, (1947).

————"Asymptotic minimax solutions of sequential point of estimation problems," *Proc. 2nd Berkeley Symp. Math. Stat. & Prob.*, 1-11, (1951).

---

*May be read as Chaudhary, F. S.

13

Non-Sampling Errors

After people have repeated a phrase a great number of times, they begin to realize it has meaning and may even be true.

H.G. Wells

13.1 INTRODUCTION

In the foregoing chapters, it was seen whatsoever good a sampling design may be and howsoever well-planned a sampling enquiry may be conducted, the sample estimate will always be subject to deviation from the parameter or the population value. Usually, these deviations depend on the variation in the values of the individual units of the population and the sample size. For example, in simple random sampling, it is well known that the sample mean $\bar{y}$ is an unbiased estimate of the population mean $\bar{Y}$ and its deviation from the mean value, usually called *sampling error* is measured by $\sigma/\sqrt{n}$ or $\sqrt{(N-n)/Nn}\, S$ depending on the sampling scheme whether with replacement or without replacement, where σ and S, having their usual meanings, are indicators of the variation. It is also well known that if all units of the population are measured or in other words, if $n=N$, the estimate will be free from sampling error. In actual practice, the plan of enquiry seldom gets implemented according to the described rules. The reasons for departure from such rules may be many, which may affect preciseness of operations in the enquiry at any stage. Consequently, the estimate based on sample will involve errors which are different from sampling errors. The errors arising at the stages of ascertainment (responses or observations) and processing of data are termed *non-sampling errors*. Generally, the amount of sampling error decreases with increase in the sample size but surprisingly it becomes otherwise in case of non-sampling error.

The sampling errors are assigned to an estimate because it is based on a 'part' from the 'whole' while non-sampling errors are assigned because there is departure from the prescribed rules of the survey, such as survey design, field work, tabulation and analysis of data, etc. This is the reason that the census results although free from sampling errors are subject to various types of non-sampling errors. Sometimes, these non-sampling errors may be so large that they affect the survey results substantially. Thus it is desirable for the survey statistician to be fully conversant with such type of errors and their contributions to the survey results so that proper care can be taken to control these errors.

Various techniques for assessing and controlling non-sampling errors have been developed by Mahalanobis (1940, 1944, 1946), Deming (1944, 1950), Birnbaum and Sirken (1950), Durbin (1954), Lahiri (1958), and Mahalanobis and Lahiri (1961). Hansen, *et al.* (1946, 1951) have examined the problems of non-sampling errors in census and sample surveys. Sukhatme and Seth (1952) have worked out some models for measurement of such errors. Singh, *et al.* (1974) have studied non-response in successive sampling designs. A good account of biases and non-sampling errors has been presented by Kish (1965) and Zarkovich (1965, 1966). In this chapter, we shall discuss some of these techniques and treatment given here will be applicable to both census and sample surveys.

13.2 SOURCES AND TYPES OF NON-SAMPLING ERRORS

We have discussed in Chapter 1 about the principal steps involved in a survey programme, which include definitions and concepts of sampling units, sampling frame, characteristics under study, objectives, methods to be used for collection of data, measurements to be made and field work, analysis of data, etc. The non-sampling errors can occur at any one or more of the stages of a survey, i.e. planning, field work, and tabulation of survey data. In this section, the implications of such errors will be discussed. For simplicity, these errors are broadly classified as follows:

Group A Errors (Non-response errors)—Errors resulting from inadequate preparation.

Group B Errors (Response errors)—Errors resulting in the stage of data collection or taking observation.

Group C Errors (Tabulation errors)—Errors resulting from data processing.

Group A Errors: These errors may be assigned mainly due to the use of faulty frame of sampling units, biased method of selection of units, inadequate schedule, etc. If the sampling frame is not updated or old frame is used on account of economy or time-saving device, it may lead to serious bias as the targeted population is not enumerated. For example, in a household survey if the old list of households prepared for the population census some years ago is used for selection of the sample, some newly added households will not form a part of the sampling frame. Similarly, a number of households already migrated will remain in the frame. Thus the use of such frames may lead either to inclusion of some units not belonging to the population or to omission of units belonging to the population. Such procedures may bring unknown bias. In some situations, a part of sampled units may refuse to respond to the questions or may be not-at-home at the time of interview. It may also lead to this type of error. It can be seen that the method will provide biased estimate. Some of the main sources assigned to these errors may be as follows:

(i) omission or duplication of units due to ambiguous definitions of locale, units, or wrong identity of units, and or inaccurate and inconsistent specification of objectives;

(ii) inaccurate methods of interview or inappropriate schedules; and

(iii) difficulties arising due to unawareness on the part of respondents or faulty methods of enumeration/data collection.

Group B Errors: These errors refers, in general, to the difference between the individual true value and the corresponding sample value irrespective of the reasons for discrepancy. For example, in an agricultural survey, a landholder reports the total area of his holdings, which amounts to 10 ha while cadastial data, assumed to be accurate, show it 11 ha. The response given by the land holder contains a response error. Depending on the type and nature of enquiry or information to be collected, these errors may be assigned to the respondents or the enumerators or both. Sometimes there may be interaction between both of them and it may inflate these errors. The measurement device or technique may be defective and may cause observational errors. Several studies in farm account surveys have shown that the cost accounting (actual physical verification) and survey results give alto-gether different data. The main sources to these errors may be assigned as under:

(i) inadequate supervision and inspection of field staff;
(ii) inadequate trained and experienced field staff; and
(iii) problems involved in data collection and other type of errors on the part of respondents.

Group C Errors: These errors can be assigned to a number of defec-tive methods of editing, coding, punching, tabulation, etc. Obviously, these methods may differ according to the techniques employed and equipments available for data processing. To these errors, the bias due to estimation procedure may also be included. As explained in Chapter 6, the ratio method of estimation provides biased estimate. This bias may be considered as a part of tabulation errors. The main sources to these errors may be assigned as the following:

(i) inadequate scrutiny of basic data;
(ii) errors in data processing operations such as coding, punching, listing, verification, etc., and
(iii) other errors committed or admitted during publication/presen-tation of results.

These sources enumerated here are not exhaustive but are listed only to illustrate some of the possible sources of errors. As discussed above, a large number of different errors may appear in a survey. However, while work on various phases of the survey is in progress it is difficult to ensure which of these errors are admitted, what is the frequency of their occurrence, and what are their effects on the results? It is parti-cularly difficult in the course of survey to make an appraisal to the net effects of integration of various types of errors. What important is that the survey statisticians and the users of survey results should be aware of the existence of non-sampling errors and make joint efforts to control them to bring to the minimum possible extent.

13.3 BIASES AND VARIABLE ERRORS

In sampling theory, the square of total error (the mean square error) of a given estimator (t) combines with the variable error (variance) and bias, which in a relationship may be written as

$$E(t - \theta)^2 = E[t - E(t)]^2 + [E(t) - \theta]^2$$
$$= V(t) + B^2(t)$$

where θ is the true value, $V(t)$ is variance of t, and $B(t)$ is the bias in accepting t as the estimator of θ.

Thus, the total error may be expressed as

$$\text{Total Error} = [V(t) + B^2(t)]^{1/2} \qquad (13.3.1)$$

It has been seen that sample values are subject to both sampling and non-sampling errors. The non-sampling errors occur because the procedures of observation (data collection) are not perfect and their contribution to the total error of the survey may be substantially large and affecting survey results adversely. It is, therefore, desirable to study carefully the causes contributing these errors.

The sampling errors arise because a part (sample) from the whole (population) is taken for observation in the survey. Suppose that a sample has been chosen under reasonably accurate conditions/methods. Even then there are chances to arrive at different estimates because of various sources of error present under usual operational conditions. Let us suppose that θ_p be the expected survey value and θ' be parametric value (estimand) which are distinguished from its true value θ for the present discussion. Thus the total error of a sample value or its departure from its true value may be splitted into four as

$$t - \theta = [t - E(t)] + [E(t) - \theta_p] + [\theta_p - \theta'] + [\theta' - \theta] \qquad (13.3.2)$$

By utilising this concept, one may separate the errors due to the sampling process from the other errors inherent in the survey designs. The sampling errors are represented by $[t - E(t)] + [E(t) - \theta_p]$; and non-sampling errors by $[\theta_p - \theta'] + [\theta' - \theta]$. Both variable errors and biases can arise either from sampling or non sampling operations. In this way, dichotomy gives two-way classification of errors; biases of sampling or non-sampling and variable errors of sampling or non-sampling. The difference of the expected value of the estimator from its true value will be termed *total bias* and it will comprise both *sampling bias* and *non-sampling bias*. Simiarly, the variable error will measure the divergence of the estimator from its expected value and it will comprise both *sampling variance* and *non-sampling variance*. Thus one may visualise that there are several potential sources of errors in each of these classes and every operation is a potential source of variable errors and biases.

First, we may consider the total bias which is the algebraic sum of all biases. Some may move in the positive direction, while others in negative, thus partially balancing each other. The deletion of one source may increase the total bias. The two components of total bias may be classified further. Sampling biases may be classified as (i) consistent sampling bias (ii) constant statistical bias, and (iii) frame biases. Similarly, non-sampling biases may be classified as (i) non-response biases, (ii) response biases, and (iii) listing biases.

Secondly, let us consider the variable errors which may include both sampling and non-sampling errors. Variable errors due to sampling may consist of any number of components which generally depends upon the sampling design used for the survey. Variable errors due to non-sampling may include three components, viz., non-response variance, response variance and processing variation. Thus, the expected value of the squared deviation of the sample value from it true value can be written as

$$E(t-\theta)^2 = (\sum_i B_i)^2 + \sum_j S_j^2/a_j n_j \qquad (13.3.3)$$

where B_i stands for the ith type of bias,

S_j^2 stands for the jth type variance,

n_j stands for the sample size, and

a_j stands for the term for the sampling design used in the survey.

The first term is the square of the combined bias, which is the algebraic sum of all bias terms. The second term represents the sum of all variance terms. In a simple geometrical representation form, all these components are shown in Fig. 13.1.

13.3.1 Various Components of Total Error in Terms of Sampling and Non-Sampling Errors and their Relationship

The right angled triangles are taken to illustrate conveniently the underlying ideas. With the two sides representing the variable errors and the bias, the hypotenuse measures the total error. In this representation, the total bias is the algebraic sum of all biases, and the variable errors include both sampling and non-sampling errors. Suitable modifications can be made by changing right angle to α if interactions are present there.

Sampling errors are shown for the sake of simplicity with two components, which generally depends on the sampling design used in the

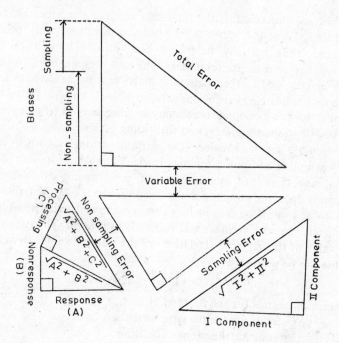

Fig. 13.1

survey. For example, the sampling variance of an estimator of the population mean based on a two-stage design would be represented by two components as

$$V(\bar{y}) = \frac{S_b^2}{n} + \frac{S_w^2}{mn}$$

where n and m are sample sizes at the first and the second stages of sampling, respectively, and S_b^2 and S_w^2 are the population values of variances at these stages.

In Fig. 13.1, these two components appear combined into hypotenuse $\sqrt{I^2+II^2}$ for both sampling errors. Similarly third component may be added by taking additional right-angle with third component as perpendicular to the hypotenuse already obtained for other components. If covariances are present, a separate covariance component may be added or subtracted depending on its sign, or by changing right angle to α, with the correlation coefficient $\rho = \cos \alpha$ in the relation. Symbolically,

$$V(\bar{y}) = I^2 + II^2 + 2\rho\, I\cdot II$$

Non-sampling variable errors may also be represented as the sums of square camponents. In the geometrical representation, three such components are shown combined into hypotenuse $(A^2 + B^2 + C^2)^{1/2}$ for non-sampling errors.

13.4 NON-SAMPLING BIAS

Bias refers to a systematic error that affects survey results with the same constant error. Separate biases are algebraically additive. Considerable separate biases in different directions, sometimes lead to an unbiased value; or on the other side, small separate biases sometimes may produce a large total bias. For example, total area under a crop as obtained in the census may be exact while the distribution of this area by sizes of holdings may bring large bias if tabulation errors affect some specified catagory such as small holdings are ignored generally by the village accountant (Patwari or Lekhpal). Different biases can be considered as a set of constants, determined by survey conditions although their values usually remain unknown. Each bias can be taken as positive or negative according as it increases or decreases the value of the estimator. Most of the biases cannot be reduced by increasing the sample size but only by improving the data collecting practices and field operations.

The magnitude of bias and its sign are not of equal importance in all the investigations. Users of results will generally be interested in the magnitude of the bias. In some cases, however, the information about the sign of the bias may become important. For illustration, let us assume that N agricultural holdings have to be enumerated in an agricultural census. However, the sampling frame has listed N' holdings only such that $N' < N$. In this case, one will have to deal with two separate biases in the census total. The first is the frame bias which can be defined as

$$B_1 = Y' - Y \qquad (13.4.1)$$

where $Y' = \sum_{i}^{N'} y_i$, $Y = \sum_{i}^{N} y_i$ and y_i is the value of the ith item in the frame.

In other words, the frame bias reflects the omissions in the census list and is independent of response errors. The other is due to response errors, which can be defined by

$$B_2 = Z' - Y' \tag{13.4.2}$$

where $Z' = \sum_{i}^{N'} (yi + di) = Y' + D$, with di and D being individual and total error, respectively.

In other words, the response bias exclusively reflects the effect of response errors. We can now define the total bias as

$$B = B_1 + B_2 \tag{13.4.3}$$

Clearly, B is the net effect and its magnitude will depend on the algebraic values of B_1 and B_2.

Generally, non-sampling biases pose problems for a systematic measurement, and affect simultaneously the population value as much as the sample values. Among the non-sampling errors of surveys, one can distinguish the biases of observation from those of non-observation. Biases of observations are assigned by obtaining and recording observations incorrectly. Biases of non-observation are assigned by failure in observation on some units or segments of the population on account of non-response or non-coverage. The former arises in the field operations and performance of observations, which comprises interview, enumeration or count, measurement, etc. Thus these are also classified as *response biases*. The latter arises due to faulty methods of data collection or estimation, non-coverage, incomplete frames, etc. These are called *processing biases*.

Biases in a census can be estimated by selecting a sample of units in a check survey with improved techniques of data collection and better measurement devices. Such surveys are called *post enumeration* surveys usually conducted after main surveys in order to check the quality of the collected data. In a sample survey, non-sampling biases can be estimated by taking a sub-sample of sample units, which are measured more carefully with accuracy and preciseness. Such surveys are called as *check surveys*. The survey is conducted in such a way that a unit by unit check or *unitary check* is always possible. If it is not possible due to cost constraints or operational considerations, a simpler procedure, termed *aggregate check*, may be used. For a check survey to be more effective, it is essential to ensure the following points:

(i) The survey should be done immediately after the main survey and the time of check should be so chosen that the purpose is achieved.

(ii) The survey should not influence the main survey and necessary steps to minimize the conditioning effect should be taken.

13.5 NON-COVERAGE: INCOMPLETE FRAMES AND MISSING UNITS

As mentioned earlier, errors of non-observation may be assigned due to failure in procuring data from some units in the target population. These may be classified as *non-coverage* and *non-response*. Non-coverage denotes failure to including some units or a group of units from the target population, which arises due to various problems relating to the sampling frame such as incomplete, outdated, rapid changes in the population, etc. Non-response refers to failure in getting response or information from some units selected from the sampled population, which may arise on account of refusal, not-at-home, unreturned or lost schedules, etc. Sukhatme (1947), Kish and Hess (1958), Zarkovich (1965) and others have discussed in details and they have given various methods for its measurement.

Generally, non-coverage refers to an error which is negative in nature. Similarly there may be positive error of *over coverage* which may arise due to the inclusion of some units which do not belong to the target population. Thus the *gross non-coverage* refers to the sum of the absolute values of non-coverage and over coverage errors. Also one may get the *net non-coverage* if algebraic sum of over coverage and non-coverage is taken together. Defective sampling frames are the source of these errors. In many populations, frames become outdated on account of continuous changes, e.g. a list of persons or households, based on a census, becomes outdated by lapse of a specified time. In many cases, frames are defective for its compilation from inexact material, e.g. list of fields growing paddy in a district is prepared by using informations available from the village accountants (Patwari or Lekhpal) records, the resulting frame will be defective and inaccurate as the survey was done by them at the time of crop sowing, which might include fields with crop failure also.

Construction of a good frame is often one of the major practical problems. Usually there is hardly a situation in practice, where the frame is available in the desired form. In almost all the populations, there is one common deficiency of incompleteness of frames and it is inevitable as most of the populations undergo some changes with passage of time. Even if one succeeds in preparing an accurate and complete frame but by the time survey work starts the frame becomes incomplete on account of dynamic nature of such populations. Yates (1949) and Goodman (1949) have considered these problems in details. Deming and Classer (1959) suggested a method for estimation of duplication in frames. Seal (1962), Hartley (1962) and Hansen, *et al.* (1963) have discussed successive and multiframes to overcome these problems. Srivastava (1977), Singh and Singh (1983), Chaudhary (1985) and others have studied problems of dynamic population and incomplete frames at length and optimum sampling plans along with suitable cost functions have been discussed.

13.6 NON-RESPONSE ERRORS

Measurement of non-response errors has always been a theme of interest in sample survey studies. Non response errors arise due to various causes right from the stage when the survey is designed and planned. The initial stages of a survey involve decisions about number of respondents, designing a schedule for collecting response data, etc. One may encounter non-response errors in surveys, which are classified with sources as below:

(i) *Not-at-homes*: This class consists of those respondents who may not be at home when enumerator calls on them. This is particularly so with the surveys when respondents are not aware with enumeration of survey work and respondent is temporarily away from the house.

(ii) *Refusal*: The class consists of those respondents who refuse to deliver information for one reason or the other or do not respond to the enumerators/questionnaires or are far away from their houses during the period of survey. In many cases, legal obligations do not exist or due to lack of publicity respondents are unaware about the conduct of the survey and, therefore, they feel it unnecessary to reply. Another reason may be the nature and the sequence of questions which affect motivation. Here all these cases are classified as refusals.

(iii) *Lost Schedules*: The class includes respondents who are not identified or not followed because it would be too expensive; or schedules which were mailed but lost or destroyed in transit. There may be some respondents who are physically or mentally incapable to respond during the survey period. In some cases, a few schedules are also found, incomplete or unusable. All these will be termed *lost schedule*.

We shall be using the term *non-response* as a common name for all such cases. Sometimes it is also referred *incomplete samples* or *missing data*. There are different ways and means to control non-response. One way of dealing with the problem of non-response is to make more efforts to collect information by taking a sub sample of units not responding in the first attempt. Another way of dealing with the problem of non-response is to estimate the probability of responding informants of their being at home at a specified point of time and weighting results with the inverse of this probability. In subsequent sections, we shall examine the effect of non-response in such cases.

13.7 TECHNIQUE FOR ADJUSTMENT OF NON-RESPONSE

A technique to deal with problem of non-response was developed by Hansen and Hurwitz (1946). Assume that the population is divided into two classes, a *response class* who responded in the first attempt and a *non-response* class who did not. Thus the total N units of the population will comprise N_1 and N_2 units, respectively, in these classes such that $N_1 + N_2 = N$. The population mean $\overline{Y}$ can be written as

$$\overline{Y} = \frac{N_1 \overline{Y}_1 + N_2 \overline{Y}_2}{N} = W_1 \overline{Y}_1 + W_2 \overline{Y}_2 \tag{13.7.1}$$

where W_1 and W_2 are the proportions of units in the response and non-response classes such that $W_1 + W_2 = 1$, and $\overline{Y}_1$ and $\overline{Y}_2$ are the population means in these classes.

Let $\bar{y}_1$ be the sample mean based on n_1 units from response class. It can be seen easily that $\bar{y}_1$ is a biased estimator of the population mean $\overline{Y}$ and its bias is given by

$$B(\bar{y}_1) = E(\bar{y}_1) - \overline{Y} = \overline{Y}_1 - \overline{Y} = W_2(\overline{Y}_1 - \overline{Y}_2) \tag{13.7.2}$$

The bias will be negligible if W_2 and $(\overline{Y}_1 - \overline{Y}_2)$ are small or for moderate values of W_2, the means of response and non-response classes do not differ significantly.

To avoid bias due to non-response, an approach which is due to Hansen and Hurwitz (1946), is to select a sub-sample of n_2 units drawn randomly, wor, from n_2 non-responding units such that $n_2 = n - n_1$, where n is the size of random sample, wor and n_1 are responding in the first attempt. Since the population mean $\bar{Y}$ is expressed in terms of unknown parameters N_1, N_2, $\bar{Y}_1$ and $\bar{Y}_2$, one should attempt to derive their unbiased estimators. It can be shown easily that unbiased estimators of N_1 and N_2 are given by

$$\hat{N}_1 = \frac{n_1}{n} N, \quad \text{and} \quad \hat{N}_2 = \frac{n_2}{n} N \qquad (13.7.3)$$

The technique given by Hansen and Hurwitz (1946) is useful in obtaining unbiased estimators and runs as follows:

(i) take a random sample, wor, of n respondents and mail a survey schedule to all of them, (ii) when the dead line of reply is over, calculate non-response, (iii) select a sub-sample, wor, of n_2' units in the non-response class and collect information by personal interview, and (iv) pool the results from both the classes to estimate the population values.

Assume that $\bar{y}_2'$ is the sub-sample mean of n_2' units. Let us define a pooled estimator of the population mean $\bar{Y}$ as

$$\bar{y}_w = \frac{1}{n} (n_1\bar{y}_1 + n_2\bar{y}_2') \qquad (13.7.4)$$

THEOREM 13.7.1 The estimator defined in relation (13.7.4) is unbiased. Its variance is given by

$$V(\bar{y}_w) = (1 - f) \frac{S^2}{n} + \frac{k - 1}{n} W_2 S_2^2 \qquad (13.7.5)$$

where S^2 is as usual, S_2^2 is the mean square in the non-response class and k is the ratio to be sub-sampled in non-response class, i.e. $k = n_2/n_2'$.

Proof:

Since $\quad E(\bar{y}_w) = E_1 E_2 (\bar{y}_w \mid n_1, n_2)$

$$= E_1 E_2 \left(\frac{n_1\bar{y}_1}{n} \mid n_1 \right) + E_1 E_2 \left(\frac{n_2\bar{y}_2'}{n} \mid n_2 \right)$$

Now $\quad E_1 E_2 \left(\frac{n_1\bar{y}_1}{n} \mid n_1 \right) = E_1 \left[\frac{n_1}{n} E_2 (\bar{y}_1 \mid n_1) \right]$

$$= E_1 \left(\frac{n_1}{n} \bar{y}_1 \right) = \frac{N_1}{N} \bar{Y}_1$$

and $\qquad E_1 E_2 \left(\dfrac{n_2 \bar{y}_2'}{n_1} \,|n_2 \right) = E_1 \left[\dfrac{n_2}{n} \, E_2 \, (\bar{y}_2'|n_2) \right]$

$$= E_1 \left(\dfrac{n_2}{n} \bar{y}_2 \right) = \dfrac{N_2}{N} \, \overline{Y}_2$$

Combining both, we get $E(\bar{y}_w) = \overline{Y}$. Hence $\bar{y}_w$ is an unbiased estimator.

If the sampling fraction of the original sample is designated by $f = n/N$ and taking $n_2/n_2' = k = 1/f_1$, we can derive sampling variance,

$$V(\bar{y}_w) = V_1 E_2 \, (\bar{y}_w) + E_1 V_2 (\bar{y}_w)$$

$$= V_1(\bar{y}) + E_1 \, [V_2(\bar{y}_w|n_1, \, n_2)]$$

Here, we have $\quad V_1(\bar{y}) = (1-f) \dfrac{S^2}{n}$ \hfill (i)

and $\qquad V_2(\bar{y}_w|n_1, n_2) = V_2 \left(\dfrac{n_2}{n} \, \bar{y}_2'|n_2 \right)$

$$= \dfrac{n_2^2}{n^2} \left(\dfrac{1}{n_2'} - \dfrac{1}{n_2} \right) s_2^2 = \dfrac{n_2}{n^2} \, (k - 1) \, s_2^2$$

where s_2^2 is the variance based on n_2 units. Hence

$$E_1 V_2(\bar{y}_w|n_1, n_2) = \dfrac{k-1}{n} E_1 \left(\dfrac{n_2}{n} \, s_2^2|n_2 \right) = \dfrac{k-1}{n} \, \dfrac{N_2}{N} \, S_2^2 \quad \text{(ii)}$$

where S_2^2 is variance of non-response class, analogous to S^2.

Adding (i) and (ii), we get the result.

COROLLARY 1 If $(k - 1) \, W_2 S_2^2 = V_s$, the result in relation (13.7.5) can be written as

$$V(\bar{y}_w) = \dfrac{1}{n}(V_s + S^2) - S^2/N \qquad (13.7.6)$$

where the term V_s/n denotes sub-sampling variance.

This shows that there is an increase in variation which is given by the second term in relation (13.7.5). If $k = 1$, the second term will vanish, which is possible only when data on all the units of non-response are collected. It may also be seen that the technique of making further inquiry in the non-response class leads to larger variance than that would have been achieved by a successful simple random sample of n units in the first attempt.

Another point of interest in Hansen and Hurwitz's technique is that it can be used with slight modifications to general interview surveys and an extension to the technique is given by El-Badry (1956). Fora-dori (1961) has designed some estimators of total along with their variances in multiphase sampling.

COROLLARY 2 If the population proportion P of a characteristic is estimated by $p = m_1/n_1$, with m_1 the number of units possessing the characteristic in n_1 responding units from a random sample of n units, wor, then the estimator p is a biased estimator and its bias is given by

$$B(p) = W_2 (P_1 - P_2) \qquad (13.7.7)$$

where W_2 is the weight of units in the non-response class, and P_1 and P_2 are the population proportions of the characteristic in response and non-response classes, respectively. Obtain an unbiased estimator of P. Also derive the variance of the estimator.

The bias can be negligible if the proportion of non-response class W_2 is very small or the difference between the proportions of units possessing the characteristic in response and non-response classes is very small.

COROLLARY 3 The bias resulting from the use of

$$s_1^2 = \sum_1^{n_1} \frac{(y_i - \bar{y}_1)^2}{(n_1 - 1)}$$

as an estimate of the population variance S^2 is given by

$$B(s_1^2) \cong W_2 (S_1^2 - S_2^2) - W_1 W_2 (\bar{Y}_1 - \bar{Y}_2)^2 \qquad (13.7.8)$$

where W_1, W_2, $\bar{Y}_1$ and $\bar{Y}_2$ are defined as previously and, S_1^2 and S_2^2 are the population values of variances in response and non-response classes, respectively, analogous to S^2.

Obviously, the bias in estimating S^2 will be negligible if both the population means and variances of response and non-response classes do not differ significantly, or the difference in the population variances of these classes are nearly W_1 times the squares of difference in the population means in both the classes. From relation (13.7.8), it can be seen that s_1^2 is an unbiased estimate of S^2 if $S_1^2 = S_2^2$ and $\bar{Y}_1 = \bar{Y}_2$. Since neither S_2^2 nor $\bar{Y}_2$ is known, it follows that the data collected in the first call cannot be used reasonably for estimating the population variance without risk of bias. Generally, the relative magnitude of the population means and variances will not be available and therefore, one should use very carefully the non-response data for estimating population values. For detailed discussion, readers are referred to Birnbaum and Sirken (1950).

In Hansen and Hurwitz technique, a sub-sample of n_2' is drawn arbitrarily from the non-response but no rationale is given for fixing this value. The usual way is to construct an appropriate cost function and relate it to the final precision of the estimator resulting from enquiry.

A simple cost of the survey may have four components (i) the overhead cost, say a, (ii) the cost of including a sample unit in the initial survey, say c, (iii) the cost of collecting information per unit in the response class, say c_1, and (iv) the cost of collecting information per unit in the non-response class, say c_2. Thus the cost function may be written as

$$C' = c_0 + cn + c_1 n_1 + c_2 n_2 \qquad (13.7.9)$$

Since C' will vary from sample to sample, we can get the average cost by

$$C = E(C') = c_0 + n\left(c + c_1 \frac{N_1}{N} + \frac{N_2}{kN} c_2\right)$$

$$= c_0 + n\left(c + c_1 W_1 + \frac{c_2}{k} W_2\right) \qquad (13.7.10)$$

A rational approach is found by minimizing the variance in relation (13.7.5) for the fixed survey cost C and obtaining the optimum values of n and k. If the desired degree of precision is specified in advance, the alternative would be to minimize C for a specified V_0 and obtain the optimum values of n and k.

THEOREM 13.7.2 The optimum values of n and k for a specified variance V_0, which minimize the cost, are given by

$$n_{\text{opt}} = \frac{S^2 + (k-1) W_2^2 S_2^2}{(V_0 + S^2/N)} \qquad (13.7.11)$$

and

$$k_{\text{opt}} = \left[\frac{c_2(S^2 - W_2 S_2^2)}{(c_1 + c_1 W_1)S_2^2}\right]^{\frac{1}{2}} \qquad (13.7.12)$$

Proof By denoting V, the variance expression in relation (13.7.5), we may define

$$\phi = C + \lambda(V - V_0)$$

where λ is Lagrange's multiplier.

Differentiating ϕ w.r.t. n and k, and equating to zero, we get k_{opt} and n_{opt} as given in the above relations.

On the basis of the quantities obtained from relation (13.7.12), the optimum sub-sampling fraction $f_1 = 1/k$ is determined and then is used in relation (13.7.11) to derive the total sample size. To start with the survey programme, a sample of that size is selected randomly, wor, and information from respondents are obtained. After identifying the non-respondents, a sub-sample is selected for personal interview in the non-response class. Unless the cost of obtaining information at the second attempt is larger than that of obtaining it at the first attempt it will be better to interview all the non-response class. If value given by relation (13.7.12) is less than 1, there will be no sub-sampling at all and the optimal procedure will be only to make new attempt to interview all the non-response class. If $S_2^2 = S^2$, we have

$$k_{opt} = \left[\frac{c_2 W_1}{c + c_1 W_1} \right]^{1/2} \qquad (13.7.13)$$

Moreover, since $\dfrac{1}{n}(S^2 + V_s)$ equals to $V_0 + \dfrac{S^2}{N}$, we can write

$$n_{opt} = \frac{S^2(1 + V_s/S^2)}{(V_0 + S^2/N)} \qquad (13.7.14)$$

If a upper bound for N_2 is known, say N_2', first a sample of size n given by (13.7.14) is drawn simply by replacing V_s by V_s' (an expression given for V_s with N_2 replaced by N_2') and k is given by

$$\left[\frac{c_2(S^2 - N_2' S_2^2/N)}{c + c_1(N - N_2')/N} \right]^{1/2}$$

Then, if n_2 do not respond, a subsample of a fraction $f_1 = 1/k$ from n_2 is drawn, where k is determined by

$$(1 - n/N) S^2 + (k - 1) n_2 S_2^2/n = nV_0 \qquad (13.7.15)$$

Kish and Hess (1959) has given a different approach to the problem of non-response. Srinath (1971) has developed a modified technique for selecting a subsample of non-response such that sub-sampling rate is varied according to the non-response rate in the sample.

EXAMPLE 13.1. In a survey, the expected response rate is 1/3, $S_2^2 = \frac{3}{4} S^2$ and it costs Re. 1.00 to include a unit in the sample, Rs. 5.00 in getting information per unit in the response class, and Rs. 9.00 in collecting information per unit in the non-response class. Obtain the optimal values of k and n so that the mean value of the population is estimated with $400/S^2$ precision. If the overhead cost is Rs. 267, determine the expected cost of the survey.

Here given that

$$W_1 = \frac{1}{3}, \ W_2 = \frac{2}{3}, \ a = 267,$$

$$c = 1, \quad c_1 = 5, \quad c_2 = 9 \text{ and } V_0 = \frac{S^2}{400}$$

Using relations (13.7.11) and (13.7.12), and substituting the values, we have

$$k_{opt} = 3/2, \text{ and accordingly } n_{opt} = 500$$

After substituting these values in the cost function given by (13.7.10), we have

$$C = 267 + 500 \left(1 + \frac{5}{3} + 9 \cdot \frac{2}{3} \cdot \frac{2}{3} \right) = \text{Rs. 3600.}$$

Hansen and Hurwitz technique loses its merits when non-response is large. Durbin (1954) observed that when $S_2^2 = S^2$, and cost of collecting data in the non-response class is much larger than that of in the response class, it will not be worthwhile to use the technique. Deming (1953), Stephan and McCarthy (1958) and Zarkovich (1966) have discussed modified techniques to specify the number of call-backs and its minimum number that is to be made on any unit.

13.8 POLITZ-SIMMONS' TECHNIQUE

An interesting plan dealing with the problem of not-at-home has been devised by Politz and Simmons (1949,1950). The aim of this technique is to adjust the biases without call-backs, which cropped up due to incomplete sample and did not distribute proportionately over the response class. The plan runs as follows : Respondents to be included in the sample are visited only once by enumerators during a specific time on five week days (excluding Saturdays & Sundays). The respondent who is found at home is asked how many times in previous five days he was at home at the specific time of interview. If the respondent says that he was at home j times, the ratio $(j + 1)/6$ is considered as an estimate of the probability of availability/inclusion of respondent in the sample. If the respondent is found to have been not-at-home, no information is collected.

Let the population consist N units and n respondents are selected by simple random sampling, wr. Assuming that p_i denotes the probability that the ith respondent is available at the time of the call, an estimator of $\bar{Y}$ is defined as

$$\bar{y}_{PS} = \frac{1}{n}\sum_i^n \frac{y_i}{p_i} \tag{13.8.1}$$

where p_i is the probability of availability of the ith respondent at the time of call. Of course, y_i is zero if the respondent is not available at the time of call.

THEOREM 13.8.1 The estimator defined in relation (13.8.1) is biased. Derive expressions for its bias and variance. Also the bias is negligible if the class of never-at-home is small.

Proof Using the information for 5 days, $p_i =(j+1)/6, j$ denoting the number of times the respondent was at home during the last five days, $j = 0, 1,...,5$. Further, let us denote p_{ij} as the probability the ith respondent will be found at home j times out of five calls, then we have $p_{ij} = \binom{5}{j} p_i^j (1 - p_i)^{5-j}$ if the respondent is available

and

$$\sum_{j=0}^5 p_{ij} = 1 - (1-p_i)^6 = 1 - q_i^6 \tag{13.8.2}$$

where

$$q_i = 1 - p_i$$

Hence, we have

$$E\left(\frac{y_i}{p_i}\bigg|i\right) = y_i \sum_{j=0}^5 \frac{6}{j+1}\binom{5}{j} p_i^j (1-p_i)^{5-j}$$

$$= \frac{y_i}{p_i}\sum_{j=0}^5 \binom{6}{j} p_i^{j+1} (1-p_i)^{6-(j+1)}$$

$$= \frac{y_i}{p_i}(1 - q_i^6) \tag{13.8.3}$$

Noting that the ith unit is selected and found at home and its probability is p_i/N, we get

$$E\left(\bar{y}_{PS}\right) = E_1 E_2(\bar{y}_{PS}) = \frac{1}{n} \sum_i^n E_1 E_2\left(\frac{y_i}{p_i}\Big| i\right)$$

$$= \frac{1}{n} \sum_i^n \sum_i^N \frac{p_i}{N} \cdot \frac{y_i}{p_i} (1 - q_i^6)$$

$$= \bar{Y} - \frac{1}{N} \sum_i^N y_i\, q_i^6 = \bar{Y} + B\left(\bar{y}_{PS}\right)$$

where $\qquad B\left(\bar{y}_{PS}\right) = -\dfrac{1}{N} \sum_i^N y_i\, q_i^6 \qquad$ (13.8.4)

Thus, the estimator $\bar{y}_{PS}$ is biased. However, the bias will be negligible if the class of non-response who are never found at home is small. The method reduces the bias due to non-response but it increases the variance of the estimator $\bar{y}_{PS}$, because unequal and estimated weights are used. To derive its sampling variance, let us have

$$E\left[\left(\frac{y_i}{p_i}\right)^2\Big| i\right] = y_i^2 \sum_{j=0}^5 \left(\frac{6}{j+1}\right)^2 p_i^j (1 - p_i)^{5-j}$$

$$= \frac{6}{p_i} A_i y_i^2 \qquad (13.8.5)$$

where $\qquad A_i = \displaystyle\sum_{j=0}^5 \frac{1}{(j+1)} \left(\frac{6}{j+1}\right) p_i^{j+1} (1-p_i)^{6-(j+1)}$

Hence,

$$V\left(\bar{y}_{PS}\right) = \frac{1}{n} \left[\frac{6}{N} \sum_i^N A_i y_i^2 - \left\{\frac{1}{N} \sum_i^N y_i (1 - q_i^6)\right\}^2\right] \qquad (13.8.6)$$

Although the variance of $\bar{y}_{PS}$ is slightly complicated and difficult to appraise without applying it to some specific populations, however, with usual approximations for a ratio estimator, one can express the results similarly. The readers are referred to Deming (1953) for detailed discussion. An estimate of variance is obtained by

$$v\left(\bar{y}_{PS}\right) = \frac{1}{n(n-1)} \sum_i^n \left(\frac{y_i}{p_i} - \bar{y}_{PS}\right)^2 \qquad (13.8.7)$$

It should be remembered that the Politz-Simmons technique was developed with an aim to avoid call-backs altogether. If call-backs are not feasible or dependable, the technique may be very efficient in terms of information obtained per unit of time. For example, if the survey is designed to collect information on the household's consumption expenditures during a specified day, call-backs after some days may not bring any useful results.

13.9 RESPONSE ERRORS

As mentioned earlier, response errors are mainly contributed by the respondents or enumerators or both. If field instructions to enumerators are not strictly followed, it may contribute substantially to these errors. Many times, the respondents may also contribute to these errors on account of several factors such as lack of understanding, ambiguous questionnaires, memory errors, untruthful reporting, deliberately incorrect and careless answers, etc. An eye estimate of crop is an excellent example of the source of response error by enumerator or respondent which will be generally influenced by his personal judgement. Response errors may be accidental or may be introduced purposely or they may arise from lack of information. It is also observed that all these errors, whether due to enumerators or respondents may have a systematic character and cannot be normally ignored. Since errors from different sources may not cancel each other their cumulative effect on the estimates will not be negligible. Several excellent analyses of sources and types of response errors are available. Deming (1944), Marks and Mauldin (1950), Marks, et al. (1953), Mahalanobis (1946), Sukhatme and Seth (1952), and Hansen, et al. (1951, 1953, 1961, 1964) have developed several important techniques for measuring and controlling response errors, particularly those arising from enumerators. Some of the succeeding sections of this chapter are devoted to an explicit formulation of a mathematical model for the response errors. An important aspect of such a formulation is to determine some basic requirements that a mathematical model should meet so that it may conform to actual survey conditions. Other important feature of all survey designs is to formulate good estimating procedures. The processes of survey planning, data collection and coding and data processing may introduce errors in survey results.

These errors may be to some extent affected by the choice of estimators and survey designs. In the present discussion, however the relationship between response errors and these estimators with various survey designs has not been considered.

13.10 RESPONSE BIAS AND RESPONSE VARIANCE

Usually large scale surveys are conducted with the help of enumerators, especially employed and trained for the job so as to get worthwhile results. Many times, some changes are made into the data by these enumerators, called *enumerator effect*, which are in fact reflections to the enumerators personality, training, education, job efficiency, etc. or in other words, enumerators introduce their *personal equations*. Definitely, the enumerator effect will vary in magnitude from one item to another and from one enumerator to another. A general model to discuss it was developed by Hansen, *et al.* (1951, 1953, 1961, 1964), and Sukhatme and Seth (1952). For simplicity, let us assume that m enumerators, selected randomly from a large pool of M enumerators, are participating in the survey work and a random sample of n units is selected from a large population of N units. The sample units are assigned at random to the m enumerators.

Let y_{ij} denote the reported value of the jth unit by the ith enumerator. A general model may be defined by

$$y_{ij} = x_j + a_i + \epsilon_{ij} \qquad (13.10.1)$$

where x_j is the true value of the jth unit, a_i is the ith enumerator effect on the response, called as *bias of response* or *systematic error*, and ϵ_{ij} is the *random component* in the jth unit by the ith enumerator. It can be seen that $E(\epsilon_{ij} \mid i, j) = 0$, $E(\epsilon_{ij}^2 \mid i, j) = S_e^2$ and $\mathrm{cov}(\epsilon_{ij}, \epsilon_{ij}') = 0$.

In addition, let n_{ij} denote the number of responses on the jth unit by the ith enumerator and assume that $n_{ij} = 1$, if the unit U_j is in the sample and zero otherwise. Thus, we shall have

$n_i. = \sum_j n_{ij}$, the number of responses collected by the ith enumerator;

$n_{.j} = \sum_i n_{ij}$, the number of responses on the jth unit;

$n.. = \sum_i \sum_j n_{ij}$, the total number of responses.

It will be assumed that each enumerator has collected an equal number of responses, $n_i. = n/m = \bar{n}$, and the number of responses (repetitions) for each unit is equal, i.e. $n._j = n/l = p$; where l units in the sample are allotted to different enumerators.

Now we consider how the response error changes when one passes from one enumerator to another, and or from one unit to another. The reported response y_{ij} is a random variate and distributed with the value x_j and a specified variance. The mean value of responses obtained by the ith enumerator on all the units of the population will be $E(y_{i_j}|i) = \bar{Y}_i$. The expected value of all the M enumerators will be $E(\bar{Y}_i) = \bar{Y}'$ and let us call it the expected survey value which may be different from the true mean value $\bar{Y}$. The difference, $\bar{Y}' - \bar{Y}$, between the expected survey value and the true mean value is called the *response bias*. Thus the total error in response may be splitted into components as

$$y_{ij} - \bar{Y} = (y_{ij} - Y_j) + (Y_j - \bar{Y}') + (\bar{Y}' - \bar{Y}) \qquad (13.10.2)$$

where Y_j is the expected value of the jth unit reported by the enumerators of the survey. It can be seen from the above relation that the total error comprises measurement error, response deviation and response bias. Here the response deviation and response bias will obviously depend upon the interview procedures, schedules and training of enumerators. Unless some procedures are deviced to control them it will not be advisable to go ahead with the survey work. In the present discussion, we shall describe some of the methods to separate out and to control various components of response errors.

Before any method of estimation is discussed, we consider the response bias, $\bar{Y}' - \bar{Y}$, which is introduced on account of the reason that the enumerators/respondents are brought into picture. The response bias may be an important component of the mean square error of the estimate if its value is large. Since it involves the true mean value of the population, $\bar{Y}$, it is not possible to measure as such from the survey. To make an estimate of response bias, one approach is to conduct a small scale study called *post-enumeration survey*, just after the census or sample survey for a comparative study of the data. The differences of the estimates based on the main survey and the post-enumeration survey can be used to estimate the response bias. On this basis, if one finds that the response bias is not significant, it will be a worthwhile attempt to reduce the other components.

Let us consider the model given by relation (13.10.1) in terms of its components. For the bias component a_i, there may be a constant bias, say $E(a_i) = a$, that affects all the units in the population, and the variable component of bias, $(a_i - a)$, which is distributed with the mean value zero and the variance S_a^2. As mentioned already, the random component of the response follows a frequency distribution with the mean value zero and the variance S_e^2. Thus, the total error in a response may be splitted into different components as

$$y_{ij} - \overline{Y} = (x_j - \overline{Y}') + (a_i - a) + a + \epsilon_{ij} \qquad (13.10.3)$$

Averaging over the sample, we have

$$\bar{y}.. - \overline{Y} = \bar{x} - \overline{Y}' + (\bar{a} - a) + a + \bar{e} \qquad (13.10.4)$$

where
$$\bar{y}.. = \frac{1}{n} \sum_i^m \sum_j^l n_{ij} (x_j + a_i + \epsilon_{ij})$$

$$= \frac{1}{n} \sum_j^l n_{.j} x_j + \frac{1}{n} \sum_i^m n_{i.} a_i + \frac{1}{n} \sum_i^m \sum_j^l n_{ij} \epsilon_{ij}$$

$$= \bar{x} + \bar{a} + \bar{\epsilon} \qquad (13.10.5)$$

It is remarkable to note that the sample mean $\bar{y}..$ is not an unbiased estimator, i.e. the estimated mean is biased unless y_i's vary in such a way that the biases compensate each other and disappear. If the measurements on all units are subject to constant bias, a, whose magnitude is unknown, then the simple random sample mean is subject to the bias, a, as

$$E(\bar{y}.. - \overline{Y}) = a \qquad (13.10.6)$$

and the bias passes undetected.

THEOREM 13.10.1 Under the sampling scheme defined as above, the variance of the estimator $\bar{y}..$ is given by

$$V(\bar{y}..) = \left(\frac{1}{l} - \frac{1}{N}\right) S_x^2 + \left(\frac{1}{m} - \frac{1}{M}\right) S_a^2 + \frac{S_e^2}{n} \qquad (13.10.7)$$

where
$$S_x^2 = \sum_i^m \sum_j^l (x_{ij} - \overline{Y}')^2 / (N - 1)$$

S_a^2 and S_e^2 have already been defined.

Proof For the variance of $\bar{y}..$, we can write

$$V(\bar{y}..) = V\left[\frac{1}{n}\sum_i^m \sum_j^l n_{ij}(x_j + a_i + \epsilon_{ij})\right]$$

$$= V\left[\frac{1}{n}\sum_i^l n_{.j}x_j + \frac{1}{n}\sum_i^m n_{i.}a_i + \frac{1}{n}\sum_i^m \sum_j^l n_{ij}a_{ij}\right]$$

Taking $n_{.j} = p$, $n_{i.} = \bar{n}$ and substituting these values, we get the result.

The first term in the right hand side of relation (13.10.7) is the usual sampling variance component and the second term is an addition due to enumerator effect, generally termed as the *response variance* component.

COROLLARY 1   If one response is available for each unit, i.e. $p = 1$ or $l = n$ the $V(\bar{y}..)$ reduces to

$$V(\bar{y}..) = \left(\frac{1}{n} - \frac{1}{N}\right)S_x^2 + \left(\frac{1}{m} - \frac{1}{M}\right)S_a^2 + \frac{S_e^2}{n} \qquad (13.10.8)$$

COROLLARY 2   In the case N and M are large, we have

$$\left.\begin{aligned}
V(\bar{y}..) &= \frac{S_x^2 + S_e^2}{n} + \frac{S_a^2}{m} \\
&= \frac{S_y^2}{n} + \left(\frac{1}{m} - \frac{1}{n}\right)S_a^2
\end{aligned}\right\} \qquad (13.10.9)$$

where $\qquad S_y^2 = S_x^2 + S_a^2 + S_e^2$

COROLLARY 3   The population proportion P is estimated by drawing a sample of n unit randomly from a population of N units. Some units were misclassified due to response errors. Assuming that response errors and units are uncorrelated, the variance of the sample proportion p is given by

$$V(p) = \frac{(1 - f)}{n}\sum_i^N \frac{(P_i - P)^2}{N - 1} + \frac{1}{nN}\sum_i^N P_i Q_i \qquad (13.10.10)$$

where P_i is the probability that the unit U_i is classified to the belonging class, being given $\sum_i^N P_i = NP$ and $Q_i = 1 - P_i$.

COROLLARY 4 If the covariance in terms of intra-enumerator correlation coefficient is defined by

$$\rho_E = \frac{E(y_{ij} - \overline{Y}_i)(\tilde{y}_{ij} - \overline{Y}_{ij})}{E(y_{ij} - \overline{Y}_i)^2}$$

an expression for $V(\tilde{y}..)$ is obtained as

$$V(\tilde{y}..) = \frac{S_y^2}{n} + [1 + (\overline{n} - 1)\rho_E]\frac{S_e^2}{n} \qquad (13.10.11)$$

This expression is due to Hansen, *et al.* (1951). It can be seen from above relation that even if the correlation is small, the contribution to the response variance will be considerable if the number of units enumerated by each enumerator is large. ρ_E is likely to be positive if each enumerator has a tendency either to over-estimate or to under-estimate consistently. If $\overline{n} = 1$, i.e. each unit is enumerated only once by a separate enumerator then the covariance term will vanish from the expression. For uncorrelated data, the second term in relation (13.10.11) will not appear and the usual variance expression will be obtained.

Another important point should also be remembered that the mean square error will comprise four components; sampling variance response variance, covariance and the squared overall bias. If the survey procedures are such that the response bias is very large relative to $V(\tilde{y}..)$, the variance term will give a misleading picture of efficiency obtained in the survey. It is the total error of the estimate, which is to be made small and it can be done by the mean square error only. In case of census, the sampling variance will be zero and so the variance and the mean square error of the census figure $\tilde{y}'$ will be given by

$$V(\tilde{y}') = [1 + (m - 1)\rho_E]\frac{S_e^2}{n} \qquad (13.10.12)$$

and

$$MSE(\tilde{y}') = V(\tilde{y}') + (\tilde{y}' - \overline{Y})^2 \qquad (13.10.13)$$

COROLLARY 5 In mail surveys, the response errors are assumed to be uncorrelated from one unit to another and the model is defined by

$$y_{ij} = x_j + \epsilon_{ij} \tag{13.10.14}$$

where
$$E(\epsilon_{ij}) = 0 \text{ and } V(\epsilon_{ij}|i) = S_{ei}^2$$

The random sample mean of n units, $\bar{y}..$ is an unbiased estimator of the population mean and its variance is given by

$$V(\bar{y}..) = (1 - f)\frac{S_y^2}{n} + \frac{1}{nN}\sum_i^N S_{ei}^2 \tag{13.10.15}$$

It should be noted that $V(\bar{y}..)$ is not equal to the usual variance but it gets inflated by a quantity depending upon the variance S_{ei}^2 of individual response errors.

COROLLARY 6 When a systematic bias of a_j is associated with the jth unit, the model is defined by

$$y_{ij} = x_j + a_j + \epsilon_{ij} = x_j' + \epsilon_{ij} \tag{13.10.16}$$

where ϵ_{ij} are assumed to be uncorrelated, and $E(a_j) = a$.

The random sample mean, $\bar{y}..$ is subject to response bias, of a, and its variance is given by

$$V(\bar{y}..) = (1 - f)\frac{S_{y'}^2}{n} + \frac{1}{nN}\sum_i^N S_{ei}^2 \tag{13.10.17}$$

COROLLARY 7 When the observed value of the unit is the true value itself, i.e. $y_{ij} = x_j$, the sample mean $\bar{y}_j$. and its variance are in conformity with the usual estimator.

13.11 ESTIMATION OF VARIANCE COMPONENTS

For estimation of different components of variance in relation (13.10.8), the reader is referred to Sukhatme and Seth (1952). Here we shall discuss some outlines, in brief, to derive estimates of various components.

THEOREM 13.11.1 If $\bar{y}_i$. is the sample estimate for the ith enumerator, an unbiased estimator of $V(\bar{y}..)$ is obtained by

$$v(\bar{y}..) = \sum_i^m (\bar{y}_i. - \bar{y}..)^2/m(m - 1) \tag{13.11.1}$$

Proof We know that

$$E(\sum_i^m \bar{y}_i^2 - n\bar{y}_{..}^2) = m[mV(\bar{y}..) + \overline{Y}'^2 - V(\bar{y}..) - \overline{Y}'^2]$$

$$= m(m-1)V(\bar{y}..)$$

Hence the result follows.

The result shows that if l independent samples are enumerated by m enumerators selected randomly with equal probability from the large pool of enumerators then an unbiased estimator of the total variance can be obtained. Before proceeding to estimate the enumerator effect in the survey data, one should apply the technique of analysis of variance, which is summarised in Table 13.11.1.

Table 13.11.1   Analysis of Variance for Enumerator Effect in Survey Data

Source of variation	Degrees of freedom	Sum of squares	Mean squares	F value
Between enumerators	$m-1$	$\sum_i (\bar{y}_{i.} - \bar{y}..)^2$	B	B/W
Within enumerators	$m(l-1)$	$\sum_i \sum_j (y_{ij} - \bar{y}_{i.})^2$	W	
Total	$(ml-1)$	$\sum_i \sum_j (y_{ij} - \bar{y}..)^2$	A	

From the analysis of variance table, it may be examined whether the enumerator effect is present or not. By the value of F with corresponding values of tabulated F for the given degrees of freedom and level of significance, one can test the significance of the difference. If there is no significant enumerator effect one may get estimate of $V(\bar{y}..)$ by dividing A by ml. If significantly different the effect does exist. An estimate of between enumerator effect component may be obtained by

$$s_{BE}^2 = B - \frac{W}{ml} = C \qquad (13.11.2)$$

Similarly, an estimate of within enumerator effect component may be estimated by

$$s_{WE}^2 = W \tag{13.11.3}$$

On this basis, the estimate of $V(\bar{y}..)$ becomes

$$s_y^2 = \frac{C}{m} + \frac{W}{mn} \tag{13.11.4}$$

Another approach which is more rigorous is explained here. Let us denote the estimates of the mean squares by

$$s_b^2 = \sum_i (\bar{y}_i. - \bar{y}..)^2/(m - 1) \tag{13.11.5}$$

(between enumerators)

$$s_w^2 = \sum_i \sum_j (y_{ij} - \bar{y}_i.)^2/m(l - 1) \tag{13.11.6}$$

(within enumerators)

and

$$s_u^2 = \sum_j (\bar{y}._j - \bar{y}..)^2/(l - 1) \tag{13.11.7}$$

(between unit means)

It can be shown easily that

$$E(s_b^2) = \frac{m(m - p)}{lp(m - 1)} S_x^2 + S_a^2 + \frac{m}{lp}S_e^2 \tag{13.11.8}$$

$$E(s_w^2) = S_x^2 + S_e^2 \tag{13.11.9}$$

$$E(s_u^2) = S_x^2 + \frac{l(m - p)}{(l - 1)mp} S_a^2 + \frac{S_e^2}{p} \tag{13.11.10}$$

By using the relations (13.11.8), (13.11.9) and (13.11.10) one may get estimates of S_x^2, S_a^2 and S_e^2. Here we may write

$$\hat{S}_a^2 = \frac{p(m - 1)(l - 1)}{lpm - pl - pm + m}\left[s_b^2 + \frac{m}{l(m - 1)} s_u^2 - \frac{m^2}{lp(m - 1)} s_w^2\right] \tag{13.11.11}$$

In practice when $p = 1$, we have

$$\hat{S}_a^2 = s_b^2 - \frac{m}{l}s_w^2 \tag{13.11.12}$$

If N and M are sufficiently large and $p = 1$, an unbiased estimator of $V(\bar{y}..)$ is given by

$$v(\bar{y}..) = \frac{s_b^2}{m} = \frac{s_u^2}{l} + \frac{(l-m)}{l(m-1)}(s_u^2 - s_w^2) \qquad (13.11.13)$$

It is quite evident from the above relation that s_u^2/l does not provide an unbiased estimator of the variance of the estimated mean but it is slightly inflated by a quantity $(l-m)(s_u^2 - s_w^2)/l(m-1)$, which may disappear only when differential biases are zero.

In case each enumerator gets information on just one unit, an unbiased estimator of $V(\bar{y}..)$ is given by $\sum_{i}^{n}(\bar{y}_i. - \bar{y}..)^2/n(n-1)$, which coincides with the estimator given in relation (2.3.9) used in simple random sampling, wr. Apparently the usual variance estimator is quite good provided each enumerator collects information from one unit only but the estimator of mean, $\bar{y}$, will be a biased one if response errors are present. In practice, an enumerator collects information from more than one unit and the estimator $\sum_{i}(\bar{y}_i. - \bar{y}..)^2/n(n-1)$ will be no more unbiased.

Further it should also be borne in mind that an unbiased estimate of the sampling variance $V(\bar{y}..)$ of $\bar{y}..$ can be obtained even in the presence of response errors but it is not possible to make so for the total mean square error.

EXAMPLE 13.2 In a survey 10 units each were enumerated by two enumerators and the following responses were noted:

Enumerator I 405 400 411 396 300 396 398 394 399 400

Enumerator II 400 400 411 392 304 395 396 391 395 395

When $N = 400$, estimate the variance of $\bar{y}..$ and its components, assuming that the correlations are non-negative.

By subtracting 390 from all the observations, we have

$$s_b^2 = 961.8, \; s_w^2 = \frac{1}{10}(1^2 + 21^2) - \frac{1}{20}\cdot 22^2 = 20.0$$

and $n = 10$; so a reasonable estimate of $V(\bar{y}..)$ is given by

$$v(\bar{y}..) = \left(\frac{1}{20} - \frac{1}{400}\right)s_b^2 = 46$$

Applying analysis of variance, we conclude that the enumerator effect is not significant and the estimate of variance obtained is valid. However, other components may be obtained with further analysis.

13.12 OPTIMUM NUMBER OF ENUMERATORS

There remains one pertinent question still unanswered what should be the number of enumerators employed in a survey? Actually this question can be replied in totality after examining the position of budget and degree of precision to be retained in the survey results. Though the precision can be increased by increasing the number of enumerators in the survey but there is a limit under budget conditions that a number beyond that one cannot employ enumerators. We shall discuss the problem of determining the optimum number of enumerator m, to be employed in the survey with the maximum number of units be assigned to each of them that the variance is minimized with a fixed budget. The simplest cost function may be defined by

$$C = c_0 + c_1 l + c_2 m \qquad (13.12.1)$$

where c_0 is the overhead cost, c_1 is the average cost of collecting information per unit and c_2 is the average cost per enumerator. Suppose the cost is fixed, say C_0. To obtain the optimum values of l and m, we consider

$$\phi = \frac{S_y^2 - S_a^2}{l} + \frac{S_a^2}{m} + \lambda \, (c_0 + c_1 l + c_2 m - C_0) \qquad (13.12.2)$$

where λ is Lagrange's multiplier.

Differentiating w.r.t. l and m, equating $\dfrac{\partial \phi}{\partial l}$ and $\dfrac{\partial \phi}{\partial m}$ to zero and solving the equations, we have

$$\left.\begin{array}{c} m = lD \\[2mm] l = \dfrac{C_0 - c_0}{c_1 + Dc_2} \end{array}\right\} \qquad (13.12.3)$$

and

where

$$D = \frac{c_1}{c_2} S_a^2 / (S_y^2 - S_a^2) \qquad (13.12.4)$$

Generally, in surveys, it is observed that c_1 is smaller than c_2 and S_a^2 is smaller than S_y^2. If the contribution of S_a^2 to variance is large relative to S_y^2, it is obvious that larger number of enumerators should be participating in the survey. However, if the number of enumerators is increased substantially, it will obviously be necessary to employ less qualified and trained enumerators.

If the cost function is formed as

$$C = c_0 + c_1 l + c_2 m + c_3 \sqrt{n} \qquad (13.12.5)$$

where c_3 is the average cost of travel per unit distance and the rest have their usual meanings. Since a sample of l units is randomly allocated to each of m enumerators it will be reasonable to assume that the cost of the survey is given by

$$C = c_0 + c_1 l + c_2 m + c_2 \sqrt{lm} \qquad (13.12.6)$$

Proceeding on similar lines as discussed above, we may derive an equation in l and m after eliminating λ, which is written as

$$2t^4 + ut^3 - vtD^2 - 2D^2 = 0 \qquad (13.12.7)$$

where $t^2 = m/l$, $u = c_3/c_2$, $v = c_3/c_1$, and D is defined as in relation (13.12.3). It can be seen that the equation has two real roots which can be obtained by iterative method.

13.13 EXTENSION TO OTHER SAMPLING DESIGNS

The theory presented in previous sections is of limited use as it is uncommon to allot units at random to the selected enumerators. In this procedure, enumerators are supposed to travel over the entire area of enquiry resulting an increase in the cost of travel. To overcome this problem, Hansen, *et al.* (1951) have recommended stratified random sampling procedure. Assume that a population of N units is divided into k strata with N_h units in the hth stratum. In the survey, a simple random sample of n units is selected from N units such that n_h units are drawn from the hth stratum and $\sum_h n_h = n$. Supoose an enumerator enumerates $\bar{n}$ units within a stratum so that $\bar{n} \, m_h = l_h$, where m_h enumerators, selected at random from M_h, are participating in the hth stratum. Let $m = \sum_h^k m_h$, the total number of enumerators participating in the survey. With this sampling design, let y_{hij} be the response obtained on the jth unit by the ith enumerator in the hth stratum. The estimators of the corresponding mean values are defined by

$$\bar{y}_{hi\cdot} = \frac{1}{\bar{n}} \sum_{j}^{\bar{n}} y_{hij} \tag{13.13.1}$$

average for the ith enumerator in the hth stratum,

$$\bar{y}_{h\cdot\cdot} = \frac{1}{nm_h} \sum_{i}^{m_h} \sum_{j}^{\bar{n}} y_{hij} = \frac{1}{m_h} \sum_{i}^{mh} \bar{y}_{hi\cdot} \tag{13.13.2}$$

average of the sample in the hth stratum,

$$\bar{y}_{\cdots} = \frac{1}{n} \sum_{h} \sum_{i} \sum_{j} y_{hij} \tag{13.13.3}$$

the estimator of the mean value.

For the sake of simplicity, S_e^2 may be assumed constant from stratum to stratum. Sticking to the notations used in the previous section, we get

$$V(\bar{y}_{\cdots}) = \sum_{h}^{k} W_h^2 V(\bar{y}_{h\cdot\cdot}) = \sum_{h}^{k} W_h \left[\frac{S_{hx}^2 + S_e^2}{l_h} + \frac{S_{ha}^2}{m_h} \right] \tag{13.13.4}$$

where S_{hx}^2, S_{ha}^2 are analogous values in the hth stratum.

An unbiased estimator of $V(\bar{y}_{\cdots})$ can be obtained by

$$v(\bar{y}_{\cdots}) = \frac{1}{m(n-1)} \sum_{h} \frac{n_h - 1}{m_h - 1} \sum_{i}^{mk} (\bar{y}_{hi\cdot} - \bar{y}_{\cdots})^2$$

$$+ \frac{1}{n(\bar{n}-1)} \sum_{h} n_h (\bar{y}_{h\cdot\cdot} - \bar{y}_{\cdots})^2 \tag{13.13.5}$$

If the contribution of the enumerator effect to the variance is to be shown separately, we can write

$$v(\bar{y}_{\cdots}) = \frac{1}{n(n-1)} \sum_{h} \sum_{i} \sum_{j} (\bar{y}_{hij} - \bar{y}_{\cdots})^2 + \frac{n-m}{m(n-1)} s_{yE}^2 \tag{13.13.6}$$

where
$$s_{yE}^2 = \frac{1}{m} \sum_h \frac{m_h}{m_h - 1} \sum_i (\bar{y}_{hi\cdot} - \bar{y}_{h\cdot\cdot})^2$$

$$- \frac{1}{n(\bar{n} - 1)} \sum_h \sum_i \sum_j (y_{hij} - \bar{y}_{hi\cdot})^2$$

It should be noted that the term s_{yE}^2 is due to the enumerator effect on response and interaction between enumerators and respondents. If the observation is of such a type which provides a good chance to include a large enumerator effect than the contribution of the term s_{yE}^2 to the variance of $\bar{y}_{\cdot\cdot\cdot}$ may be quite substantial. It may be possible to reduce this contribution significantly by training and adequately supervising the enumerators. From relation (13.10.11) it may be seen that if cost were not a factor, maximum efficiency with the sampling design would be achieved by assigning one unit to each enumerator.

Numerous applications of these methods, called *replicated sampling*, are discussed by Deming (1960). Replicated sub-samples are obtained when two or more subsamples are taken from the population by the same method of selection. For example, each stratum might consist of two random samples, each assigned to different enumerators who would be required to cover the whole stratum instead of half the stratum. Thus every stratum may provide 1 *df* for $V(\bar{y}_{\cdot\cdot\cdot})$. More complex estimates based on realistic assumptions are given by Fellegi (1964). If E_1 and E_2 denote the two enumerators and S_1 and S_2, the two replicated subsamples, comparisons of $(E_1 S_1)$ with $(E_2 S_1)$ and $(E_1 S_2)$ with $(E_2 S_2)$ provide the replicated measurements, while comparisons of $(E_1 S_1)$ with $(E_2 S_2)$ and $(E_1 S_2)$ with $(E_2 S_1)$ give the interpenetration measurements. Thus one may obtain estimates of the simple response variances, the correlated component, the total response variance. Hansen, Hurwitz and Bershad (1961) called S_a^2/n the simple response variance, and term $(n-1)\rho_E S_a^2/n$ as the correlated component of the total response variance $[1 + (n-1)\rho_E] S_a^2/n$. An index of inconsistency, analogous to the quantity $(1 - \phi)$, where ϕ is the coefficient or reliability, is defined by Pritzker and Hanson (1962). The quantity $\hat{I} = n(b + c)/[(a + b)(c + d) + (a + c)(b + d)]$ is an index which measures inconsistency in a 2×2 table with set of values (y_{i1}, y_{i2}) reported by two enumerators (For details, see Problem 13.9).

Number of responses

Enumerator I

		A	α	Total
	A	a	c	$a + c$
Enumerator				
II	α	b	d	$b + d$
Total		$a + b$	$c + d$	n

Assuming binominal trials with 1 and 0 values of the variate, one may find that $(b + c)$ is the number of units where inconsistency in reporting of the responses has occurred. Interpretations of these comparisons are, of course, doubtful. For discussions of merits and demerits of various variates of such measurements, the readers are referred to Bailar and Dalenius (1969). Singh, *et al.* (1974) have studied problem of non-response in successive sampling designs. Koch (1973) has given a general decomposition of the mean square error in multivariate sample surveys while Singh and Singh (1983) have demonstrated its application in a bivariate population.

13.14 TABULATION ERRORS

Processing of data in a large scale survey or census involves scrutiny of data, sorting of schedules, listing, coding, punching, etc. When processing of data starts, a large number of errors is possibly to enter at these stages and even well-trained personnel may make errors. Generally tabulation errors are assigned to omission, duplication, misclassification, erroneous enumeration. Operations such as listing, coding, punching, transcribing, etc. are potential sources of these errors. Tabulation errors are subject to control through verification, consistency checks, etc. Built-in checks and cross-checks on computations are necessary for reducing these errors. Suitable sampling methods may be used to assess and control these tabulation errors at various stages. The main point to be considered is how much control in terms of verification, consistency and checks is required to achieve in terms of efficiency or precision.

The first and foremost job to be done before tabulation of data is to make a thorough scrutiny of the material and detect out listing errors, inconsistencies and other defects. The second job is to finalise the number, the shape and size of the tables that are to be included in the final report so that the main findings are presented. A detailed discussion for the control of these errors is given by Zarkovich (1966). We shall present here, in brief, some outlines with description of some designs used for control of listing errors.

Compact Clustering Design—For detailed discussion of such designs the reader is referred to Hansen, *et al.* (1953). In this design, compact clusters of contiguous sampling units or compact areas are made up so that comparison of the results obtained in survey and check will offer information about omissions and duplications. Since area segments are conceived here as compact clusters in tabulated data, this means that compact clusters shall make possible to provide a measure of listing errors. Let us take $u_{ij} = 1$, if the jth unit of the ith cluster is listed in the census and zero otherwise. Similarly, $v_{ij} = 1$, if the jth unit is listed in the check survey and zero otherwise. The listing bias is defined as

$$b_{ij} = u_{ij} - v_{ij} \qquad (13.14.1)$$

With this, we shall group the units of the ith cluster in three classes, (i) units listed in both the census and the check, for these units there is *agreement* between both the surveys, i.e. $b_{ij} = 0$; (ii) units listed in the survey and not in the check, for these units there is *erroneous inclusion*, i.e. $b_{ij} = 1$; and (iii) units listed in the check and not in the survey, for these units there is omission, i.e. $b_{ij} = -1$. Thus the total bias is given by

$$B = U - V = \sum_i \sum_j b_{ij} = \sum_i^M B_i \qquad (13.14.2)$$

Assuming that the population consists of M clusters and out of these a sample of m clusters is selected in the check survey, an estimate of B is obtained by

$$\hat{B} = \frac{M}{m} \sum_i^m B_i \qquad (13.14.3)$$

where B_i is obtained by grouping the units of the ith cluster. Summing up the totals in (ii) and (iii) we can get B_i. After totalling over the sampled clusters, $\hat{B}$ can be obtained. Calculating its variance, we have

$$V(\hat{B}) = \frac{M^2}{m} V(B_i)$$

$$= \frac{M^2}{m} \frac{1}{(M-1)} \left[\sum_i^M B_i^2 - \frac{1}{M} (\sum_i^M B_i)^2 \right] \qquad (13.14.4)$$

If the second term in relation (13.14.4) is negligible, we can write

$$V(\hat{B}) = \frac{M^2}{m} \frac{1}{(M-1)} \sum_i^M B_i^2$$

$$\cong \frac{M}{m} \sum_i^M B_i^2 \qquad (13.14.5)$$

An estimate of $V(\hat{B})$ is obtained by

$$v(\hat{B}) = \frac{M^2}{m(m-1)} \left[\sum_i^m B_i^2 - \frac{1}{m} (\sum_i^m B_i)^2 \right] \qquad (13.14.6)$$

In case, the second term is small, it can be approximated by

$$v(\hat{B}) = \frac{M^2}{m(m-1)} \sum_i^m B_i^2 \qquad (13.14.7)$$

Before estimating the bias component V, we should examine significance of $\hat{B}$ by defining

$$t = \frac{\hat{B}}{\sqrt{v(\hat{B})}} \qquad (13.14.8)$$

If the value is statistically significant, the listing bias is considered significant and there is a reason to worry about it, and in that case one should also get an estimate of the bias component V by using the relation $\hat{V} = \hat{U} - \hat{B}$. In the present discussion, the difference estimator is used. On similar lines one may use ratio estimator also for estimating these errors.

Extended Compact Clustering Design—In some situations, it may be that the primary sampling units contain a large number of elementary units and relisting of all the units becomes costly operation. For example, dwelling units or villages are clusters of households, and households are clusters of persons. So clusters will comprise persons (elementary units) within a household or village. Thus the compact clustering design discussed earlier may not suffice the purpose. In this case the design is modified slightly. A sample of extended compact clusters which consist of primary sampling units is first selected and within selected clusters a check is carried out. This will provide an estimate of bias due to listing errors in primary units. Similarly an estimate of bias due to these errors in elementary units can also be derived by a check of elementary units within the selected clusters. A combined estimate is obtained by pooling both of the components. In practice, these checks of primary units and elementary units are arranged in one single field operation. Whenever an enumerator enumerates the clusters of primary units he will also enumerate the elementary units in the check. Retaining the previous notations, it is assumed that the quantities U_{ij} and V_{ij} are available after the check survey has been conducted. Let U_{ij} denote the total number of elementary units listed in the census to the check with analogous meaning. If U and V are two bias components, we can write an estimator

$$\hat{B} = \hat{U} + \hat{V} \qquad (13.14.9)$$

where $\hat{U}$ is an estimator of the bias of errors in listing clusters and $\hat{V}$ is an estimator of the bias of errors in listing the elementary units. The estimator $\hat{V}$ is obtained by

$$\hat{V} = \frac{M}{m} \sum_i^m \frac{N_i}{n_i} \sum_j^{ni} B_{ij} = \frac{M}{m} \sum_i^m \frac{N_i}{n_i} B_i' \qquad (13.14.10)$$

where N_i and n_i are the numbers of primary units in the population and the sample in the ith cluster, $B_{ij} = U_{ij} - V_{ij}$, such that $\sum_j B_{ij} = B_i$. Writing $B_i' = \frac{N_i}{n_i} B_i$ and $\bar{B} = \frac{1}{m} \sum_i^m B_i'$, we have variance of $\hat{V}$

$$V(\hat{V}) = \frac{M^2}{m} \sum_i^M V(B_i') = \frac{M^2}{m(M-1)} \left[\sum_i^M B_i'^2 - \frac{1}{M} (\sum_i^M B_i')^2 \right] \qquad (13.14.11)$$

An unbiased estimate of $V(\hat{V})$ is given by

$$v(\hat{V}) = \frac{M^2}{m(m-1)} \left[\sum_i^m B_i'^2 - \frac{1}{m} (\sum_i^m B_i')^2 \right] \qquad (13.14.12)$$

In addition, all the primary units which should not have been enumerated in the census were also counted. After the check survey is over, we have

U_i = The total number elementary unit enumerated in all the primary units listed in the census in the ith cluster.

V_i = The corresponding total in the check for the ith cluster.

$W_i = U_i - V_i$, hence $\hat{U}$ may be estimated as in relation (13.14.3), and its variance may be obtained similarly by a relation as in (13.14.4).

After totalling the components $\hat{U}$ and $\hat{V}$, one may get the estimator $\hat{B}$ in relation (13.14.9). As regards the variance of $\hat{B}$, we have

$$V(\hat{B}) = V(\hat{U}) + V(\hat{V}) + 2 \operatorname{cov}(\hat{U}), \hat{V} \qquad (13.14.13)$$

The covariance term vanishes, if the components $\hat{U}$ and $\hat{V}$ are estimated independently by taking a separate check for the elementary units. For a detailed discussion the readers are referred to Zarkovich (1966) wherein these are dealt in at length.

SET OF PROBLEMS

13.1 What are non-sampling errors? How do you distinguish between sampling and non-sampling errors? What methods would you suggest to assess and control non-sampling errors?

13.2 Discuss the problem of non-response in surveys. Show how to obtain an unbiased estimator of the mean of the whole population by resurveying a pre-assigned fraction f_1, of the non respondents. Derive variance of the estimator. By taking a simple cost function, determine the optimum value of f_1.

13.3 What is meant by non-response in sample surveys? What are the effects of non-response on the estimates? What are the various ways of dealing with non-response in sample surveys with a view to control their bad effects?

13.4 In a mail enquiry, a random sample of n units was taken with equal probability and with replacement, out of which n_1 units responded in the first call. For making call-backs to collect the required information, a sample with sampling fraction f_1 was drawn from the non respondent class n_2 ($= n - n_1$) with equal probability and with replacement. Assuming that in the second call all responded, derive an unbiased estimator of the population mean $\bar{y}$. Derive its variance. Find the optimum values of n and f_1 which would minimize the variance for the fixed cost C_0, the cost function is given by $C = nc_1 + (1 - p)nf_1c_2$, where c_1 is the cost per unit for the first call, c_2 is the cost per unit for the call-backs, and p is the proportion of units responding in the first call in the population.

13.5 If there is a constant measurement bias of a and the response is defined by $y_i = u_i + a$, where y_i, is the true values of the ith unit, $i = 1, 2 \ldots, N$. Prove that the mean value based on a simple random sample, wor, will subject to the constant bias of a, what would be the effect of this bias on the variance of $\bar{y}$ and on the variance of estimator? If measurements on an ancillary variate x are also subject to a constant bias of k, discuss its effect on these estimators $\frac{\bar{y}}{\bar{x}} \bar{X}$, $\bar{y} \frac{\bar{x}}{\bar{X}}$, $\bar{y} + c (\bar{X} \quad \bar{x})$ and $\bar{y} + b (\bar{X} - \bar{x})$, where c is a constant and b is the least square estimate of the regression of y on x.

13.6 Suppose the response follows the model $y_{ij} = x_i + a_j + E_{ij}$ where y_{ij} denotes the response reported by the jth enumerator for the ith unit, $i = 1, 2, \ldots, N; j = 1, 2, \ldots, M$, x_i is the true value of the ith unit, a_j is the bias of the jth enumerator, and ϵ_{ij} is the random error with then mean value 0 and variance σ_ϵ^2. Assuming that a random sample of n units is drawn from the population and these are assigned at random to m enumerators selected from M enumerators such that each enumerator enumerate only l units and each sample unit is observed by the same number p of enumerators. Derive bias and variance of the sample mean. Also obtain an unbiased estimator of variance and write assumptions for it.

13.7 In exercise 13.6, the cost of survey is represented by $C = c_1 n + c_2 m + c_3(nm)^{1/2}$, find the optimum number of enumerators for which the variance of the sample mean will be minimum when cost is fixed. Also discuss the case when $c_3 = 0$.

13.8 In a population, P_{ij} denotes the proportion of units belonging to class (i,j), i being 1 if the unit is available for interview and 0 otherwise, and j being 1 if the unit will answer "Yes" and 0 if "NO". To estimate the proportion responding "Yes" in the population, $P_{11} + P_{01}$, a sample of n units is drawn at random without replacement and p is the proportion of units who answer "Yes" to the question out of the units available for interview. Derive an expression of the bias of p and obtain conservative bounds for this bias.

(Birnbaum and Sirken, 1950)

13.9 In order to estimate P, the population proportion belonging to class A, a random sample of n units is drawn with replacement and each unit is classified to the class A, or α (not A) on the basis of observation which is subject to error. The same units are enumerated by two enumerators independently and the results are presented as follows:

Enumerator I

		A	α	Total
Enumerator	A	a	c	$a + c$
II	α	b	d	$b + d$
	Total	$a+b$	$c \div d$	n

If p_{ij}, q_{ij} $(i = 1, 2)$ $(j=1, 2)$ denote the probabilities that a unit belonging to class A or α is correctly classified or incorrectly classified, respectively, by the enumerators. Thus the probability that a unit selected at random is incorrectly classified by enumerator I is $\bar{q}_1 = Pq_{11} + (1 - P) q_{12}$ and the corresponding probability for enumerator II is $\bar{q}_2 = Pq_{21} + (1-P)q_{22}$, show that the estimator

$$\hat{P} = \frac{a}{n} + \frac{b+c}{2n}$$

is a biased estimator of P and its relative bias is given by

$$\text{Rel. Bias} (\hat{P}) = \frac{\bar{q} - (q_{11} + q_{21})E(\hat{P})}{E(P) - q}$$

Where $\bar{q} = (\bar{q}_1 + \bar{q}_2)/2$. Also obtain conservative bounds for bias in $\hat{P}$.

13.10 From a population comprising N respondents, a sample of n respondents is taken at random and j calls are made on each sample respondent. Responses on the study variate y were collected from the respondents who were available atleast once. If p_i is the probability that a specified respondent would be at home, the probability that he will be available at least once in j calls is given by $1 - q_i^j$, where $q_i = 1 - p_i$. Show that the expected value of the estimated mean would be

$$\sum_i^N yi(1 - q_i^j)/N$$

13.11 Considering a generalised model for given by $y_{ijk} = x_j + a_i + b_{ij} + \epsilon_{ijk}$ where x_j is the true value of the characteristic on the jth unit, a_i is the bias of the ith enumerator, b_{ij} is the interaction of the ith enumerator with the jth unit and ϵ_{ijk} is the random error component, show that the expected value and variance of the sample mean are given by

$$E(\bar{y}...) = \bar{y} + a$$

$$V(y...) = \left[\frac{\sum\limits_j n_{ij}^2}{n^2} - \frac{1}{N}\right] S_x^2 + \left[\frac{\sum\limits_i n_{i\cdot}^2}{n^2} - \frac{1}{m}\right] S_a^2$$

$$+ \frac{\sum\sum\limits_{ij} n_{ij}^2}{n^2} \sigma_{bi}^2 + \frac{S_\epsilon^2}{n}$$

where $\sigma_{bi}^2 = E(b_{ij}^2/i)$ and the rest are as usual.

(Sukhatme and Seth, 1952)

13.12 In a mail survey, the response rate is expected 30 per cent. If the degree of precision desired is $4000/S^2$ with no non-response, cost of mailing a questionnaire is Re 0.25. and the cost of processing a completed questionnaire is Re. 0.30. How many questionnaires should be mailed out and what fraction of non-response should be interviewed, if the cost of a personal interview be Rs. 3.20?

REFERENCES

Bailar, B.A. and T. Dalenius, "Estimating responses variance components of the US Bureau of the Census Model," *Sankhya*, **31B**, 341-360, (1969).

Birnbaum, Z. W. and M. G. Sirken, "Bias due to non-availability in sampling surveys," *J. Amer. Statist, Assoc.*, **49**, 91-111, (1950a).

———, "On the total error due to non-interview and to sampling," *Int. J. Opinion and Altitude Res.*, **4**, 179-191, (1950b)

Chaudhary, Narendra, *On the use of outdated frame in successive sampling*, M.Sc. Thesis (Unpublished) New Delhi, (1985).

Deming, W.E., "On Errors in Surveys," *Amer. Socio. Rev.*, **9**, 359-369, (1944).

———, *Some Theory of Sampling*, Wiley & Sons, New York, (1950).

———, "On a probability mechanism to attain an economic balance between the resultant errors of non response and the bias of non response," *J. Amer. Statist. Assoc.*, **48**, 743-772, (1955).

———, *Sample Design in Business Research*, Wiley & Sons, New York, 1960.

———and G.J. Classer, 'On problem of matching lists in sample' *J. Amer Statist. Assoc.*, **54**, 403-415, (1959).

Durbin, J., "Non-response and call-backs in surveys," *Bull. Int. Statist. Inst.*, **34**. 72-86 (1954).

El-Badry, M.A., "A Sampling procedure for mailed questionnaires," *J. Amer. Statist. Assoc.*, **51**, 209-227, (1956).

Foradori, G. T., "Some non response sampling theory for two stage designs," Mimeo. *297, North Carolina State College*, Raleigh, (1961).

Fellegi, I., "Response variance and its estimation," *J. Amer. Statist. Assoc.*, **59**, 1016-1041, (1964).

Goodmen, L. A., "On estimation of number of classes in a population," *Ann. Math. Statist.* **20**, 572-579, (1940).

Hartley, H.O., "Multiple frame surveys," *Proc. Soc. Statist. Sec. Amer. Statist. Assoc.*, **57**, 203-206, (1962).

Hansen, M.H. and W.N. Hurwitz, "The problem of non response in Sample Surveys," *J. Amer. Statist. Assoc.*, **41**, 517-529, (1946).

———and M. Gurney, "Problem and methods of the sample survey of business," *J. Amer. Statist. Assoc.*, **41**, 173-189, (1946).

———, E.S. Marks and W.P. Mauldin, "Response errors in Surveys," *J. Amer. Statist. Assoc.*, **46**, 147-190, (1951).

———and W.G. Madow, *Sample Survey Methods and Theory*, Wiley & Sons, New York, (1953).

———and M. Bershad, "Measurement errors in censuses and surveys," *Bull. Int. Statist. Inst.*, **38**, 359-374, (1961).

———and T.B. Jabine, "Use of" imperfect tests for probability sampling at US Bureau of Census," *Bull. Int. Statist. Inst.* **40**, 497-517, (1963).

———and L. Pritzker, "Estimation and interpretation of gross differences and simple response variance," Paper presented in Mahalanobis birthday celebration lectures, (1964).

Kish, L., *Survey Sampling*, Wiley & Sons, New York, (1965).

———and I. Hess, "On non coverage of sample dwellings," *J. Amer. Statist. Assoc.* **53**, 509-524, (1958).

————, "A replacement procedure reducing the bias of non response," *Amer. Statistician*, **13** (4), 17-19, (1959).

Koch, G., "An alternative approach to multi-variate response errors model for sample survey data with applications to estimators involving subclass means", *J. Amer. Statist. Assoc.*, **68**, 906-913, (1973).

Lahiri, D.B., "Observation on the use of interpenetrating samples in India," *Bull. Int. Statist, Inst.*, **36**, 144-152, (1958).

Mahalanobis, P.C., "A sample survey of the acreage under jute in West Bengal," *Sankhya* **4**, 511-530, (1946).

————, "On large scale sample surveys," *Phill. Trans. R. Statist. Soc.*, **231B**, 329-451, (1944).

————, "Recent experiments in statistical sampling in the Indian Statistical Institute," *J.R. Statist. Soc.*, **109**, 326-378, (1946).

————and D.B. Lahiri, "Analysis of errors in censuses and Surveys with special reference to experience in India," *Bull. Int. Statist. Inst.*, **38**, 401-433, (1961).

Marks, E.S. and W.P. Mauldin, "Problems of responses in enumerative surveys," *Amer. Socio. Rev.*, **15**, 649-657, (1950a).

————, "Response errors in census research," *J. Amer. Statist. Assoc.*, **45**, 424-438, (1950b)

————and H. Nisselon, "Post-enumeration survey of 1950 census; A case history of survey Designs", *J. Amer. Statist. Assoc.*, **48**, 220-243, (1953).

Pritzker, L. and R. Hanson, "Measurement errors in 1960 Census of population," *Proc. Soc. Statist. Sec., Amer. Statist, Assoc.*, 80-89, (1962).

Politz, A. and W. Simmons, "An attempt to get the "not-at-homes", into the sample without call-backs," *J. Amer. Statist. Assoc.*, **44**. 9-31, (1949).

————, "Note on "An attempt to get the not-at-homes into the sample without call-backs" *J. Amer. Statist. Assoc.*, **45**, 136-137, (1950).

Seal, K.C., "Use of out dated frames in large scale surveys", *Bull. Cal. Statist. Assoc.*, **11**,68-84, (1950).

Singh, D., Singh, R. and Singh, Padam, "A study of non response in successive sampling", *J. Ind. Soc. Agr. Statist.*, **26**, 37-41, (1974).

Singh, D. and R. Singh, "Sampling with partial enumeration from bivariate population", *J. Statist. Planning and Inference*, **7**, 343-351, (1983).

Sukhatme, P.V., "The problem of plot size in large-scale surveys, *J. Amer. Statist. Assoc.*, **42**, 297-310, (1947).

————and G.R. Seth, "Non sampling errors in surveys" *J. Ind. Soc. Agr. Statist.*, **4**, 5-41, (1952).

Srinath, K.P., "Multiphase in nonresponse problems", *J. Amer. Statist. Assoc.*, **16**, 583-586, (1971).

Srivastava, A.C., *Sampling from dynamic population*, Ph.D. Thesis, (Unpublished), Delhi, (1977).

Stephan, F.F. and P.J. McCarthy, *Sampling opinions*, Wiley & Sons, New York, (1958).

Yates, F., *Sampling Methods for Censuses and Surveys*, Charles Griffin, London, (1949).

Zarkovich, S.S., *Sampling Methods and Census*, F.A.O., Rome, (1965).

————, *Quality of Statistical Data*, F.A.O., Rome, (1966).

Author Index

Subject Index